Maths
The Basics

Functional
Skills Edition

June Haighton
Bridget Philips
Veronica Thomas
Deborah Holder

Text © J Haighton, D Holder, B Phillips, V Thomas 2004, 2013
Original illustrations © Nelson Thornes Ltd 2004, 2013

First published by Nelson Thornes 2004

This edition published by:
Nelson Thornes Ltd
Delta Place
27 Bath Road
CHELTENHAM
GL53 7TH
United Kingdom

13 14 15 16 17 / 10 9 8 7 6 5 4 3 2 1

A catalogue record for this book is available from the British Library

ISBN: 978 1 4085 2112 0

Illustrations and page make-up by Pantek Media, Maidstone

Printed and bound in Spain by GraphyCems

Contents

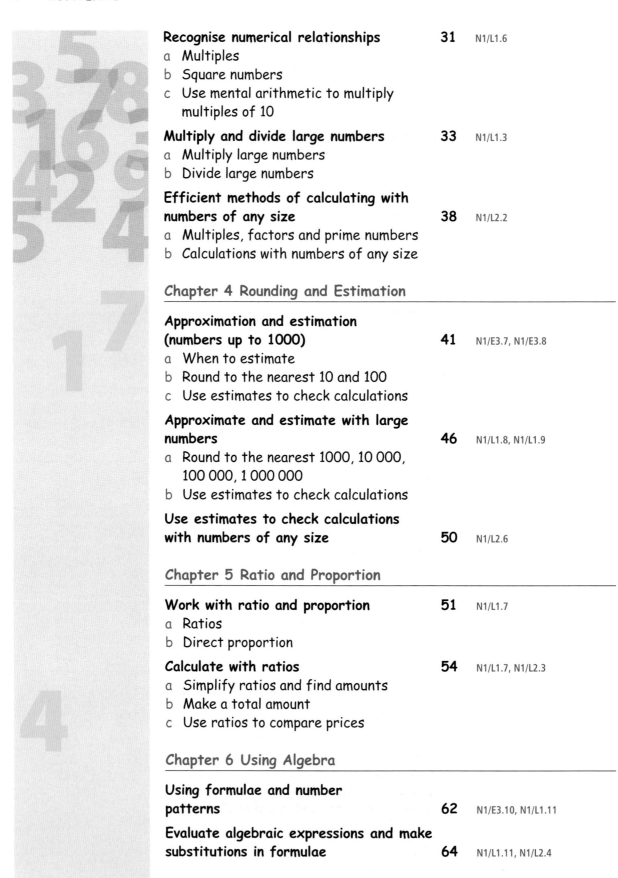

2 Measures, Shape and Space

3 Handling Data

Introduction

About this book

This book is divided into the three main sections of the Adult Numeracy Core Curriculum:

- **Number**
- **Measures, shape and space**
- **Handling data**

The level of the work ranges from Entry level 3 (E3) through Level 1 (L1) to Level 2 (L2). Entry levels 1 and 2 are covered in separate Worksheet Packs.

The curriculum elements covered in each topic are given in the contents list and at the beginning of each topic. Some topics include more than one level of work. These topics begin with work at the lowest level (usually E3) and progress to the higher levels. The icons shown here **L1** **L2** indicate where the level of difficulty of the work increases.

How to use this book

This depends on what you are aiming for, but in all cases it is recommended that you work through the levels you need, starting with E3. If you already have some knowledge of a topic, you may be able to miss out some parts, but always look through the work to make sure.

Is your main aim to improve your understanding of a particular mathematical topic? If so, look for the topic in the Contents list and work through from E3 to the level you need.

Is your main aim to prepare for a Functional Mathematics qualification? Functional Mathematics can be taken at all of the levels included in this book – the higher the level, the more difficult the real-life problems that you will be asked to solve. Usually you can use a calculator to work things out, but you might also need to check your answer without a calculator. When you have finished the working you need to be able to explain what your answer means.

If you are preparing for a qualification at Entry level 3, work through all the topics labelled E3. For Level 1 work through all the topics labelled E3 and L1. For Level 2 work through all of the topics in this book.

To accompany the book, you will find FREE additional Functional Skills resources on our Planet Vocational website, at www.planetvocational.co.uk/mtbfs. For each level there are 5 sets of resources linked to chapters in 'Maths The Basics'. Each set consists of a worked example which includes a range of questions set in a real-life context and some worksheet practice questions for you to try. There is also a Mixed Practice worksheet with questions that involve topics from all parts of the book.

Number

1 Number Value

Numbers up to 1000

a Place value

There are three hundred and sixty-five days in a year.

The number 365 has

3	Hundreds
6	Tens
5	Units

```
H T U
3 0 0
  6 0
    5
─────
3 6 5
```
← 3 Hundreds, 6 Tens and 5 Units

The 3, 6 and 5 are called digits.

Practice

1 Write the following numbers in figures.
Example: Six hundred and seventy
Use columns like this on squared paper.

H	T	U
6	7	0

a) One hundred and sixty-five
b) Fifty-three
c) Two hundred and twelve
d) Four hundred and three
e) Nine hundred and ninety
f) Sixty-eight
g) Eight hundred and twenty
h) Seven
i) Two hundred and seventeen
j) Nine hundred and two

2 Write the following numbers in words. Then write the value of each digit.
Example: 245 is **two hundred and forty-five**
245 has **2 hundreds, 4 tens, 5 units.**

	H T U		H T U		H T U		H T U		H T U
a)	2 3 6	b)	1 0 5	c)	6 1 7	d)	8 2	e)	7 5 0
f)	5 0 0	g)	1 2 3	h)	9 0	i)	9 3 8	j)	3 4 4

3 Copy this table.

Planet	Number of days to go round the sun	Planet	Approximate number of years to go round the sun
Earth		Jupiter	
Mercury		Neptune	
Mars		Saturn	
Venus		Uranus	

Read the information below.
Write the coloured numbers as figures in your table.

> It takes the **Earth** three hundred and sixty-five days to go round the sun.
>
> It takes **Mercury** just eighty-eight days, **Mars** six hundred and eighty-seven days and **Venus** nearly two hundred and twenty-five days.
>
> The other planets take a lot longer.
>
> **Jupiter** takes nearly twelve years.
>
> **Neptune** takes nearly one hundred and sixty-five years, **Saturn** nearly thirty years and **Uranus** eighty-four years.

4 List the planets in order according to how long they take to go round the sun. Start with the planet that takes the **least** time.

5 This table shows the times for five dwarf planets to go around the sun.

Ceres	Eris	Haumea	Makemake	Pluto
5 years	557 years	285 years	310 years	248 years

Write each time in words.

Activity

Pick three **different** digits. (Choose from 0, 1, 2, 3, 4, 5, 6, 7, 8, 9.)
See how many different numbers you can make from them.
Put the numbers you have made in order, starting with the **smallest**.
Try this again but with 2 digits the same and the other one different.
What happens if all of the digits are the same?
Write a rule for making the **largest** possible number with 3 digits.
Write a rule for making the **smallest** possible number with 3 digits.

b Count in 10s

If you add or subtract 10, the Tens digit changes. The Units digit doesn't change.

Adding:

9 +	19 +	119 +	Subtracting:	18 –	58 –	158 –
10	10	10		10	10	10
19	29	129		8	48	148

The Hundreds digit may change.
It changes if adding 10 takes you up into the hundred above.

91 +	99 +	191 +	399 +
10	10	10	10
101	109	201	409

It changes if subtracting 10 takes you down into the hundred below.

106 –	103 –	206 –	503 –
10	10	10	10
96	93	196	493

Practice

1 Copy each list. Fill in the missing numbers by adding 10.

a)
11	21	31		51	61	71	81		101	111		131	141

b)
441	451		471	481	491		511	521		541	551	561	

c)
383	393		413	423	433		453	463	473		493	503	

2 Copy each list. Fill in the missing numbers by subtracting 10.

a)
195	185	175		155	145	135		115	105	95		75	65

b)
596	586		566	556		536	526	516		496	486	476	

c)
139	129		109		89	79	69	59		39	29	19	

c Count in 100s

If you add or subtract 100, the Hundreds digit changes. The Tens and Units digits don't change.

Adding:

234 +	Subtracting:	586 –
100		100
334		486

Practice

1 Copy each list. Fill in the missing numbers by adding 100.

a)
25	125		325	425		625		825	

b)
63		263		463	563		763		963

2 Copy each list. Fill in the missing numbers by subtracting 100.

a)
982	882		682		482	382		182	82

b)
907		707		507	407		207	107	

d Odd and even numbers

An **even** number is any number that **you can divide exactly by 2**.
All the answers in the 2 times table are even numbers.
If you count up in 2s, all the numbers you get are even numbers.
All the other counting numbers are **odd** – they **cannot be divided exactly by 2**.

Practice

1 a) Copy and complete the following table.

$1 \times 2 =$		$6 \times 2 =$		$11 \times 2 =$		$16 \times 2 =$	
$2 \times 2 =$		$7 \times 2 =$		$12 \times 2 =$		$17 \times 2 =$	
$3 \times 2 =$		$8 \times 2 =$		$13 \times 2 =$		$18 \times 2 =$	
$4 \times 2 =$		$9 \times 2 =$		$14 \times 2 =$		$19 \times 2 =$	
$5 \times 2 =$		$10 \times 2 =$		$15 \times 2 =$		$20 \times 2 =$	

b) Complete the first sentence in the box.

> ▶ All **even numbers** end in _, _, _, _ or _.
> ▶ All **odd numbers** end in 1, 3, 5, 7 or 9.

2 List the numbers below that are **even**.

8	80	23	180	17	280	15	172	501	754
466	123	99	300	256	312	89	202	11	125

3 List the numbers below that are **odd**.

3	209	148	77	774	20	99	403	127	501
700	162	713	680	1000	14	19	471	345	176

4 Say whether each statement is *True*, *False* or *Maybe true*.
Give reasons for each answer.

a) If the houses in a street are numbered so **all the even numbers are on one side** and **all the odd numbers are on the other side**, numbers 32 and 180 will be on the **same** side of the street.

b) If there are 435 people working at a factory, there could be an equal number of men and women.

c) If 500 boxes need packing, the work can be shared equally between two people.

d) If a car park with 1000 spaces is on 2 levels, there are an equal number of spaces on each level.

Large numbers

N1/L1.1

a Place value

In 1851 the population of London was about two million, six hundred and eighty-five thousand.

pop 2 685 000

The number 2 685 000 has:

2	Millions	2 000 000
6	Hundred Thousands	600 000
8	Ten Thousands	80 000
5	Thousands	5 000
0	Hundreds	000
0	Tens	00
0	Units	0
		2 685 000

To read large numbers split the digits into groups of three (starting at the end).

Example   7 302 064 is:

| 7 | 302 | 064 |
| seven million, | three hundred and two thousand | and sixty-four |

and

3 020 500 is three million, twenty thousand, five hundred.

	Thousands					
M	H	T	U	H	T	U
7	3	0	2	0	6	4
3	0	2	0	5	0	0

Practice

1 Copy the headings onto squared paper.
Write these numbers as figures in the correct columns.

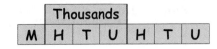

a) Two thousand

b) Thirty-two thousand

c) One hundred and sixty-five thousand, two hundred

    d) One hundred and nine thousand, eight hundred and seventy
    e) Four million, two hundred thousand
    f) Ninety thousand, seven hundred and fifty
    g) Eight thousand and twenty-five
    h) One hundred and nineteen thousand
    i) Six million, two hundred and ninety-nine thousand
    j) Nine million, three hundred and six

2 Write each number in words.

		Thousands					
	M	H	T	U	H	T	U
a)			4	5	0	0	0
c)	6	7	0	0	0	0	0
e)			7	0	7	5	0
g)			5	0	6	5	
i)			4	0	3	5	2
k)	5	0	9	0	4	0	0

		Thousands					
	M	H	T	U	H	T	U
b)		3	7	5	0	0	0
d)		2	0	5	0	0	0
f)	1	2	0	0	0	0	0
h)			2	0	3	0	6
j)	9	6	7	0	0	8	0
l)	3	5	0	3	2	4	1

3 List the numbers in question 2 in order of size, starting with the **smallest**.

4 a) Copy the table.
    b) Read the information below.
        Write the coloured numbers as
        figures in your table.

Year	Estimated population of London
1100	
1300	
1500	
1600	
1801	
1851	
1939	
2000	

In **1100** the population of London was about twenty-five thousand.
By the year **2000** it was about seven million, six hundred and forty thousand.
Between these two dates the population both increased and decreased.
By **1300** the population could have been as large as one hundred thousand.
Due to the black death it fell to about fifty thousand by **1500**.
Research shows that by **1600** the population had reached two hundred thousand.
The first census in **1801** recorded a population of one million, one hundred and seventeen thousand.
The population doubled by **1851** reaching two million, six hundred and eighty-five thousand.
At the beginning of the Second World War in **1939** London's population reached eight million, seven hundred thousand.

c) Use your table to answer the following questions.
- **i)** Between which years did the population fall?
- **ii)** In which year was the population the highest?

b Greater than > and less than <

Symbols can be used to show that one number is greater (larger) than or less (smaller) than another.

> means **greater than**   < means **less than**

5 is greater than 4.   This can be written as 5 > 4.
100 is greater than 50.   This can be written as 100 > 50.
6 is less than 9.   This can be written as 6 < 9.
25 is less than 50.   This can be written as 25 < 50.

> ▶ The larger number always goes at the open end of the symbol.

Practice

1 Rewrite each of these using < or >.
- **a)** 1 is less than 3
- **b)** 7 is greater than 5
- **c)** 9 is greater than 4
- **d)** 0 is less than 1

2 Copy these. Write < or > between each pair of numbers.
- **a)** 2   8
- **b)** 6   1
- **c)** 18   13
- **d)** 19   23
- **e)** 99   101
- **f)** 170   159

3 Rewrite each of these using < or >.
- **a)** 269 is less than 270
- **b)** 15 000 is greater than 2500
- **c)** 9900 is greater than 7500
- **d)** 50 000 is less than 75 000
- **e)** 9 is less than 900
- **f)** 10 500 is greater than 10 499
- **g)** 99 900 is less than 100 000
- **h)** 4 010 000 is greater than 4 009 999

Positive and negative numbers in practical contexts   N1/L1.2

Positive numbers are **greater than** 0.
Negative numbers are **less than** 0.

Negative numbers are sometimes used to describe temperature.
For example, −2°C is 2 degrees **colder** than 0°C.
Note that, −6°C is **colder** than −2°C.

Negative numbers are also used to describe overdrawn bank accounts.
For example, a balance of −£25 means that £25 is **owed** to the bank.

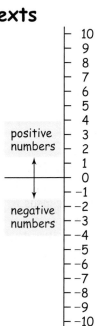

Practice

1 In each part, put the temperatures in order.
   Start with the **coldest**.

   a) −5°C   8°C   4°C   0°C   −2°C   1°C
   b) 10°C   −8°C   −10°C   9°C   −7°C   6°C
   c) 0°C   12°C   −15°C   20°C   −18°C   −10°C

2 A supermarket chain needs to store food at the following temperatures to meet health and
   safety regulations:

Chilled food	Between 0°C and 8°C
Frozen vegetables	−18°C or less
Ice cream	−20°C or less

   The temperatures below were recorded at different supermarkets.
   Write down the temperatures that do not meet health and safety regulations.

   a) Chilled food   5°C   9°C   7°C   −1°C   3°C   2°C
   b) Frozen vegetables   −20°C   −16°C   −19°C   −17°C   −21°C   −14°C
   c) Ice cream   −18°C   −21°C   −19°C   −23°C   −17°C   −24°C

3 Four bank accounts have these balances:

Account name	M Patel	J Robinson	R Dale	C Yeung
Balance (£)	255	−365	−50	75

   a) Which account has the most money in it?
   b) Which account is the most overdrawn?

4 An account holder has arranged an overdraft of £200 with his bank. If he owes any more
   than this he has to pay a fee. When the balance of the account is zero the account holder
   writes two cheques, one for £125 and one for £100.
   Will he have to pay a fee?

Compare numbers of any size   N1/L2.1

A **billion** is one thousand million.   In figures **1 billion = 1 000 000 000**
The distance from the earth to the sun is about 150 billion metres (150 000 000 000 m).

Four billion, five hundred and thirty million
is written like this:

	Millions			Thousands					
B	H	T	U	H	T	U	H	T	U
4	5	3	0	0	0	0	0	0	0

Very large numbers can be written neatly using decimals.

> **Examples**
>
> 1 500 000 (one million, five hundred thousand) is **1.5 million**.
> 3 450 000 000 (three billion, four hundred and fifty million) is **3.45 billion**.
> (Put the decimal point after the whole millions or billions.)

Practice

1 Copy the headings from the previous page. Write these numbers in full.

a) 2.5 million   b) 2.56 million   c) 1.7 billion

d) 1.85 billion   e) 30 million   f) 300 million

g) 1.3 billion   h) 0.5 billion   i) 0.25 million

2 Write these numbers in millions or billions, using decimals where necessary.

a) 2 500 000   b) 5 600 000   c) 4 750 000

d) 3 800 000 000   e) 2 950 000 000   f) 1 050 000 000

g) 20 000 000   h) 600 000 000   i) 34 500 000

3 The table gives the biggest lottery wins during the first four years it was played.
Write each amount in full and put them in order of size, starting with the **largest**.

Year	Largest win
1994	£17.9 million
1995	£22.6 million
1996	£14 million
1997	£14.1 million

4 A company's turnover (the amount of business it does) and profit (the amount of money it makes) are often large numbers.
A **negative profit** means the company made a **loss**.

Use the table to answer the questions.

a) Which company had:
  i)   the greatest turnover
  ii)   the smallest turnover
  iii) the greatest profit?

b) i)   Which companies made a loss?
  ii)   Which company made the greatest loss?

Company	Turnover £	Profit £
A	2.6 billion	1.5 million
B	12 billion	45.6 million
C	20 million	−99 000
D	45 million	60 000
E	8 billion	−1.5 million
F	1.2 billion	125 000
G	0.9 billion	−25 000

Number

2 Addition and Subtraction

Three-digit numbers

N1/E3.2, N1/E3.3

a Add in columns

Line up the numbers in columns to add Units to Units, Tens to Tens, etc.

Example	332 + 265

```
H T U
3 3 2 +
2 6 5
─────
5 9 7
```

▶ Words for adding:
   total, altogether, sum, plus

Add the Units, then the Tens, then the Hundreds.

If the Units total is 10 or more, carry 10 to the Tens column as 1 Ten to be added to the other Tens.

Example	336 + 45

```
H T U
3 3 6 +
  4 5
─────
3 8 1
  1
```

6 + 5 = 11
so 1 Ten is carried over to the Tens column.
Then 3 + 4 + 1 makes a total of 8 in the Tens column.

▶ It doesn't matter where you put the number you carry – some people write it at the top and some at the bottom. Take care to add it to the correct column.

If the total of the Tens column is 10 or more, carry 10 Tens to the Hundreds column as 1 Hundred.

Example	765 + 165

```
H T U
7 6 5 +
1 6 5
─────
9 3 0
1 1
```

5 + 5 = 10 so 1 Ten is carried over to the Tens column
Then 6 + 6 + 1 makes a total of 13 in the Tens column.
3 Tens go in the Tens column and 1 Hundred (10 Tens) is carried over to the Hundreds column and added to the 7 and 1.

Practice

1 Copy and complete these.

a) 234 +
145
_____

b) 530 +
249
_____

c) 403 +
_93
_____

d) 188 +
301
_____

e) 415 +
365
_____

f) 528 +
144
_____

g) 269 +
125
_____

h) 458 +
329
_____

i) 160 +
340
_____

j) 193 +
125
_____

k) 675 +
271
_____

l) 194 +
111
_____

m) 199 +
201
_____

n) 834 +
_97
_____

o) 363 +
258
_____

p) 255 +
355
_____

You can add more than 2 numbers – line them all up in columns.

| **Example** | 476 men, 195 women and 72 children attend a football match. How many people is this altogether? |

$$\begin{array}{r} 4\ 7\ 6\ + \\ 1\ 9\ 5 \\ 7\ 2 \\ \hline 7\ 4\ 3 \\ {}_{2\ \ 1} \end{array}$$

In the Tens column:
7 + 9 + 7 + 1 = 24
Put 4 in the Tens column
Carry the 2
Add 4 + 1 + 2 in the
Hundreds column

In the Units column:
6 + 5 + 2 = 13
Put 3 here
Carry 1 ten

Total
= **743 people**

---

2 Line up the Hundreds, Tens and Units in columns, then add the numbers.

a) 890 + 76 + 23   **b)** 99 + 109 + 264   **c)** 189 + 7 + 45

3 An insurance company has 258 full-time employees and 83 part-time employees. How many employees do they have altogether?

4 A secretary photocopies a letter for every child in a school. There are 208 boys and 232 girls. How many photocopies are needed?

5 A lorry travels 162 miles from Dover to Peterborough, then 188 miles from Peterborough to Liverpool. How far does the lorry travel altogether?

6 In a stocktake, shop staff count 42 boxes of disks on display and 440 in the stock room. How many boxes of disks are there altogether?

7 A customer buys a computer for £699, a scanner for £89 and a dust cover for £7. How much does the customer pay altogether?

8 A holiday company charges £879 for a flight and accommodation, plus a £9 booking fee and £38 for insurance. What is the total cost of the holiday?

9 A swimming pool sells 45 adult tickets, 162 child tickets and 8 infant tickets in an afternoon. How many tickets is this altogether?

> ▶ Now use a calculator to check your answers.

b Add in your head

Examples

$63 + 24$	Add the **Tens** and **Units** separately in your head. $6 + 2 = 8$, $3 + 4 = 7$ giving **87**

$253 + 145$ Add the Hundreds, Tens and Units separately.
$2 + 1 = 3$, $5 + 4 = 9$, $3 + 5 = 8$, giving **398**.

> ▶ Try different methods.

$71 + 27$ Count on in tens, then add the units:

71   81   91   **98**

  $+10$ $+10$ $+7$

$(27 = 10 + 10 + 7)$

$234 + 45$ 234 244 254 264 274 **279**

  $+10$ $+10$ $+10$ $+10$ $+5$

$(45 = 10 + 10 + 10 + 10 + 5)$

$156 + 45$ Break up the number to be added. Here break 45 into 40, 4 and 1:
$156 + 40 = 196$ $+4 = 200$ $+1 = $**201**

$199 + 76$ Round to a near number (here 200) then adjust the answer:
$200 + 76 = 276$ then subtract $1 = $**275**

Practice

Experiment with these.

1 a) $35 + 22$ **b)** $41 + 16$ **c)** $32 + 37$ **d)** $29 + 14$
   e) $57 + 15$ **f)** $38 + 25$ **g)** $49 + 36$ **h)** $19 + 79$

2 a) $216 + 25$ **b)** $159 + 22$ **c)** $199 + 199$ **d)** $154 + 233$
   e) $145 + 63$ **f)** $129 + 56$ **g)** $250 + 156$ **h)** $123 + 123$
   i) $206 + 156$ **j)** $149 + 152$ **k)** $312 + 256$ **l)** $209 + 329$
   m) $515 + 59$ **n)** $267 + 104$ **o)** $311 + 190$ **p)** $297 + 196$

c Subtract in columns

Line up the numbers so Units are subtracted from Units, Tens from Tens, etc.

Example	365 − 45

```
H T U
3 6 5 −
  4 5
─────
3 2 0
```

> ▸ **Words for subtracting:**
>   **take away, difference, minus**

Subtract the Units, then the Tens and then the Hundreds.

Sometimes changes are needed to make subtracting possible.

Method 1

Example	382 − 165

```
H T U
  7 1
3 8̸ 2 −
1 6 5
─────
2 1 7
```

Look at the Units – you cannot subtract 5 from 2.
So take one 10 from the Tens column and put it with the 2 to make 12 Units. This leaves 7 Tens.
(The 382 was changed into 3 Hundreds, 7 Tens and 12 Units.)

Example	512 − 291

```
H T U
4 1
5̸ 1 2 −
2 9 1
─────
2 2 1
```

Start with the Units.
Now look at the Tens – you cannot subtract 9 from 1
So take a Hundred and put it in the Tens column as 10 Tens.
You can now subtract the 9 Tens from 11 Tens.
(The 512 was changed into 4 Hundreds, 11 Tens and 2 Units.)

Example	306 − 159

```
H T U
2 9 1
3̸ 0̸ 6 −
1 5 9
─────
1 4 7
```

Start with the Units. You cannot subtract 9 from 6.

You cannot take a Ten because there aren't any!
Take a Hundred. Put it in the Tens column as 10 Tens.
Now take a Ten and put it with the 6 to make 16 Units.
This leaves 9 Tens.
(The 306 was changed into 2 Hundreds, 9 Tens and 16 Units.)

> ▸ **Some people use a different method called 'paying back'.**
> **If the 'paying back' method below looks more familiar, use it instead.**

Method 2 (Paying Back)

Example   382 – 165

```
H T U
    1 1
3 8 2 –
  7
1 6 5
2 1 7
```

Start with the **Units**. You cannot subtract 5 from 2.

Make it 12 and 'pay this back' by adding 1 to the **Tens** column of the number you are subtracting.
(This increases both numbers by 10, the difference stays the same.)

Example   512 – 291

```
H T U
  1
5 1 2 –
3
2 9 1
2 2 1
```

Start with the **Units**. Then the **Tens** – you cannot subtract 9 from 1. Make it 11 and 'pay this back' by adding 1 to the Hundreds column of the number you are subtracting.

Example   306 – 159

```
H T U
  1 1
3 0 6 –
2 6
1 5 9
1 4 7
```

Start with the **Units**. You cannot subtract 9 from 6.

Make it into 16 and 'pay this back' by adding 1 to the **Tens** column of the number you are subtracting, making it 6.
You cannot subtract 6 from 0. Make it 10 and 'pay this back' by adding 1 to the **Hundreds** column of the number you are subtracting.

Whichever way you subtract, you can check your answer by adding.

For example if   403 –   then   226 +
                  177   177
                  226   403

> Adding takes you back to the **number you started with**. Adding is called the **inverse** of subtracting.

Practice

1 Copy and complete these, then check the answers.

a) 465 –
    254
    ____

b) 927 –
    606
    ____

c) 763 –
    563
    ____

d) 670 –
    155
    ____

> Not all of these sums need changes.

e) 832 –
    523
    ____

f) 276 –
     49
    ____

g) 300 –
    230
    ____

h) 201 –
    170
    ____

i) 518 –
   365
   ___

j) 330 –
   131
   ___

k) 523 –
   344
   ___

l) 721 –
   654
   ___

m) 500 –
   255
   ___

n) 206 –
   137
   ___

o) 901 –
   602
   ___

p) 700 –
   199
   ___

q) 404 –
    75
   ___

r) 600 –
   301
   ___

▶ When subtracting remember to put the larger number at the top.

2 Line up the Hundreds, Tens and Units in columns, then do the subtractions.

a) 465 – 78   b) 503 – 406   c) 783 – 85

3 175 envelopes are used from a box of 500.
How many are left in the box?

4 A television costing £399 is reduced by £75 in a sale.
What is the sale price?

SALE

£399 reduced by £75

5 A company employs 412 men and 285 women.
How many more men than women does it employ?

6 A travel company charges £529 for a holiday in April
and £635 for the same holiday in July. What is the difference in price?

7 The journey from London to Edinburgh is 413 miles. I have already driven 250 miles of this journey. How far have I left to go?

d Subtract in your head

Examples

▶ Try different methods.

Start with the smaller number.

53 – 29   Count on to the next ten, then count on in tens:

29 + 1 = 30   + 20 = 50   + 3 = 53   Answer **24**

230 – 175   175 + 5 = 180   + 10 = 190   + 10 = 200   + 30 = 230   Answer **55**

Subtract a near number, then adjust:

71 – 19   71 – 20 = 51   + 1 = **52**

234 – 45   234 – 44 = 190   – 1 = **189**

199 – 172   Break the number up: 199 – 100 = 99   – 70 = 29   – 2 = **27**

Practice

Experiment with these.

1 a) 42 – 19   **b)** 54 – 38   **c)** 75 – 47   **d)** 81 – 25
   e) 56 – 37   **f)** 89 – 47   **g)** 90 – 53   **h)** 63 – 29

2 a) 200 – 51   **b)** 249 – 50   **c)** 183 – 34   **d)** 832 – 210
   e) 600 – 199   **f)** 350 – 145   **g)** 460 – 254   **h)** 321 – 122
   i) 459 – 133   **j)** 650 – 160   **k)** 679 – 522   **l)** 880 – 181
   m) 301 – 102 **n)** 479 – 180 **o)** 900 – 191 **p)** 786 – 492

e Use addition and subtraction

Try these questions. You will need to decide whether to add or subtract.
Look out for words like 'total' and 'altogether' – they usually mean that you need to add.
Words like 'difference' and 'how many more' usually mean that you need to take away.

1 Paul spends £56 on a jacket and £38 on trousers.
How much does he spend altogether?

2 A pencil costs 29p. A pen costs 82p.
What is the difference in price?

3 There are fifty-four people on a coach trip. Thirty-eight of these are children.
How many are adults?

4 An estate agent sells forty-three houses and eighteen flats in one month.
How many is this in total?

5 Meera earns £576 basic pay and £165 commission.
What is her total pay?

6 A laptop costing £430 is reduced by £45 in a sale.
What is the sale price?

7 Ahmed uses 148 sheets of paper from a pack of 500.
How many sheets of paper are left?

£45 off!
Usual Price
£430

8 A company employs 362 men and 178 women.
   a) Find the total number of employees.
   b) How many more men than women are there?

9 In one day a card shop sells 357 birthday cards, 109 anniversary cards and 216 other
cards. How many cards do they sell altogether?

10 Sally earns £175. She buys a dress for £39, a bag for £24 and
shoes for £35.
How much does she have left?

Now use a calculator to
check all your answers.

Add and subtract large numbers

N1/L1.3

With large numbers take even more care to line up the digits in columns.
Leave a space between the thousands and hundreds.

Examples

Add 12 535 and 8 036.

```
  1 2   5 3 5 +
      8   0 3 6
  2 0   5 7 1
      1       1
```

Subtract 1 659 from 75 000.

```
        4  9  9  9  1
  7  5   0  0  0 −
      1   6  5  9
  7  3   3  4  1
```

> You can check using the inverse operation (i.e. doing the opposite), see page 14, or by using a calculator.

Practice

1   Write the numbers in columns before adding or subtracting them.
    Check each answer.

a) 50 360 + 1 999   b) 2 652 + 36 853   c) 945 + 27 650
d) 136 500 + 23 999   e) 3 565 000 + 175 000   f) 6 500 000 + 27 500
g) 325 156 − 13 067   h) 250 000 − 37 525   i) 1 675 000 − 149 500
j) 902 856 − 7 580   k) 500 000 − 136 750   l) 301 000 − 4 650

2   It is 11 934 miles from London to Melbourne by air.
    How many miles will you travel if you fly there and back?

3   A couple earn £28 000 and £19 762.
    What is their joint income?

4   A live concert is also shown on the television. There are 12 000 people at the concert and
    1 320 000 people watch it on television. How many people watch the concert altogether?

5   The diagram shows the mileage of a car at the start and
    end of a journey. How far has the car travelled?

    Start   End
    4 8 5 7 6   5 0 3 1 0

6   In its first year of trading a company has a turnover of
    £78 600. In the second year its turnover is £1 265 000. By how much has the turnover
    increased?

7   A charity is trying to raise £1 000 000. It raises £45 650 from charity shop sales and
    £19 500 from donations.

    a) Altogether how much money has the charity raised so far?
    b) How much more does the charity need to raise?

8   In an election 659 208 people voted for Party A, 330 629 for Party B and 78 344 for
    Party C.

    a) How many people voted altogether?
    b) How many more people voted for Party A than either of the other parties?

Add and subtract numbers of any size

You may have to work with positive and negative numbers or very large numbers.

Example   Two national lottery prizes were not claimed in one month.
One prize was £700 000 and the other was £1.04 million.
What was the total?

Work in millions   or   write the numbers in full.
   0.7 +   700 000 +
   1.04   1 040 000
£1.74 million   1 740 000 = **£1.74 million**

Example   The table gives the temperature at midnight in some European cities.

How much warmer is it in London than in Berlin?

City	Temperature (°C)
Berlin	−7.2
Brussels	−2.5
Dublin	2.3
London	−1.8
Paris	3.1

The difference in temperature is

   7.2 −
   1.8
   5.4

London is 5.4°C warmer than Berlin.

┬ 0
├ −1.8 (London)
│ difference in temperature
┴ −7.2 (Berlin)

Example   In its first year of trading a company makes a profit of £50 000.
In its second year the company makes a profit of
−£25 000 (i.e. a loss of £25 000).
By how much has the profit fallen
in the second year?

The answer is **£75 000**.
(£50 000 down to 0 and then a further £25 000
down to −£25 000)

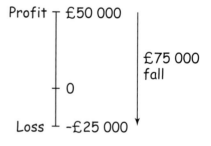

Profit ┬ £50 000
│
│ £75 000 fall
┼ 0
│
Loss ┴ −£25 000

Practice

1   Use the city temperatures given in the second Example on page 18.
    Find the difference in temperature between:
    a) London and Paris
    b) Berlin and Brussels
    c) Dublin and Brussels
    d) London and Brussels.

2   Calculate the following. You may need to write some numbers in full first.
    a) 3 million + 0.5 million
    b) 6.5 million − 3.9 million
    c) £1.2 million − £86 500
    d) £15 billion − £8.5 billion
    e) £2 billion + £250 million
    f) £3.2 billion − £790 million

3   In one national lottery game the prize for 6 balls was £2.35 million and the prize for
    5 balls and the bonus ball was £160 000.
    What is the difference between these prizes?

4   The table below shows the profits made by a company's outlets.

Outlet	Profit for year (£ millions)				
	2009	2010	2011	2012	Total
A	0.5	1.2	1.3	1.7	
B	0.9	0.1	−0.7	−1.1	
C	2.6	2.4	2.8	2.9	
D	−2.5	−1.7	−0.7	0.5	
Total	1.5	2	2.7	4	

Remember:
positive means **profit**,
negative means **loss**.

a) Calculate the total profit for each outlet.
    Check your answer by adding the totals for each year together. The result should agree
    with your total for the four outlets.
b) Use the table to answer the following questions.
    i)   What is the difference between Outlet A's profits in 2011 and 2012?
    ii) By how much did Outlet B's profits fall between 2010 and 2011?
    iii) By how much did Outlet D's profits increase between 2011 and 2012?
    iv) What is the difference between Outlet A and Outlet B's total profit for the
         4 years?
    v)   What is the difference between Outlet C and Outlet D's total profit for the
         4 years?
    vi) What is the difference between Outlet B and Outlet D's total profit for the
         4 years? Explain your answer.

Number

3 Multiplication and Division

Recall multiplication facts

N1/E3.5

Use multiplication facts that you know to help you work out others.

Example 3×9

Do you know the answer?
Does it help if you turn it round? 9×3
or you may know that $3 \times 10 = 30$
Take off one 3 $3 \times 9 = 30 - 3 = \mathbf{27}$

▶ Words for multiplying:
times, product, lots of

Example 6×6

Do you know the answer?
If not, you may know that $5 \times 6 = 6 \times 5 = 30$
Add an extra 6 $6 \times 6 = 30 + 6 = \mathbf{36}$

Practice

1 Answer the following.
   Use the tables you know to help you with any that you don't know.

   a) 3×5 b) 8×10 c) 7×2 d) 4×4

   e) 5×5 f) 6×3 g) 4×9 h) 6×7

   i) 9×9 j) 4×8 k) 6×8 l) 7×9

   m) 8×8 n) 4×7 o) 7×7 p) 9×6

   q) Each box holds six eggs.
      How many eggs are there in eight boxes?

   r) Jane saves £9 each week.
      How much does she save in four weeks?

   s) How many days are there in six weeks?

All the tables up to 10×10 are shown in this multiplication grid.

×	1	2	3	4	5	6	7	8	9	10
1	1	2	3	4	5	6	7	8	9	10
2	2	4	6	8	10	12	14	16	18	20
3	3	6	9	12	15	18	21	24	27	30
4	4	8	12	16	20	24	28	32	36	40
5	5	10	15	20	25	30	35	40	45	50
6	6	12	18	24	30	36	42	48	54	60
7	7	14	21	28	35	42	49	56	63	70
8	8	16	24	32	40	48	56	64	72	80
9	9	18	27	36	45	54	63	72	81	90
10	10	20	30	40	50	60	70	80	90	100

Look at the arrows.

They show that
$$3 \times 5 = 15$$

Use the grid to check all your answers to question 1.

The numbers in this row are called **multiples of 8**.

The numbers in this column are the **multiples of 3**.

2 **a) i)** What digits do the multiples of 2 end in?

ii) Which of the following do you think are multiples of 2?

23   30   50   75   86   99   162   204   318

b) i) What digits do the multiples of 5 end in?

ii) Which of the following do you think are multiples of 5?

65   79   80   100   115   123   125   203

c) What do you notice if you add the digits in each of the multiples of 3?
(e.g. $3, 6, 9, 1+2=3, 1+5=6, 1+8=9.$)

d) What do you notice about the multiples of 9?

Hint:
Look down column 3 in the grid.

You can use your tables to help you multiply larger numbers.

When you multiply by 10, the numbers move one column to the left.
This is because each column is worth 10 times more than the last.

H T U   H T U
$3 \times 10 =$ 3 0
$1\ 2 \times 10 =$ 1 2 0

When a whole number moves left,
it leaves a space in the units column.

Fill this with a 0.

Example   Calculate 4×30

Multiplying 4×30 is the same as multiplying $4 \times 3 \times 10$
$4 \times 3 = 12 \times 10 = \mathbf{120}$

3 Multiply these. Use the number grid to help you if necessary.

a) 3×20	**b)** 4×50	**c)** 6×60	**d)** 5×70
e) 3×90	**f)** 4×70	**g)** 6×90	**h)** 8×90
i) 8×80	**j)** 9×70	**k)** 8×60	**l)** 9×90

Multiplication methods

N1/E3.4

Example 21×3

You can break up the 21 into $20 + 1$, then multiply each of these by 3
There is more than one way to write this out.
Like this:

$$\begin{array}{c|c|c} \times & 20 & 1 \\ \hline 3 & 60 & 3 \end{array} = \textbf{63}$$

$3 \times 20 = 60$ and $3 \times 1 = 3$
Add the 60 and the $3 = 63$

Note: $3 \times 2 = 6$
so $3 \times 20 = 60$ (10 times as big)

or like this:

$$\begin{array}{r} 2\ 1\ \times \\ 3 \\ \hline \textbf{6\ 3} \end{array}$$

$3 \times 1 = 3$ (or $1 \times 3 = 3$)
Put 3 in the **Units** column
$3 \times 2 = 6$ (or $2 \times 3 = 6$)
Put 6 in the **Tens** column

Example 45×7

$$\begin{array}{c|c|c} \times & 40 & 5 \\ \hline 7 & 280 & 35 \end{array} = \textbf{315}$$

$7 \times 40 = 280$ and $7 \times 5 = 35$
Add the 280 and the $35 = 315$

or

$$\begin{array}{r} 4\ 5\ \times \\ 7 \\ \hline \textbf{3\ 1\ 5} \\ {\scriptstyle 3} \end{array}$$

$7 \times 5 = 35$ (or $5 \times 7 = 35$)
Put 5 in the **Units** column
Carry the 3
$7 \times 4 = 28$ (or $4 \times 7 = 28$)
Add the 3 making 31

It doesn't matter which method you use. Using boxes shows how the calculation works.
The other way does it all at once.

Use the method that is more familiar, or try both methods and see which one you prefer.

Practice

1 Multiply these. Use the multiplication grid to help you if necessary.

a) 24×2 b) 31×3 c) 44×2 d) 42×3

e) 35×3 f) 43×4 g) 65×4 h) 75×6

i) 39×6 j) 56×8 k) 76×4 l) 89×6

m) 88×7 n) 49×9 o) 97×5 p) 93×9

2 There are 12 curtain rings in each pack. How many are there in 5 packs?

3 There are 25 paper plates in each pack. How many are there in 3 packs?

4 A film lasts for 3 hours. How many minutes in this? (1 hour = 60 minutes)

5 A worker earns £9 an hour. How much does the worker earn for a 37 hour week?

6 A hotel has 7 floors. Each floor has 23 bedrooms.
   How many bedrooms does the hotel have altogether?

7 An office assistant makes 9 photocopies of a 42 page report.
   How many pages is this altogether?

8 A commuter travels 85 miles each day to get to work and back.
   Altogether how many miles does the commuter travel in 5 days?

9 The instructions on a packet of rice say,
   'Allow 75 grams per person.'
   How many grams do you need for 8 people?

> ❭ Now use a calculator to check your answers.

Division   N1/E3.6

a Different ways

You can think of division in different ways:

> ❭ Words for dividing:
> share, goes into

- the opposite of multiplication
- how many times a number goes into another
- how many times you can take one number away from another.

Example $12 \div 4$

'How many 4s are there in 12?'

$3 \times 4 = 12$ so $12 \div 4 = 3$ or   4 into 12 goes **3 times**

If 12 stamps are shared equally between 4 people, they each get **3**.

You can use times tables or the multiplication grid to calculate this.

To use a multiplication grid for $12 \div 4$
go down from the 4 until you reach 12,
then go across to find the answer **3**.

Look at the multiplication grid on page 21

You can also think of $12 \div 4$ as how many times can 4 be taken away from 12.
$12 - 4 = 8$ (once) $8 - 4 = 4$ (twice) $4 - 4 = 0$ (**three** times)

Example $13 \div 4$

'How many 4s are there in 13?'

$3 \times 4 = 12$ (not quite enough) and $4 \times 4 = 16$ (too much).
The closest you can get is **3 with one left over**.

You can write this as **3 r 1**.

So $13 \div 4 = 3$ **remainder 1**.

Sometimes the remainder can be broken down further.
For example, a bill of £13 shared between 4 people is £3 each and £1 left over.
The £1 can be broken down into 25p each.
However it is not always possible to do this. For example, if you are sharing a book of 13 stamps you cannot divide the remaining stamp.

Practice

1 Calculate these. (Some have remainders.)

a) $25 \div 5$	**b)** $50 \div 10$	**c)** $24 \div 6$	**d)** $28 \div 4$
e) $32 \div 8$	**f)** $45 \div 9$	**g)** $63 \div 7$	**h)** $49 \div 7$
i) $13 \div 2$	**j)** $42 \div 10$	**k)** $37 \div 6$	**l)** $28 \div 5$
m) $22 \div 3$	**n)** $51 \div 8$	**o)** $60 \div 9$	**p)** $59 \div 7$

2 A £20 taxi fare is shared equally between 4 people. How much does each person pay?

3 A swimmer gets £80 sponsorship money for swimming 10 lengths.
   How much was she sponsored per length?

4 A packet of 5 pencils costs 45p. How much is this per pencil?

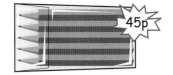

5 48 tins are arranged in 6 rows. How many tins are in each row?

6 A bag of 30 sweets is shared between 4 children. If all the children have to have the same number of sweets, what is the maximum number of sweets each child can have?

7 28 people want to go to a football match. Each car holds 5 people.
How many cars do they need?

8 How many teams can be made for 5-a-side football if there are 29 players?

9 40 people are invited to a wedding reception. Each table seats 6 people.
How many tables are needed?

10 An office orders 50 filing trays. They are shared equally between the 8 office clerks.
How many filing trays are left over?

b Standard method

When the number to be divided is too large, it is not possible to use times tables or a multiplication grid.
You can use a method that breaks the number into Tens and Units.

Example	$96 \div 3$

Re-write the sum $3\overline{)96}$ Divide 3 into 9 first: $9 \div 3 = 3$
Put the 3 above the 9

$\dfrac{3}{3\overline{)96}}$ Then divide 3 into 6: $6 \div 3 = 2$
Put the 2 above the 6

$\dfrac{32}{3\overline{)96}}$ $96 \div 3 = \mathbf{32}$

Sometimes the numbers do not divide exactly.

Example	$57 \div 2$

$2\overline{)57}$ $5 \div 2 = 2$ remainder 1
$\dfrac{2}{2\overline{)5^17}}$ Put the 2 above the 5 and the 1 in front of the 7.
The 1 is 1 Ten so this makes 17.
Divide 2 into 17,
$17 \div 2 = 8$ remainder 1.

$\dfrac{28 \text{ r}1}{2\overline{)5^17}}$ Put the 8 above the 17 then write down the remainder.
$57 \div 2 = \mathbf{28 \ r \ 1}$

Practice

1 Copy and complete these divisions.

a) $2\overline{)48}$ b) $3\overline{)99}$ c) $2\overline{)80}$ d) $5\overline{)60}$

e) $5\overline{)75}$ f) $4\overline{)92}$ g) $6\overline{)78}$ h) $4\overline{)76}$

i) $3\overline{)87}$ j) $3\overline{)88}$ k) $5\overline{)92}$ l) $7\overline{)86}$

m) $8\overline{)99}$ n) $4\overline{)70}$ o) $3\overline{)95}$ p) $6\overline{)89}$

2 How many 5 millilitre doses of medicine can be taken from a 90 millilitre bottle?

3 A doctor allows 5 minutes for each patient.
How many patients can the doctor see in 1 hour (60 minutes)?

4 Four people divide a £68 restaurant bill equally. How much does each person pay?

5 A child's school trip costs £75. The trip is paid for in 3 equal payments.
How much is each payment?

6 A 70 metre length of tape is cut into 5 equal pieces. How long is each piece?

7 75 plants are packed into trays of 6.

     a) How many trays are filled?
     b) How many plants are left over?

8 It is recommended that there is 1 adult to look after every
3 children under the age of 2 in a nursery. How many adults
are needed to look after 40 children under the age of 2?

9 A box of 50 pens is shared equally between 4 people.
How many pens are left over?

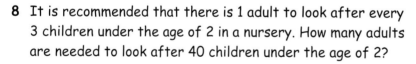

Multiplication and division are **inverse** operations. You can check the result of a multiplication by dividing. If the result of a division is exact, you can check it by multiplying.

In the first example in this section we found that $96 \div 3 = 32$.
To check this, work out 32×3. The answer is 96.
Note that if there is a remainder, you multiply then add the remainder.
The second example showed that $57 \div 2 = 28$ r 1.
The check for the second example is $28 \times 2 = 56 + 1 = 57$.

$$\begin{array}{r} 32 \\ 3\overline{)96} \end{array} \qquad \begin{array}{r} 32\times \\ 3 \\ \hline 96 \end{array}$$

$$\begin{array}{r} 28 \text{ r } 1 \\ 2\overline{)5^17} \end{array} \qquad \begin{array}{r} 28\times \\ 2 \\ \hline 56 \\ \hline 1 \end{array}$$

Check your answers to question **1** by multiplying.

Multiply and divide whole numbers by 10, 100 and 1000   N1/L1.4

a Multiply by 10, 100 and 1000 in your head

Multiplying (and dividing) by 10, 100 and 1000 in your head is useful when you need to estimate answers or work with large numbers.

Example 36×10

When you **multiply by 10** the numbers move **one column to the left**.

 See the example on page 21.

Th	H	T	U
		3	6

$\times 10 =$

Th	H	T	U
	3	6	0

$36 \times 10 = 360$

Multiplying by 100 is the same as multiplying by 10, then 10 again.
The numbers move **two columns to the left**. With a whole number this leaves two spaces to be filled by zeros.

Example 75×100

Th	H	T	U
		7	5

$\times 100 =$

Th	H	T	U
7	5	0	0

$75 \times 100 = 7500$

Multiplying by 1000 is the same as multiplying by 10, then 10, then 10 again.
The numbers move **three columns to the left**. With a whole number, this leaves three spaces to be filled by zero.

Example 540×1000

Thousands			H	T	U
H	T	U			
			5	4	0

$\times 1000 =$

Thousands			H	T	U
H	T	U			
5	4	0	0	0	0

$540 \times 1000 = 540\,000$

Practice

1 Calculate these in your head.

a) 13×10 b) 75×10 c) 136×10 d) 202×10

e) 90×10 f) 190×10 g) 12×100 h) 37×100

i) 129×100 j) 120×100 k) 303×100 l) 23×1000

m) 70×1000 n) 904×1000 o) 750×1000 p) 3581×1000

2 How many pence would you get for a £5 note? (£1 = 100p)

3 How many 10p coins would you get for a £20 note? (£1 = 10 × 10p)

4 There are 10 millimetres in 1 centimetre.
   How many millimetres are there in 66 centimetres?

5 There are 100 centimetres in one metre. How many are there in 15 metres?

6 There are 100 envelopes in one box. How many are there in 25 boxes?

7 A cashier has forty-five £10 notes in the till. How much is this?

8 There are 10 disks in one box. How many disks are there in 50 boxes?

9 100 people each donate £25 to charity. How much money is this altogether?

10 There are 1000 millimetres in 1 metre. How many are there in 12 metres?

11 There are 1000 leaflets in one pack. How many leaflets are there in 250 packs?

12 One kilogram equals 1000 grams. How many grams are equal to 78 kilograms?

b Divide by 10, 100 and 1000 in your head

When you divide by 10, 100 or 1000 the numbers move to the right.

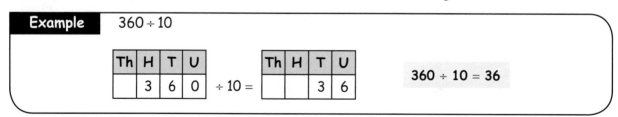

| Example | $360 \div 10$ |

Th	H	T	U
	3	6	0

÷ 10 =

Th	H	T	U
		3	6

$360 \div 10 = 36$

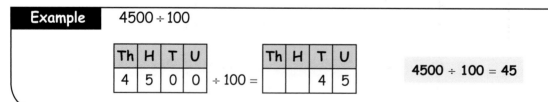

| Example | $4500 \div 100$ |

Th	H	T	U
4	5	0	0

÷ 100 =

Th	H	T	U
		4	5

$4500 \div 100 = 45$

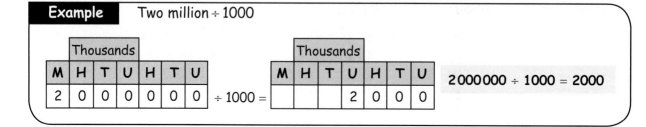

| Example | Two million ÷ 1000 |

	Thousands					
M	H	T	U	H	T	U
2	0	0	0	0	0	0

÷ 1000 =

		Thousands				
M	H	T	U	H	T	U
			2	0	0	0

$2\,000\,000 \div 1000 = 2000$

If the number does not end in zero, the answer is a decimal.

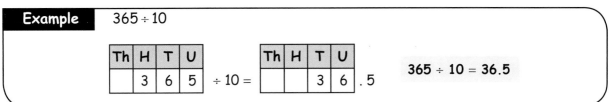

Example $365 \div 10$

Th	H	T	U
	3	6	5

$\div 10 =$

Th	H	T	U	.
		3	6	. 5

$365 \div 10 = 36.5$

Example $4523 \div 100$

Th	H	T	U
4	5	2	3

$\div 100 =$

Th	H	T	U	.
		4	5	. 23

$4523 \div 100 = 45.23$

Example $738 \div 1000$

H	T	U
7	3	8

$\div 1000 =$

H	T	U	.
		0	. 738

$738 \div 1000 = 0.738$

0 needed to emphasise the position of the decimal point

If you need an exact answer, give the decimal form (see page 115).
If you do not need an exact answer, you can round the number (see page 116).

Practice

1 Calculate these.

a) $460 \div 10$ b) $4600 \div 10$ c) $4600 \div 100$ d) $46\,000 \div 1000$

e) $530 \div 10$ f) $5300 \div 10$ g) $5300 \div 100$ h) $53\,000 \div 100$

i) $9900 \div 10$ j) $750 \div 10$ k) $57\,000 \div 1000$ l) $3800 \div 100$

m) $64\,000 \div 1000$ n) $2300 \div 1000$ o) $96\,200 \div 1000$ p) $275 \div 100$

2 A lottery win of £12 000 is shared between 100 people.
How much does each person get?

3 A printer charges £15 for 100 colour copies.
What is the cost per copy? (£1 = 100p)

4 There are 10 millimetres in 1 centimetre.
What is 720 millimetres in centimetres?

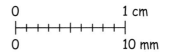

5 There are 100 centimetres in 1 metre.
   What is 24 000 centimetres in metres?

6 Paperclips are packed in boxes of 100.
   How many boxes would 275 000 paperclips fill?

7 Candles are sold in boxes of 10. How many boxes will 3500 candles fill?

8 100 workers in a call centre each answer the same number of calls.
   How many does each worker answer out of a total of 17 600 calls?

9 About one out of ten people are left-handed.
   How many people out of 150 are likely to be left-handed?

10 There are 1000 metres in 1 kilometre.
   a) What is 18 000 metres in kilometres?
   b) What is 3500 metres in kilometres?
   c) What is 675 metres in kilometres?

Use multiplication and division   N1/E3.5, N1/E3.6, N1/L1.5

Try to do these in your head.
In some questions you will need to decide whether to multiply or divide.

Practice

1 Copy and complete these by putting the missing number in each box.
   a) $4 \times 7 = \boxed{}$ b) $5 \times 6 = \boxed{}$ c) $5 \times \boxed{} = 50$
   d) $6 \times \boxed{} = 36$ e) $\boxed{} \times 8 = 32$ f) $9 \times 7 = \boxed{}$
   g) $9 \times \boxed{} = 81$ h) $7 \times \boxed{} = 42$ i) $\boxed{} \times 7 = 56$
   j) $\boxed{} \times 8 = 64$ k) $4 \times \boxed{} = 36$ l) $8 \times 6 = \boxed{}$

2 How many 2 pence coins make 20p?

3 There are 6 eggs in a box.
   How many boxes do you need to get 54 eggs?

4 A £72 bill is shared equally between 8 people.
   How much does each person pay?

5 One book costs £9. How much do 7 of these books cost?

6 One box holds 3 light bulbs. How many boxes do I need to buy to
   get 27 light bulbs?

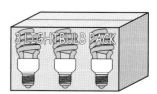

7 I save £8 each week. How many weeks will it take me to save £64?

8 I work 8 hours each day, for 5 days a week.
   How many hours is this altogether?

9 I earn £63 for 7 hours work. How much is this per hour?

Recognise numerical relationships   N1/L1.6

a Multiples

A **multiple** of a number can be divided **exactly** by that number.

Example	18 is a multiple of 2, 3, 6 and 9 because $2 \times 9 = 18$ and $3 \times 6 = 18$

Practice

1 a) Which of these are multiples of 4?   7   12   23   32   37   40
   b) Which of these are multiples of 7?   21   29   36   42   64   70
   c) Which of these are multiples of both 3 **and** 4?
       6   12   17   20   21   24
   d) Which of these are multiples of both 2 **and** 5?
       4   10   15   18   25   30

2 Write down all the multiples below 100 of:

   a) 6   b) 8   c) 9   d) 11   e) 20.

3 Write down all the multiples below 1000 of:

   a) 50   b) 200   c) 250.

4 Write down five multiples of 1000.

b Square numbers

A **square** number is made when you multiply any whole number by itself.

Example	25 is a square number, as $5 \times 5 = 25$

Practice

1   Copy out the multiplication grid on page 21. Shade the square numbers.

2   Which of these are square numbers?

    6   9   20   24   25   32   40   49   54   62   81   100

c Use mental arithmetic to multiply multiples of 10

This is useful for estimation.

Example	5×60

$5 \times 60 = 5 \times 6 \times 10$ ←   **Think of 60 as 6 × 10.**
$\quad\quad\quad = 30 \times 10 = \mathbf{300}$

Example	30×40

Think of 30 as 3 × 10.

$30 \times 40 = 3 \times 4 \times 10 \times 10$

Think of 40 as 4 × 10.

$3 \times 4 = 12 \quad\quad \times 10 = 120 \quad\quad \times 10 = 1200 \quad\quad$ So $30 \times 40 = \mathbf{1200}$

Practice

1   Work these out in your head.

    a) 3×30 **b)** 4×60 **c)** 7×20 **d)** 9×50
    e) 8×60 **f)** 3×90 **g)** 20×70 **h)** 40×50
    i) 60×70 **j)** 70×50 **k)** 60×60 **l)** 90×80

2   Five people pay £20 each for concert tickets. How much is this altogether?

3   A typist estimates there are 20 words in a line and 40 lines on a page.
    Approximately how many words are there on a page?

4   It takes 30 payments of £40 each to repay a loan.
    How much is paid altogether?

5   50 teams of 6 people take part in a quiz.
    How many people is this altogether?

6 About 80 people per day visit an exhibition.
Roughly how many people visit the exhibition in a month of 30 days?

7 A bag of potatoes weighs 30 kilograms.
What is the total weight of 90 bags?

8 30 coaches each take 50 people to a rugby match.
How many people is this?

9 How many seconds are there in an hour?
(1 minute = 60 seconds, 1 hour = 60 minutes)

Multiply and divide large numbers

N1/L1.3

a Multiply large numbers

Break down the numbers so that the **Tens** and **Units** of each number are multiplied together.
There are different ways of doing this. Use the method that seems most familiar or
straightforward to you.

Example 32×41

Grid method   This works out the **Tens** and **Units** separately.

×	30	2
40	1200	80
1	30	2

1280 +
  32

1312

Separate the **Tens** and **Units**.
Multiply 40×30 and 40×2.
Multiply 1×30 and 1×2.
Add the answers across then down.

Traditional method   This works out 32×1 then 32×40, then adds them.

```
    3 2 ×
    4 1
    ---
    3 2
  1 2 8 0
  -------
  1 3 1 2
```

Multiply $1 \times 2 = 2$ Put the 2 in the **Units** column.
Multiply $1 \times 3 = 3$ Put the 3 in the **Tens** column.
                                   (You have now multiplied 32×1.)
Put a zero in the **Units** first (because you are now multiplying tens).
Multiply $4 \times 2 = 8$ Put the 8 in the **Tens** column.
Multiply $4 \times 3 = 12$ Put this down on the left.
                                   (You have now multiplied 32×40.)
Add together the 32 and the 1280 = **1312**

If you prefer, you can start by multiplying 32×40.
The first line in the answer would be 1280 with the 32 underneath.

Example 245×63

Grid method

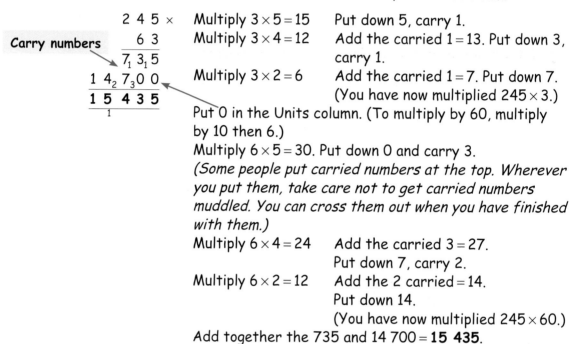

$\times$	200	40	5
60	12 000	2400	300
3	600	120	15

14 700 +
735
15 435
1

Separate the **Hundreds, Tens** and **Units**.
Multiply 60×200, 60×40 and 60×5.
Multiply 3×200, 3×40 and 3×5.
Add the answers across then down.

Traditional method   This works out 245×3 then 245×60, then adds them.

Carry numbers

2 4 5 $\times$
6 3
7₁ 3₁ 5
1 4₂ 7₃ 0 0
1 5 4 3 5
1

Multiply $3 \times 5 = 15$ Put down 5, carry 1.
Multiply $3 \times 4 = 12$ Add the carried $1 = 13$. Put down 3, carry 1.
Multiply $3 \times 2 = 6$ Add the carried $1 = 7$. Put down 7.
(You have now multiplied 245×3.)
Put 0 in the Units column. (To multiply by 60, multiply by 10 then 6.)
Multiply $6 \times 5 = 30$. Put down 0 and carry 3.
(Some people put carried numbers at the top. Wherever you put them, take care not to get carried numbers muddled. You can cross them out when you have finished with them.)
Multiply $6 \times 4 = 24$ Add the carried $3 = 27$. Put down 7, carry 2.
Multiply $6 \times 2 = 12$ Add the 2 carried $= 14$. Put down 14.
(You have now multiplied 245×60.)
Add together the 735 and 14 700 $=$ **15 435**.

If you find the grid method and traditional method difficult, try this last method:

Example 245×63

Lattice method

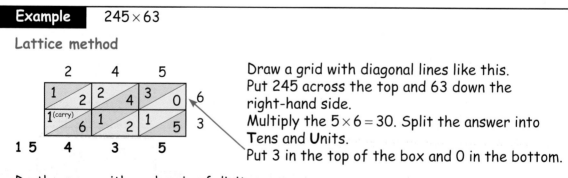

Draw a grid with diagonal lines like this.
Put 245 across the top and 63 down the right-hand side.
Multiply the $5 \times 6 = 30$. Split the answer into Tens and Units.
Put 3 in the top of the box and 0 in the bottom.

Do the same with each pair of digits,
e.g. $4 \times 6 = 24$ Put 2 in the top of the box and 4 in the bottom.
To find the answer you now need to add together the numbers in the diagonal lines.
Start with the bottom right hand corner.

$$
\begin{aligned}
5 &= 5 \\
0+1+2 &= 3 \\
3+4+1+6 &= 14 \text{ (carry the 1)} \\
2+2+1\text{(carried)} &= 5 \\
1 &= 1
\end{aligned}
$$

giving the answer **15 435**.

▸ Note: each method can be extended for any number of digits and decimal numbers.

Practice

1   Work these out without a calculator, then use a calculator to check.

a) 41×25	**b)** 60×32	**c)** 52×43	**d)** 45×62
e) 38×56	**f)** 29×43	**g)** 56×78	**h)** 57×44
i) 36×70	**j)** 45×67	**k)** 54×45	**l)** 63×49
m) 123×35	**n)** 632×47	**o)** 225×57	**p)** 342×80
q) 512×63	**r)** 309×42	**s)** 708×96	**t)** 573×48

2   A caterer charges £16 per head. How much is the charge for 45 people?

3   There are 48 tins in each tray. How many are there in 35 trays?

4   A coach company charges £27 a ticket for a trip. 42 people buy tickets.
    How much money is this altogether?

5   A school has 36 classes. Each class has 31 children.
    How many children are there at the school?

6   A concert hall has 85 rows of 48 seats. What is the total number of seats?

7   A lorry driver travels about 350 miles a day.
    Approximately how many miles does he travel in 20 days?

8   Jade makes a New Year's resolution to cycle 25 kilometres
    every day. If she keeps to her resolution, how many
    kilometres will she cycle in the year (365 days)?

9   Roughly how many days old is a man on his 85th birthday?
    Assume that there are 365 days in every year.

b Divide large numbers

You can use the standard method to divide large numbers as well as small numbers.

 See page 25.

Example $9396 \div 4$

$$
\begin{array}{r}
2\ 3\ 4\ 9 \\
4\overline{)9\ {}^1 3\ {}^1 9\ {}^3 6}
\end{array}
$$

$9 \div 4 = 2$ rem 1   Put 2 above the 9.
               Carry 1 to make the 3 into 13.
$13 \div 4 = 3$ rem 1   Put 3 above the 3.
               Carry 1 to make the 9 into 19.
$19 \div 4 = 4$ rem 3   Put the 4 above the 9.
               Carry 3 to make the 6 into 36.
$36 \div 4 = 9$ Put the 9 above the 6.

The answer for $9396 \div 4$ is **2349**.

Example $18\ 053 \div 9$

$$
\begin{array}{r}
2\ 0\ 0\ 5\ \text{r}\ 8 \\
9\overline{)1\ 8\ 0\ {}^1 5\ {}^5 3}
\end{array}
$$

cannot do $1 \div 9$ so carry the 1 to make the 8 into 18.
$18 \div 9 = 2$ Put the 2 above the 8.
$0 \div 9 = 0$ Put the 0 above the 0.
cannot do $5 \div 9$ so put 0 above the 5 and carry the 5 to make 3 into 53.
$53 \div 9 = 5$ rem 8   Put the 5 above the 3.
               Write down the remainder.

The answer for $18\ 053 \div 9$ is **2005 remainder 8**.

Practice

1 Work these out.

  a) $5910 \div 6$ **b)** $3240 \div 8$ **c)** $7929 \div 9$

  d) $14\ 608 \div 4$ **e)** $29\ 470 \div 7$ **f)** $34\ 125 \div 5$

  g) $14\ 508 \div 8$ **h)** $62\ 500 \div 6$ **i)** $25\ 127 \div 3$

  j) $48\ 950 \div 6$ **k)** $50\ 000 \div 9$ **l)** $80\ 225 \div 4$

2 A market research company employs 8 people to carry out 2000 interviews. How many people will they each interview if they share the 2000 equally?

3 A bill for £2025 is paid in 9 equal payments. How much is each payment?

4 Tickets for an outdoor event cost £6 each. Altogether £19 326 is taken in ticket sales. How many tickets are sold?

5 At a factory 34 000 ice lollies are packed into boxes of 8.
How many boxes are filled?

6 26 000 millilitres of perfume is packed as 5 millilitre samples.
How many samples are there?

7 Six people inherit £135 000 to be shared equally between them.
How much does each person inherit?

8 £250 000 is divided equally between 4 charities.
How much does each charity get?

It is more difficult to divide by a large number. You may need to try out estimates at each stage. Set the working out in either of the ways shown below.

Examples $648 \div 18$

$$
\begin{array}{r}
3\ 6 \\
{}^{10} \\
18\overline{)6\ 4\ 8}
\end{array}
$$

How many 18s in 64?
There are three 20s in 60 so 3 is a good guess.
$18 \times 3 = 54$
There are 10 left.

$$
\begin{array}{r}
3\ 6 \\
18\overline{)6\ 4\ 8} \\
5\ 4 \\
\hline
1\ 0\ 8 \\
1\ 0\ 8 \\
\hline
0
\end{array}
$$

Rough working

$$
\begin{array}{r}
18\ \times \\
3 \\
\hline
54 \\
{}_2
\end{array}
\qquad
\begin{array}{r}
18\ \times \\
6 \\
\hline
108 \\
{}_4
\end{array}
$$
Answer
$= 36$

Bring the 8 down – this is neater than squeezing the 10 that are left in front of the 8.
There are six 18s in 108.

---

9 Work these out. Check your answers on a calculator.

a) $882 \div 21$ b) $988 \div 19$ c) $675 \div 15$ d) $925 \div 25$

e) $377 \div 29$ f) $800 \div 32$ g) $1170 \div 18$ h) $2312 \div 17$

10 Calculate the monthly pay if the annual pay is: **1 year = 12 months so divide by 12.**

a) £24 480   b) £15 000   c) £35 160   d) £42 984.

Efficient methods of calculating with numbers of any size N1/L2.2

a Multiples, factors and prime numbers

A **multiple** of a number can be divided **exactly** by that number.
e.g.   6, 21 and 30 are all multiples of 3.

A **factor** of a number is something that **divides exactly into it**.
e.g.   2, 3, 4 and 6 are all factors of 12.

A **prime** number has **two factors** – itself and 1.
e.g.   2, 5, 7 and 11 are all prime numbers.

Note: 1 is **not** a prime number. It has only one factor – itself.

There are some quick ways of checking for factors and multiples.

To check whether 2 is a factor of a number, look at the last digit.
If it is **even**, then **2 is a factor** (i.e. the number is a multiple of 2).
So 2 is a factor of numbers like 74, 348, 74 256, etc.

To check whether 5 is a factor of a number, look at the last digit.
If it is **0 or 5**, then **5 is a factor** (i.e. the number is a multiple of 5).
So 5 is a factor of numbers like 70, 345, 74 250, etc.

To check whether 3 is a factor of a number, add all the digits.
If the total is **divisible by 3**, then **3 is a factor**
(i.e. the number is a multiple of 3).

Example	12 345

$1+2+3+4+5=15$ then $1+5=6$
6 is divisible by 3 so 3 is a factor of 12 345
(and 12 345 is a multiple of 3).

Keep adding the digits together.

This method also works for 9 (but not other numbers).

Example	13 860

$1+3+8+6+0=18$ then $1+8=9$
so 9 is a factor of 13 860 (and 13 860 is a multiple of 9).

Practice

1   Which of the numbers listed below are multiples of:

    a) 2   **b)** 5   **c)** 3   **d)** 9?

    98   127   165   243   364   539   720   945   1603   25 245

2 **a)** What is the smallest number that is a multiple of both 3 **and** 5?
   b) Find the smallest number that is a multiple of both 4 **and** 6.
   c) What is the largest number that is a factor of both 18 **and** 30?
   d) Find the largest number that is a factor of both 21 **and** 70.

3 **a)** Find the numbers that are factors of 12 and also multiples of 3.
   b) Find 2 factors of 18 that are also factors of 81.

4 **a)** Find a prime number greater than 3 that is a factor of 30.
   b) What is the next prime number after 23?

5 Find a number that is a factor of 15, 171 and 411.

b Calculations with numbers of any size

The standard methods for multiplying and dividing are given earlier. For large numbers it is often quicker and more accurate to use a calculator, but sometimes numbers are too large to fit on the calculator display. It is also useful to be able to do rough checks to make sure the calculator is giving the right answer. Try to use mental arithmetic whenever you can – everything gets easier with practice! Look out for short-cuts.

Example   A £1 500 000 lottery win is shared equally between 30 people. How much does each person get?

To divide by 30, divide by 10 then 3
$$1\,500\,000 \div 10 = 150\,000$$
$$15 \div 3 = 5 \qquad \text{So the answer is } \textbf{£50\,000.}$$

Example   Four charities each receive £125 000. How much is this in total?

$$2 \times 125 = 250 \quad 250 \times 2 = 500$$
so $4 \times 125 \qquad = 500$
and $4 \times £125\,000 = £500\,000$

To multiply by 4, double then double again.

Practice

Work these out. Use mental arithmetic when the numbers are easy enough.

1 How much is each payment if a debt of £3600 is paid off in:

   a) 12 equal payments   **b)** 40 equal payments
   c) 100 equal payments?

2   Each month a company sends out 3500 customer account statements.
    How many statements does it send out in a year (12 months)?

3   A mortgage is paid off at £5000 a year for 30 years.
    How much is paid in total?

4   A local paper sells an estimated 32 000 papers.
    On average four people read each paper.
    How many people read the paper altogether?

5   A company doubles its original advertising budget of
    £750 000.
    What is its budget now?

6   The total salary bill for 8 company directors is £2.4 million.
    Each director earns the same amount.
    What is each salary? (Give the answer in full.)

7   A builder sells 20 flats for £145 000 each.
    How much is this in total?

8   Approximately 1.3 million tourists visit a country each year.
    About how many tourists will visit the country during the next 5 years?

9   A company announces spending of £1.5 million over the next 4 years.
    How much is this per year?

10 The table shows the profits made by the three sections
    of a company. The total profit is shared equally between
    two partners.
    How much do they each receive?

Section	Profits
A	£2.46 million
B	−£3.28 million
C	£5.14 million

Number

4 Rounding and Estimation

Approximation and estimation (numbers up to 1000) N1/E3.7, N1/E3.8

a When to estimate

Estimation is used to work out an **approximate** answer.
This is useful if you want to:

- do a quick rough calculation
- check that an answer to a calculation is sensible
- find an approximate figure when you cannot calculate something exactly.

Estimation is **not** used to find an accurate answer.

Example   A shopper has 95p left in his purse.
Can he afford to buy 3 pencils costing 28p each?

28p is nearly 30p. $3 \times 30p = 90p$.
Yes, he can afford the 3 pencils.

Example   A shopper buys 3 items costing £49 each. The shop charges £196.
Is this approximately correct?

£49 is nearly £50. $3 \times £50 = £150$.
No, the shop has charged too much.

Example   A businessman needs to arrive at an airport **at least** 2 hours before his
flight departs at 11:30 am. He has to drive 50 miles to the airport.
What time should he leave home?

There is no single correct answer to this problem.
The businessman needs to use his knowledge and experience to work it out.
He might estimate that the journey will take about **2 hours** due to traffic.
He might also allow **another hour** to park and find the check in.
If so, he will estimate that he needs to leave home at 6:30 am to arrive at 9:30 am.

Practice

1 In which of these situations would you use estimation?

a) To work out how much food to buy for a party.
b) To calculate an employee's pay.
c) To work out how much spending money to take on holiday.
d) To measure a dose of medicine.
e) To measure a piece of glass to fit a frame.
f) To work out how far it is from London to Edinburgh.
g) To work out how long it will take to get to the cinema.
h) To count the number of children on a school trip.
i) To work out how much paint to buy to decorate a room.
j) To calculate how much change to give.

b Round to the nearest 10 and 100

Rounding numbers to the **nearest 10** gives whole numbers that end in **zero** (e.g. **10, 20, 30**). This is useful because numbers like 50 and 90 are much easier to work with than numbers like 49 or 93.

Examples   Round 74 to the nearest 10

70 71 72 73 **74** 75 76 77 78 79 80

74 lies between 70 and 80. It is nearer to 70, so it is rounded down to **70**.

Round 238 to the nearest 10.

230 231 232 233 234 235 236 237 **238** 239 240

238 lies between 230 and 240. It is nearer to 240, so it is rounded up to **240**.

Round 196 to the nearest 10.

190 191 192 193 194 195 **196** 197 198 199 200

196 is between 190 and 200. It is nearer to 200, so it is rounded up to **200**.

Round 485 to the nearest 10.

480 481 482 483 484 **485** 486 487 488 489 490

485 lies between 480 and 490. It is in the middle.
In maths we use the rule:
If it is in the middle, you should **round up**, in this case to **490**.

▶ **To round to the nearest 10:**
If the Units digit is **less than 5, round down** (e.g. 28**3** = 280 to the nearest 10)
If the Units digit is **5 or more, round up** (e.g. 28**7** = 290 to the nearest 10)
The answers always end in 0.

280 281 282 283 284 285 286 287 288 289 290

Practice

1 Round each number to the nearest 10.

a) 37	b) 15	c) 22	d) 92
e) 137	f) 215	g) 422	h) 992
i) 412	j) 99	k) 699	l) 313
m) 242	n) 175	o) 368	p) 707
q) 365	r) 811	s) 901	t) 275

Large numbers are often rounded to the nearest 100.
Deciding whether to round to the nearest 10 or 100 depends on the size of the number
and how accurate you want the answer to be.

Example   Round 370 to the nearest 100.

300 310 320 330 340 350 360 **370** 380 390 400

370 lies between 300 and 400. It is nearer to 400 so it is rounded up to **400**.

▶ Any whole number from 350 to 399 would be rounded up to 400.
So numbers like 351, 365, 382 and 399 are all 400 to the nearest 100.

Example   Round 720 to the nearest 100.

700 710 **720** 730 740 750 760 770 780 790 800

720 lies between 700 and 800. It is nearer to 700 so it is rounded down to **700**.

▶ Any whole number from 700 to 749 would be rounded down to 700.
So numbers like 706, 721, 736 and 749 are all 700 to the nearest 100.

Example   Round 550 to the nearest 100.

500 510 520 530 540 **550** 560 570 580 590 600

550 lies between 500 and 600. It is in the middle.
Remember the rule in maths is: **if it is in the middle, round up**, here to **600**.

> **To round to the nearest 100:**
> If the **Tens** digit is **less than 5, round down** (e.g. 427 = 400 to the nearest 100)
> If the **Tens** digit is **5 or more, round up**   (e.g. 483 = 500 to the nearest 100)
> The answers always end in two 0s.
>
> 400 410 420 430 440 450 460 470 480 490 500
>
>   427   483

2 Round each number to the nearest 100.

a) 170	**b)** 350	**c)** 620	**d)** 880
e) 175	**f)** 357	**g)** 629	**h)** 851
i) 383	**j)** 199	**k)** 750	**l)** 749
m) 851	**n)** 999	**o)** 235	**p)** 428
q) 857	**r)** 810	**s)** 250	**t)** 59

c Use estimates to check calculations

You can check whether the answer to a calculation is sensible by rounding the numbers and estimating the answer.

Example   A student calculates that if one CD costs £18 then 3 will cost £34.

A rough check shows this cannot be correct:
£18 rounded to the nearest £10 is £20. 3 × £20 = 60. Therefore the student's answer is not sensible. (The correct answer is £54.)

Example   A saver has £912 in her bank account. She thinks that if she takes out £285 there will be £773 left in the account.

A rough check shows this cannot be correct:
£912 rounded to the nearest £100 is £900 and £285 rounded to the nearest £100 is £300.
£900 − £300 = £600. (The correct answer is £627.)

Practice

1   Use an estimate to check whether each statement is correct.

Menswear – Price list

Jumpers	£25	Jackets	£67
Shirts	£20	Coats	£85
Ties	£8	Suits	£159
Trousers	£29		

Note:
Round to numbers you
can work out easily.

a) The total cost of a jacket and trousers is £136.
b) A suit costs £92 more than a jacket.
c) A jacket and two pairs of trousers cost £125 altogether.
d) A shopper spends £87 on trousers. He has bought 2 pairs.
e) Coats are reduced by £27 in a sale. The sale price is £58.
f) A shopper buys 9 jumpers. The total cost is £225. (Hint: Work out £25 × 10)
g) A man buys 3 suits. Altogether they cost £577. (Hint: Work out £160 × 3)
h) Last year suits cost £115. The price has increased by £64.

2   The table shows the number of pairs of gloves of
different sizes sold in a shop.

Size	Number sold
Small	191
Medium	412
Large	286

a) Find the total number sold.
b) How many more medium than large pairs were sold?

Check each answer by rounding the numbers to the nearest hundred.

3   Use the information in the table below to answer the questions.
In each part you will need to carry out a series of calculations.
Use estimates to check each calculation you do.

	Cost per person, per night £	English breakfast per person £	Continental breakfast per person £
		Hotel – Price List	
Adult	55	8	6
Child	38	5	3

Find the cost for:

a) 2 adults and 1 child, staying 2 nights with English breakfast
    (*Note: Work out the cost for 1 night with breakfast, then double it*)
b) 1 adult and 3 children, staying 4 nights with Continental breakfast
c) 2 adults and 4 children, staying 3 nights with Continental breakfast.

Approximate and estimate with large numbers   N1/L1.8, N1/L1.9

a Round to the nearest 1000, 10 000, 100 000, 1 000 000

The method for rounding larger numbers is similar to that for smaller numbers.
For example:

To round a number to the nearest thousand:

- Identify which digit is in the **Thousands** column.

- Look at the **Hundreds** digit to the right – this is the 'deciding digit'.
  If it is **below 5, round down** so the Thousands digit stays the same.

- If the Hundreds digit is **5 or above, round up** so the Thousands digit increases by one.

- After the thousands digit write 000 (because a number rounded to the nearest **Thousand** will have no Hundreds, Tens or Units).

Examples

a) **To round 3 257 432 to the nearest 1000**

Thousand digit   Deciding digit

3 25<u>7</u> 432

4 is **less than 5**, so **round down**
The thousands digit, 7, stays the same.

3 257 432 to the nearest thousand is **3 257 000**.

To the nearest **Thousand** means there are no Hundreds, Tens or Units.
The number must end in 000.

b) **To round 3 257 432 to the nearest 10 000**

Ten Thousands digit   Deciding digit

3 2<u>5</u>7 432

7 is **more than 5**, so **round up**
The ten thousands digit increases by 1.

3 257 432 to the nearest ten thousand is **3 260 000**.

To the nearest **Ten Thousand** means there are no Thousands, Hundreds, Tens or Units.
The number must end in 0000.

c) **To round 3 257 432 to the nearest 100 000**

Hundred Thousands digit   Deciding digit

3 <u>2</u>57 432

When the deciding digit is 5, **round up**.
The hundred thousands digit increases by 1.

3 257 432 to the nearest hundred thousand is **3 300 000**.

A number rounded to the nearest Hundred Thousand ends in 00 000.

d) To round 3 257 432 to the nearest 1 000 000

Millions digit   Deciding digit

3 256 432   2 is **less than 5**, so **round down**.

3 256 432 to the nearest million is **3 000 000**.

A number rounded to the nearest **Million** ends in 000 000.

Practice

1 Round each number to the nearest 1000.

a) 4562   b) 6500   c) 27 421   d) 30 950
e) 255 756   f) 105 220   g) 951 021   h) 1 365 500

2 Round each number to the nearest 10 000.

a) 29 000   b) 52 000   c) 145 000   d) 209 000
e) 765 900   f) 231 900   g) 684 625   h) 1 909 000

3 Round each number to the nearest 100 000.

a) 480 000   b) 320 000   c) 1 450 000   d) 4 608 000
e) 1 659 000 f) 2 783 540 g) 1 999 000   h) 12 832 359

4 Round the following numbers to the nearest 1 000 000.

a) 2 990 000   b) 3 050 000   c) 6 500 000
d) 15 650 000   e) 4 459 999   f) 1 755 000

Newspapers often use numbers that have been rounded.
They are easier to read and have more impact.

Examples

24 989 fans attend football match!

or

25 000 fans attend football match!

The second headline is more likely to grab the reader's attention.

5 Round the numbers in the following headlines.

a)
```
2 987 000 unemployed
```

b)
```
THE NEWS
Company announces 7018 job losses
```

c)
```
12 756 people attend charity event
```

d)
```
£1 987 649 lottery win
```

e)
```
4721 new jobs created
```

f)
```
3 005 021 copies of record sold
```

g)
```
£21 095 000 profits for company
```

h)
```
278 650 people to benefit from new laser treatment
```

Activity

See how many numbers you can find in newspapers that have been rounded.

---

b Use estimates to check calculations

You can use an estimate to check whether the answer to a calculation is about the right size (sometimes said to be 'of **the correct order**').

Example

	Year 1	Year 2
Company profits	£145 270	£277 650

The chairman of the company says,
'In the second year company profits rose by over £132 000.'
Is this likely to be true?

To the nearest £10 000:   £145 270 is approximately £150 000
                           £277 600 is approximately £280 000

$280 - 150 = 130$ so £280 000 − £150 000 = £130 000
The answer is about the right size and the statement is **likely to be true**.

Practice

1   The table below gives the number of employees in the four main sectors of industry in the North East and South East regions of England in one year.

	Number of people employed			
	Manufacturing	Wholesale and retail	Real estate	Health and social work
North East	175 489	167 224	94 564	131 628
South East	491 800	656 808	621 068	390 613

Source: Census 2001 Key Statistics for Health Areas in England and Wales

Use estimates to decide whether each statement is true or false.

a) Approximately 300 000 more people were employed in manufacturing in the South East than the North East.   True or False?

b) Approximately twice as many people were employed in real estate in the South East as in the North East.   True or False?

c) Nearly three times as many people were employed in health and social work in the South East as in the North East.   True or False?

d) Nearly 200 000 more people were employed in wholesale and retail than real estate in the North East.   True or False?

e) Roughly 35 000 more people were employed in wholesale and retail than real estate in the South East.   True or False?

f) Over 150 000 more people were employed in wholesale and retail than manufacturing in the South East.   True or False?

g) Approximately 20 000 more people were employed in manufacturing than wholesale and retail in the North East.   True or False?

You will need to carry out a series of calculations with large numbers to solve the following problem. Use estimation to check your answers at each stage.

---

2   The money taken at a charity concert is listed below in the income table.
    The expenditure table shows the costs of putting on the concert.

Income	Total £
2976 tickets @ £25 each	
2204 tickets @ £12 each	
1008 programmes @ £3 each	
592 souvenir mugs @ £8 each	
Total	

Expenditure	Total £
Hire of venue and insurance	17 750
Publicity and printing	4899
8 technicians @ £489 each	
50 security staff @ £175 each	
Performers (donated time)	0
Total	

a) Copy and complete the tables to find the total income and expenditure.

b) Calculate the difference between the total income and expenditure to find out how much money was raised.

Use estimates to check calculations with numbers of any size

N1/L2.6

Use estimates to check whether answers are of an appropriate size, i.e. 'of the correct order' even when using a calculator.

Example   The table gives UK cinema admissions (no. of tickets sold). How many more admissions were there in 2010 than 2000?

2000	142.5 million
2010	169.2 million

$169.2 - 142.5 = 26.7$ million $= \mathbf{26\,700\,000}$

$$\begin{array}{r} 169.2 - \\ 142.5 \\ \hline 26.7 \end{array}$$

Check 170 million $-$ 140 million $=$ 30 million
This agrees as 26 700 000 rounded to the nearest ten million is 30 million.

Practice

1   The table gives the number of people who visited cinemas in each year shown.

   a) Use a calculator to find the increase or decrease in cinema admissions between consecutive years shown in the table.
   Give your answers in full.

   b) Use estimation to check that your answers are of the right order.

Cinema admissions (millions)	
1935	912.3
1945	1585.0
1955	1181.8
1965	326.8
1975	116.3
1985	72.0
1995	114.6
2005	164.6

Source: www.cinemauk.org.uk

2   The table gives the wins in the national lottery.

   a) Use a calculator to find the total prize money.

   b) Use estimates to check your answer to **a)**.

Prize	No. of winners
£1 194 994	3
£55 153	20
£1 156	596
£45	33 480
£10	579 471

Number

5 Ratio and Proportion

Work with ratio and proportion

N1/L1.7

a Ratios

When you dilute drinks or chemicals or follow a recipe you need to mix ingredients in the correct **ratio** to get the right result. The ratio is often given in terms of 'parts'. The size of a 'part' depends on how much of the mixture you need.

Example The instructions on a bottle of plant food say,
'Dilute 1 part plant food to 3 parts water'.

This means you need to use **3 times as much water as plant food** to get the correct strength.
You can mix up

a large amount

| 1 part plant food | 3 parts water (1 × 3) | 4 parts of diluted plant food (1 × 4) |

or

a small amount

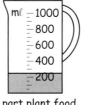

| 1 part plant food (250 mℓ) | 3 parts water (250 × 3 = 750 mℓ) | 4 parts of diluted plant food (250 × 4 = 1000 mℓ) |

depending on how much you need.

As long as the ratio of plant food to water stays the same the mixture will have the same strength or dilution.

Example Instructions for mixing pink paint say, 'Mix 1 part red with 4 parts white.'

a) How much white paint should you mix with 500 mℓ of red paint?
b) How much red paint should you mix with 500 mℓ of white paint?

a) Red paint   1 part = 500 mℓ
    White paint   4 parts = 500 × 4 = **2000 mℓ** or **2 litres**   1 litre = 1000 mℓ

b) White paint   4 parts = 500 mℓ
    Red paint   1 part = 500 ÷ 4 = **125 mℓ**

Practice

1 a) How much water should you mix with 50 mℓ of orange squash?
   b) How much water should you mix with 100 mℓ of orange squash?
   c) If you mix the correct amount of water with 200 mℓ of
      orange squash, how much diluted squash will you have?

Dilute 1
part orange squash
with 4 parts water

2 a) How much water should you mix with 250 mℓ of shampoo?
   b) How much water should you mix with 500 mℓ of shampoo?
   c) If the correct amount of water is mixed with 800 mℓ of
      shampoo, will it fit into a 5 litre (5000 mℓ) bucket?

Mix one part
shampoo with six
parts warm water

3

For **standard** mortar
mix 1 part cement
with 5 parts sand.
For **strong** mortar
mix 1 part cement
with 3 parts sand

		Cement	Sand
Standard mortar		5 kg	
			50 kg
		15 kg	
Strong mortar		3 kg	
		12 kg	
			60 kg

Hint:
For sand **multiply** by 5 (or 3)
For cement **divide** by 5 (or 3)

Copy and complete the table by giving the missing quantities for sand and cement.

b Direct proportion

Sometimes instructions give the amount of each ingredient rather than using 'parts'. If you
need more (or less) of the mixture then you need to increase (or decrease) the quantity of
each ingredient **in proportion**.

For example: to **double** the amount, **multiply everything by 2**
                to make **five** times as much, **multiply everything by 5**
                to make **half** the amount, **divide everything by 2**
                to make a **third** of the amount, **divide everything by 3**.

Example

Mushroom Soup (serves 4)
225 g mushrooms   1 onion   1 potato
710 mℓ milk   50 g butter   5 mℓ thyme
sprig of parsley   salt & pepper to taste

How much of each ingredient do you need to make mushroom
soup for 8 people?

For twice as many people, you need double the amount of soup.
You need to double the amount of each ingredient.

450 g mushrooms   **2 onions**   **2 potatoes**   **1420 mℓ milk**
100 g butter   **10 mℓ thyme**   **2 sprigs of parsley** **salt & pepper to taste**

Example   The instructions on a bottle of floor cleaner say,
'Pour 2 capfuls (30 mℓ) into a bucket with 6 litres of water.'

a) How much floor cleaner should you use with 3 litres of water?
b) How much floor cleaner should you use with 1 litre of water?

a) The amount of water has been halved (divided by 2), so the amount of floor
cleaner should be halved. This means you need **1 capful (15 mℓ)** of floor cleaner.

b) This is a third of the amount of water in part a.
You should use a third of the amount of floor cleaner. $15 \div 3 = 5$
This means you need **5 mℓ (a third of a capful)** of floor cleaner.

Practice

1 a) Rewrite the naan bread recipe to make:
     i)   12 breads
     ii) 3 breads.

   b) How much milk, flour and
      yogurt do you need to make 2
      breads?

Naan Bread Ingredients	
150 mℓ milk	2 teaspoons caster sugar
450 g plain flour	2 teaspoons dried yeast
150 mℓ plain yogurt	½ teaspoon salt
1 egg	1 teaspoon baking powder
	2 tablespoons vegetable oil

Makes 6 breads

2 Look at the instructions on the
   jar of gravy granules.
   How many teaspoons of gravy granules
   do you need to make:

   a) 560 mℓ (1 pint) of gravy

   b) 140 mℓ ($\frac{1}{4}$ pint) of gravy

   c) 70 mℓ ($\frac{1}{8}$ pint) of gravy?

Gravy granules
Mix 4 teaspoons of
granules with 280 mℓ
($\frac{1}{2}$ pint) of water

3 Look at the instructions on the box of plaster.
   a) How much water should you mix with these amounts of plaster?
      i)   1250 g
      ii) 500 g   Note: 1 litre = 1000 mℓ
      iii) 250 g

   b) How much water and plaster should you mix together to cover
      i)   2 square metres
      ii) 4 square metres?

1 litre of water
mixed with 2500 g
of plaster will
will cover 1 square
metre

Calculate with ratios

a Simplify ratios and find amounts

The ratio '1 part plant food to 3 parts water' can be written as 1 : 3.

This form is useful for simplifying ratios with large numbers and units of measurement.

> **To simplify a ratio, divide both sides by the same number.**
> This is like simplifying fractions – see pages 92-93.

Suppose a recipe for pastry uses 100 g of butter and 200 g flour.

The ratio of butter to flour is 100 : 200.

In this case, you can divide by 100

(or by 10 then 10 again, or by 5 then 20).

The ratio of butter to flour is 1 : 2.

This means there is twice as much flour as butter.

When the units are the same you do not need to write them in the ratio.

1 : 2

This ratio can be applied to any quantity.
The table shows some possibilities.

Butter	Flour
50 g	100 g
75 g	150 g
200 g	400 g

Example

A builder mixes 150 kg of sand with 50 kg of cement to make concrete.

a) Write the ratio of sand to cement in its simplest form.

b) The builder needs more concrete.
   (i) How much sand should he mix with 600 kg of cement?
   (ii) How much cement should he mix with 750 kg of sand?

a)   The ratio of sand to cement is 150 : 50.

To simplify this, divide each number by 50 (or divide by 10, then by 5).

The ratio of sand to cement is 3 : 1.

b)   Using the ratio of sand to cement 3 : 1
   (i) If 1 part cement = 600 kg,
        then 3 parts sand = 600 × 3 = 1800 kg
   (ii) If 3 parts sand = 750 kg,
        then 1 part cement = 750 ÷ 3 = 250 kg

Note:
This means the builder uses 3 times as much sand as cement.
The order is very important.
1 : 3 would mean 3 times as much cement as sand.

Sometimes ratios are more difficult.

| Example | The recipe for a children's cocktail says, 'Mix 75 mℓ of strawberry syrup with 100 mℓ of mango juice and 150 mℓ of apple juice.' |

a) Write this as a ratio in its simplest form.
b) How much mango juice and apple juice should you mix with 120 mℓ of strawberry syrup?

strawberry   mango   apple

a) The ratio of the ingredients is   75 : 100 : 150.
   To simplify this, divide each number by 25 (or 5 then 5 again).
   The ratio of strawberry syrup to mango juice to apple juice is   3 : 4 : 6.

b) Using the ratio 3 : 4 : 6   if 3 parts strawberry syrup = 120 mℓ
                                     then 1 part = 120 ÷ 3 = 40 mℓ
                                     4 parts mango juice = 4 × 40 = **160 mℓ**
                                     6 parts apple juice = 6 × 40 = **240 mℓ**

> **Find one part first, then multiply to find each amount.**

Practice

1 Simplify the ratio in each statement.
   a) The ratio of men to women in a sports club is 80 : 40.
   b) The ratio of management to staff in a factory is 30 : 150.
   c) In a batch of light bulbs the ratio of good to faulty is 7000 : 500.
   d) The ratio of adults to children is 15 : 75.
   e) The ratio of home to away supporters at a match is 24 000 : 8000.

Note:
Keep dividing
until you have the
smallest possible
numbers.

2 A painter mixes 2 litres of red paint with 8 litres of yellow paint to make orange.
   a) Write the ratio of red paint to yellow paint in its simplest form.
   b) The painter needs more of the same shade of orange paint.
      i)   How much yellow paint should he mix with 3 litres of red?
      ii) How much red paint should he mix with 400 millilitres of yellow?

3 A gardener makes potting compost by mixing 100 litres of soil with 20 litres of peat.
   a) Write the ratio of soil to peat in its simplest form.
   b) The gardener needs more compost.
      i)   How much soil should he mix with 50 litres of peat?
      ii) How much peat should he mix with 300 litres of soil?

4 Simplify the ratio in each statement.
   a) The ratio of men to women at a party is 20 : 30.
   b) The ratio of helpers to children is 12 : 54.
   c) The ratio of games won, drawn and lost is 12 : 3 : 18.
   d) The ratio of A, B and C exam grades is 30 : 45 : 75.
   e) The ratio of votes for 3 candidates is 132 000 : 72 000 : 40 000.

5 a) Look at the instructions for coffee machines.
    i) Write the ratio of descaler to water in its simplest form.
    ii) How much water should you mix with 100 mℓ of descaler?

  b) Look at the instructions for kettles.
    i) Write the ratio of descaler to water in its simplest form.
    ii) How much water should you mix with 400 mℓ of descaler?

> For coffee machines mix 125 mℓ of descaler with 500 mℓ of water.
>
> For kettles mix 500 mℓ of descaler with 750 mℓ of water.

6 a) Use the information for velvet green paint.
    i) Write the quantities of white, moss and leaf green paint as a ratio in its simplest form.
    ii) How much moss and leaf green paint should you mix with 80 mℓ of white paint?
    iii) How much white and leaf green paint should you mix with 1800 mℓ of moss paint?

  b) Use the information for sepia tint paint.
    i) Write the quantities of white, beige and chocolate brown paint as a ratio in its simplest form.
    ii) How much beige and chocolate brown paint should you mix with 80 mℓ of white paint?
    iii) How much white and chocolate brown paint should you mix with 180 mℓ of beige paint?
    iv) How much white and beige paint should you mix with 750 mℓ of chocolate brown paint?

Mix and Match Paints
Velvet Green
To make 1 litre (1000 mℓ) mix:
50 mℓ White
450 mℓ Moss
500 mℓ Leaf Green
Sepia Tint
To make 1 litre (1000 mℓ) mix:
200 mℓ White
300 mℓ Beige
500 mℓ Chocolate Brown

> ▶ Quantities must be in the **same units** before you can simplify the ratio.

Example   The instructions on a bottle of cleaning fluid say, 'Mix 50 mℓ of cleaning fluid with 1 litre of water.'

        1 litre = 1000 mℓ

How much water should you mix with 80 mℓ of cleaning fluid?

The ratio of cleaning fluid to water is 50 mℓ : 1 litre.
It is easier to convert the larger units into the smaller ones.

> **See page 212 for metric units.**

The ratio of cleaning fluid to water = 50 : 1000 = 1 : 20.

> **Divide both sides by 50, or by 10 then 5.**

The amount of cleaning fluid = 1 part = 80 mℓ.
The amount of water = 20 parts = 80 mℓ × 20 = **1 600 mℓ (or 1.6 litres).**

Example | The scale of a map is 4 cm : 1 km.

Write this as a ratio in the form 1 : *n*.

We need to convert 1 km to cm.
1 km = 1000 m and 1 m = 100 cm so 1 km = 1000 × 100 = 100 000 cm
The ratio = 4 : 100 000
Divide both sides by 4:   Ratio = **1 : 25 000**

7 Write each ratio in its simplest form.
   a) cream : milk = 40 mℓ : 1 litre
   b) biscuits : butter = 1 kg : 250 g
   c) gate : fence = 90 cm : 18 m
   d) present : postage = £12 : £1.50
   e) plan : actual distance = 6 cm : 3 m
   f) cost : VAT = £4 : 80p

Conversions	
1 litre = 1000 mℓ	1 m = 100 cm
1 kg = 1000 g	1 m = 1000 mm
1 tonne = 1000 kg	1 km = 1000 m
1 lb = 16 oz	

8 Look at the bottle of dye.
   a) Write the ratio of dye to water in its simplest form.
   b) How much dye do you need to mix with 5 litres of water?
   c) How much water do you need to mix with 50 mℓ of dye?

DYE

Mix 200 mℓ
with 1 litre
of water

9 A food shop sells different sized packs of fruit and nuts (in the
   same proportions).
   A medium pack contains 500 g of fruit and 1 kg of nuts.
   a) Write the ratio of fruit to nuts in its simplest form.
   b) A small pack contains 200 g of fruit. What weight of nuts does it contain?
   c) A giant pack contains $2\frac{1}{2}$ kg of nuts. What weight of fruit does it contain?

10 A plan of a kitchen has a scale of 20 mm : 1 m.
   a) Write this ratio in the form 1 : *n*.
   b) A kitchen unit is 600 mm wide. What is its width on the plan?
   c) The length of the kitchen is 90 mm on the plan.
   What is the actual length of the kitchen?

11 Write each ratio in its simplest form.
   a) lawn food : water = 30 mℓ : 1 litre
   b) chocolate : toffee = 1 kg : 800 g
   c) lace trim : ribbon = 25 cm : 3.5 m
   d) card : stamp = £1.95 : 60p
   e) map : actual distance = 6 cm : 3 km
   f) cost : VAT = £42 : £8.40
   g) cement : lime : sand = 500 kg : 1 tonne : 4.5 tonnes
   h) flour : butter : sugar : dried fruit = 1 lb : 8 oz : 8 oz : 12 oz

12 Look at the bottle of weedkiller.
   a) Write the ratio of weedkiller and water in its simplest form.
   b) How much water is needed with 40 mℓ of weedkiller?
   c) How much weedkiller is needed with 2400 mℓ of water?

Weed Killer

Mix 50 mℓ Weed Killer with 4 litres water

13 A pet shop makes 'Small Pet Food Mix' by mixing $1\frac{1}{2}$ kg cereal, 400 g nuts and 200 g seeds.

   a) Write these quantities in the same units as a ratio, then simplify it.   (Note: 1 kg = 1000 g)
   b) The mixture is always made with the ingredients in the same proportion.
      i)   What quantity of cereal and nuts are mixed with 20 g of seeds?
      ii) What quantity of nuts and seeds are mixed with 600 g of cereal?

For more practice with scale diagrams and maps, see pages 187–189.

14 The scale on a map is 1 : 50 000.

   a) What does 1 cm on the map represent? Give your answer in metres.
   b) What does 2 cm on the map represent? Give your answer in kilometres.

b Make a total amount

Sometimes you need to find the quantity of each ingredient to make a particular total amount.

Example

How much anti-freeze and how much water do you need to make 1 litre (1000 mℓ) of diluted anti-freeze?

The ratio 1 : 3 means
1 part anti-freeze + 3 parts water
= 4 parts diluted anti-freeze.

> Add the parts to find the total number of parts.

Anti-Freeze

Dilute with water in ratio 1 : 3

These 4 parts need to make 1000 mℓ so 1 part = 1000 ÷ 4 = 250 mℓ.

> Divide to find one part.

Amount of anti-freeze = 1 part = **250 mℓ.**
Amount of water = 3 parts = 3 × 250 = **750 mℓ.**

Check your answer by adding:
250 mℓ anti-freeze + 750 mℓ water
= 1000 mℓ diluted anti-freeze.

> The ratio 1:3 means that $\frac{1}{4}$ of the diluted anti-freeze will be anti-freeze and $\frac{3}{4}$ will be water.

Practice

1   Sue makes necklaces using black and white beads in the ratio 1 : 2.
    She wants to make a necklace with a total of 45 beads.
    How many black and how many white beads does she need?

2   A food company mixes oats and fruit in the ratio 3 : 1 to make breakfast cereal.
    Work out the weight of oats and the weight of the fruit in:

    a) 400 g of breakfast cereal
    b) 1 kg of breakfast cereal (Remember 1 kg = 1000 g)

3   Rowan mixes white and brown flour in the ratio 4 : 1 to make bread.
    How much white flour and how much brown flour does he mix to make a total of:

    a) 750 g of flour
    b) 1 kg of flour (Remember 1 kg = 1000 g)

4   The instructions on a bottle of squash say,
    'Dilute one part orange squash with four parts water.'
    How much squash and how much water do you need to make

    a) 1 litre of diluted squash (Remember 1 litre = 1000 mℓ)
    b) $1\frac{1}{2}$ litres of diluted squash?

5   Look at the information for 'Sunny Day' paint.
    How much of each colour of paint do I need to make:

    a) 5 litres of 'Sunny Day' paint
    b) 1 litre of 'Sunny Day' paint?

> **Mix and Match Paints**
> To make **Sunny Day,** mix
> 1 part Tangerine paint,
> 1 part White paint and
> 3 parts Buttercup paint

6   A health shop mixes ingredients to make bags of muesli.
    The amounts for a 1 kg bag are: 600 g cereal, 250 g fruit,
    150 g nuts.

    a) Write this as a ratio in its simplest form.
    b) Calculate how much of each ingredient is needed to make
        i)   800 g of muesli
        ii) 1800 g of muesli.

7   The population of a town consists of 15 000 Afro-Caribbean people, 45 000 Asian people
    and 150 000 white people.

    a) Write this information as a ratio and simplify it.
    b) A new company in the town plans to employ 420 people. If the number of employees
        from these ethnic groups is in the same ratio as the population, how many employees
        will be from each group?

8 Ahmed, Ben and Cilla invest £4000, £10 000 and £16 000 in a new business.
They agree to share any profits in the ratio of their investments.

a) Write the investments as a ratio in its simplest form.
b) In its first year the company makes a profit of £45 000.
How much does each person get?
c) Ben gets £25 000 from the second year's profits.
What was the total profit for the second year?
d) Cilla gets £100 000 from the third year's profits.
What was the total profit for the third year?
e) Find the total amount received by each person during these 3 years.

L2 c Use ratios to compare prices

You can compare prices by finding the value of a 'part'.

Example

The 200 g jar of coffee costs £5.50
The 300 g jar costs £6.99. Which is better value?

The ratio of the weights of the jars is $200 : 300 = 2 : 3$.

The smaller jar has 2 parts and the larger jar has 3 parts

For the smaller jar: Cost of 1 part $= £5.50 \div 2 = £2.25$

For the larger jar: Cost of 1 part $= £6.99 \div 3 = £2.33$

The **larger** jar is the better value.

> **Find the cost of 1 part for each jar.**
> **(This is the cost per 100 g in this case.)**

COFFEE 200 g

COFFEE 300 g

£5.50   £6.99

There are other ways to work out the best value. In the example you could find the cost per 100 g for the smaller jar, then multiply by 3 and compare the result with the cost of the larger jar. You could use this way as a check.

Practice

1 In each case find which is the better value.

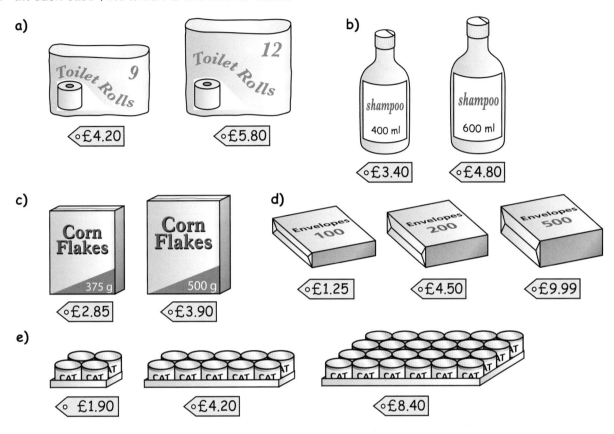

a)

Toilet Rolls 9 — ○£4.20
Toilet Rolls 12 — ○£5.80

b)

shampoo 400 ml — ○£3.40
shampoo 600 ml — ○£4.80

c)

Corn Flakes 375 g — ○£2.85
Corn Flakes 500 g — ○£3.90

d)

Envelopes 100 — ○£1.25
Envelopes 200 — ○£4.50
Envelopes 500 — ○£9.99

e)

○ £1.90   ○£4.20   ○£8.40

2 A builders' merchant delivers loose sand at £27.50 for 0.5 tonne, £52.50 for 1 tonne or £240 for 5 tonnes. The builders' merchant also sells bags of sand at £2.75 for a 25 kg bag and £4.50 for a 40 kg bag.
A local store charges £2.50 for a 5 kg bag of sand.

   a) Calculate the price per tonne in each case. (Remember 1 tonne = 1000 kg)
   b) A builder buys 0.5 tonnes of loose sand from the builders' merchant but only uses 0.3 tonnes. He sells the rest at the same rate as he paid for it.
   How much money does he get from the sale?
   c) What would a 5 kg bag cost if it was charged at the same rate as 5 tonnes of loose sand?

Number

6 Using Algebra

Use formulae and number patterns

N1/E3.10, N1/L1.11

Look at the table. It shows the cost of
hiring a boat for different lengths of time.

Number of hours	1	2	3	4
Cost (£)	7	14	21	28

Do you recognise the number pattern in the bottom row?
These numbers are **multiples** of 7.
This is because the boat costs £7 per hour to hire.

You can work out the cost for any number of hours using the **formula**:
    Cost = £7 × number of hours

> A formula gives a relationship between quantities.

Use the formula to find the cost for 10 hours:
    Cost = £7 × 10 = £70

You can check this by continuing the number pattern:
    7   14   21   28   35   42   49   56   63   70

Example

A baker says it takes 15 minutes for his oven to heat up,
then 20 minutes to bake each batch
of cakes.

a) Copy and complete the table.

Number of batches	1	2	3	4
Total				

b) Describe the pattern in the
   numbers in the bottom row of
   the table.

c) Write down a formula that works out the total time for any number of batches.

d) Use your formula to find the time needed for 8 batches.

e) Use a different method to check your answer to part d).

a) For 1 batch, total time = 15 + 20 = 35 minutes

For 2 batches, total time = 15 + 20 + 20
You can write this as 15 + 20 × 2 = 55 minutes

> You need to multiply before adding.
> Only the 20 is multiplied by 2.

For 3 batches, total time = 15 + 20 + 20 + 20
= 15 + 20 × 3 = 75 minutes
For 4 batches, total time = 15 + 20 + 20 + 20 + 20
= 15 + 20 × 4 = 95 minutes

Number of batches	1	2	3	4
Total time (minutes)	35	55	75	95

+ 20   + 20   + 20

b) Each number in the bottom row is 20 more than the previous number.

c) The formula that works out the total time for any number of batches is:
Total time = 15 + 20 × number of batches

d) For 8 batches: total time = 15 + 20 × 8 (remember to multiply before adding)
= 15 + 160 = 175 minutes

e) To check this, continue the pattern in the table: 35, 55, 75, 95, 115, 135, 155, **175**

Practice

1 The table shows how much it costs to buy different numbers of coffee mugs.

Number of mugs	1	2	3	4
Cost (£)	4	8	12	16

a) Describe the pattern in numbers in the bottom row of the table.

b) Write down a formula that works out the cost for any number of mugs.

c) Use your formula to find the cost of 6 mugs.

d) Use a different method to check your answer to part c).

2 A group of friends hire a mini-bus. They share the £60 cost equally between them.

a) Copy and complete the table to show how the amount each person pays depends on the number of friends in the group.

Number of friends in the group	2	3	4	5
Amount each person pays (£)				

b) Write down a formula that works out the amount each person pays for any number of people.

c) Use your formula to find the amount each person pays when there are 12 in the group.

d) Multiply your answer to part c) by 12 to check your answer.

3 A garden centre charges £12 per metre for fencing and £30 for a gate.

    a) Copy and complete the table to show the total cost for different lengths of fencing and a gate.

Length of fencing (metres)	3	4	5	6
Total cost of the fencing and gate (£)	66			

    b) Write down a formula that works out the total cost for different lengths of fencing.

    c) Use your formula to find the total cost of 10 metres of fencing and a gate.

    d) Use a different method to check your answer to part c).

4 Ian is saving £40 per week for a holiday that costs £280.

The formula that works out how much Ian still needs to save is:

Amount Ian still needs to save = £280 − £40 × number of weeks that Ian has been saving

    a) Copy and complete the table.

Number of weeks that Ian has been saving	1	2	3	4
Amount still to be saved (£)	240			

    b) Describe the pattern in the numbers in the bottom row of the table.

    c) Use the formula to find how much Ian still needs to save after 7 weeks. Explain what your answer means.

    d) Use a different method to check your answer to part c).

5 An electrician charges £42 for a call-out plus £25 for each hour he works on the job.

    a) Draw a table that shows how much it costs for jobs that take 1 hour, 2 hours, 3 hours and 4 hours.

    b) Describe the pattern in the numbers in your table.

    c) Write down a formula that works out the total cost for jobs that take different amounts of time.

    d) Use your formula to find the total cost for a job that takes 8 hours.

    e) Use a different method to check your answer to d).

Evaluate algebraic expressions and make substitutions in formulae
N1/L1.11, N1/L2.4

A **formula** can be written in words or in letters as an algebraic expression.

Algebraic formulae are used in many areas such as science, engineering and finance and also for working things out on a spreadsheet.

When someone hires a church hall, they pay £20 per hour plus a fixed administration fee of £10.

For 2 hours the total charge is $2 \times £20 + £10 = £50$.

For 5 hours the total charge is $5 \times £20 + £10 = £110$.

The formula in words is for total charge is:

Total charge = number of hours × £20 + £10

You can write this formula more neatly using letters. If you use 'C' for the total charge in £s and 'n' for 'the number of hours', the formula becomes:

$$C = n \times 20 + 10$$

When using letters, we usually leave out the $\times$ for multiplication to avoid confusing it with the letter x. We also write the number in '$n \times 20$' first, so the formula becomes:

$$C = 20n + 10$$

20n means 20 × n

You can use the formula to work out the charge for hiring the hall for other times.
To find the charge for 8 hours, **substitute** 8 for n and work out:

$$C = 20 \times 8 + 10 = 160 + 10 = \textbf{£170}$$

Practice

1 **a)** Hiring a hotel conference room for a day costs £15 per delegate plus
a fixed charge of £75. Write this as a formula:
   i)   in words
   ii) in letters, using T for the total cost in £s and d for the number of delegates.

   b) Use your formula to calculate the cost of a day's hire for:
   i) 10 delegates   **ii)** 20 delegates   **iii)** 50 delegates.

2 A cookery book gives this formula for the time to roast a joint of meat:
Time in minutes = number of kilograms × 50 minutes + 25 minutes extra

   a) Write this as a formula in letters, using C for the total cooking time in minutes and w for the weight of the joint in kilograms.
   b) Use your formula to find the cooking time for a joint of meat weighing:
   i) 2 kg   **ii)** 2.5 kg   **iii)** 0.8 kg.

3 A taxi company charges £1.60 a mile plus a £2 call-out fee.

   a) Write this as a formula using F for the total fare in £s and
   m for the number of miles.
   b) Use your formula to calculate the total fare for a journey of:
   i) 5 miles   **ii)** 12 miles   **iii)** 17.5 miles.

4 Each week a sales assistant earns £7.50 for each customer account he opens plus his basic pay of £280.

   a) Write this as a formula using P for the total pay in £s and a for the number of customer accounts.
   b) Use your formula to find how much he earns in a week when he opens:
   i) 4 accounts   **ii)** 10 accounts   **iii)** 15 accounts.

5 A firm uses this formula to work out the cost of producing Christmas cards:

$$C = 0.05n + 30$$

where C is the total cost in £s and n is the number of cards produced.

a) Find the cost of producing
   i) 200 cards   **ii)** 1000 cards   **iii)** 8000 cards.
b) What does the number 0.05 in the formula represent?
c) What happens to the overall cost per card if more cards are produced?

 A church charges £20 an hour for the hire of its hall, £5 an hour for the hire of its kitchen plus a fixed administration charge of £10.
What is the total charge for hiring the hall and kitchen for 3 hours?

The correct answer is £85. Did you get this? How did you work it out?
One way is to add the two charges per hour together first: £20 + £5 = £25
Then multiply by 3 to find the charge for 3 hours: £25 × 3 = £75
Then add the £10 administration charge: £75 + £10 = £85

This arithmetic can be written as one expression: $3 \times (£20 + £5) + £10$
The brackets show the part that is worked out first.

When you do arithmetic it is essential to add, subtract, multiply or divide the numbers in the correct order. The mathematical convention that you must follow is often called **BODMAS**:

Work out brackets first	then ÷ or ×	+ and − last	Note
B	**O D M**	**A S**	'Other' includes squares, see **page 31**.
brackets	other divide multiply	add subtract	

For example, $8 + 12 \div 4 = 8 + 3 = 11$ (**not** $20 \div 4 = 5$). **Divide** before **adding**.

Scientific calculators are programmed to do calculations in the right order.
Take care – many other calculators are not! Check these on your calculator:

Examples

$3 + 2 \times 3 = 9$	(multiply 2 × 3 first)	$3 \times (3 + 2) = 15$	(add 3 + 2 first)
$15 - 9 \div 3 = 12$	(divide 9 by 3 first)	$(15 - 9) \div 3 = 2$	(take 9 from 15 first)

---

6 Calculate these without a calculator – take care to use the right order.

a) $5 + 2 \times 7$ **b)** $10 - 2 \times 5$ **c)** $6 \times 3 + 2$
d) $10 + 12 \div 2$ **e)** $20 - 8 \div 2$ **f)** $(20 - 8) \div 2$
g) $2 \times (5 + 7)$ **h)** $2 \times (3 + 7 + 1)$ **i)** $5 \times 4 - (6 - 3)$

Check your answers on your calculator.

BODMAS also applies to formulae.

Example

When a car accelerates smoothly the distance in metres it travels is given by

$$d = \frac{t(u+v)}{2}$$

where u metres per second is its starting speed, v metres per second is its final speed and t seconds is the time taken.

The brackets mean you should add u and v first. Then multiply by t because this is next to the brackets (remember this means multiply). Finally divide by 2.

If a car accelerates from 8 metres per second to 12 metres per second in 5 seconds, the distance it travels is:

This means ÷ 2 after working out the top.

$$d = \frac{5(8+12)}{2} = \frac{5 \times 20}{2} = \frac{100}{2} = \textbf{50 metres}$$

---

7 Use the formula in the above example to calculate the distance d when:

   a) $t=10$, $u=0$ and $v=10$ **b)** $t=3$, $u=7$ and $v=11$.

8 A demonstrator earns £9 an hour plus £5 for every sewing machine she sells.

   a) Write this as a formula. Use P to represent the total pay in £s, t to represent the number of hours worked and n to represent the number of machines sold.

   b) Use your formula to calculate her total pay if she:
      i)   works 8 hours and sells 3 machines
      ii) works 15 hours and sells 5 machines
      iii) works 40 hours and sells 12 machines.

9 A building society uses this formula to calculate the maximum loan it allows a couple to have: $M = 3(A + B)$

M is the maximum loan and A and B are the couple's annual incomes in £s.

Calculate the maximum loan the couple can have when:

   a) $A=£15\,000$ and $B=£12\,000$ **b)** $A=£20\,000$ and $B=£18\,000$
   c) $A=£35\,000$ and $B=£30\,000$ **d)** $A=£54\,700$ and $B=£38\,250$.

10 Use the formula $V = IR$, where $V=$ voltage, $I=$ current in amps and $R=$ resistance in ohms, to find V when $I=2.4$ and $R=50$.

11 Use the formula $s = \frac{d}{t}$, where $s=$ average speed, $d=$ distance and $t=$ time, to find s when:

   a) $d=150$ miles and $t=2.5$ hours   **b)** $d=6$ km and $t=15$ minutes.

12 Skiing tuition costs a fixed charge of £10 for insurance plus £28 per hour for the instructor and £E per hour for equipment hire (depends on the equipment).

    a) Write this as a formula, using C to represent the total cost in £s and n to represent the number of hours.

    b) Use your formula to calculate the cost for:
        i) 3 hours when $E=5$ **ii)** 5 hours when $E=7$ **iii)** $7\frac{1}{2}$ hours when $E=10.50$.

13 The formula for calculating the area of a trapezium is:

$$A = \frac{h(a+b)}{2}$$ (Assume all the lengths are in cm.)

> For more area and volume formulae, see pages 240-241 and 243-246.

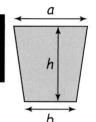

Find the area of a trapezium when:

    a) $a=6$, $b=3$ and $h=4$ **b)** $a=10$, $b=7$ and $h=8$
    c) $a=4.5$, $b=2.5$ and $h=3$ **d)** $a=25$, $b=13$ and $h=15$.

14 Use the formula $C = \dfrac{5(F-32)}{9}$ where $C=$ temperature in °Celsius and $F=$ temperature in °Fahrenheit to find C when:

    a) $F=212$ **b)** $F=77$.

Activity

You can use formulae in a spreadsheet programme to carry out calculations.
If you have access to a computer with spreadsheet software, try this example, then experiment with other numbers and situations.

Example An employee's time sheet shows the hours she has worked:
        Monday 6,   Tuesday 7,   Wednesday 8,
        Thursday 7,   Friday 5

She earns £8 an hour. Follow the instructions to find her wage for the week.

- Enter the days of the week in cells A1 to E1.
- Enter the number of hours in cells A2 to E2.
- In cell F2 enter: =8*SUM(A2:E2) and press enter.

(Note the symbol * means multiply, so this formula means $8 \times$ the sum of the amounts in cells A2 to E2.)
Cell F2 should show **£264**, the amount the employee has earned.

Number

7 Mixed Operations and Calculator Practice

Solve problems

When you use maths to solve real-life problems, take care to choose the correct operations
($+, -, \times$ or $\div$). Always **check that your answer is sensible**, especially if you use a calculator, as
it is easy to press the wrong button.

If you do enter a wrong number, you may be able to correct it by pressing a CE (Clear
Entry) button. This deletes the last number you entered, without clearing the whole sum.
A C (Clear) button, often in red, clears everything. However, calculators vary and often use
other notation such as DEL or AC. Check how your calculator works.

| Example | A householder buys a new table for £499. |

a) She pays £280 deposit. How much is left to pay?
b) She pays the rest in 6 equal instalments. How much is each instalment?

Work it out on paper   then check on your calculator
a) For the amount left to pay   Press
    work out 499 – 280   4 9 9 − 2 8 0 =

$$\begin{array}{r} 499 - \\ 280 \\ \hline \textbf{£219} \end{array}$$

The calculator will give **219**

b) To find an instalment   Carry on the calculation   **See page 108**
    work out 219 ÷ 6   ÷ 6 =   **for money in**
$$\begin{array}{r} 3\,6.5 \\ 6\overline{)21^39.^30} \end{array}$$ The calculator will give 36.5   **decimals.**

      This means **£36.50** ⟵——— **To show the pence add a zero at the end.**

Practice

Work these out on paper, then check with a calculator.

1 Jars of jam are packed in boxes of 72.
   How many jars of jam are there in 8 boxes?

2 A prize of £936 is shared equally between six people.
   How much do they get each?

3 There are 196 men, 218 women and 372 children staying in a hotel.
   a) How many people is this altogether?
   b) How many more women than men are there?

4 A householder pays £65 cash plus 6 instalments of £78 for a new cooker.
   a) What is the total of the instalments?
   b) How much does he pay altogether?

5 Two friends book a holiday. The total cost is £994.
   a) They pay a deposit of £99. How much is left to pay altogether?
   b) How much has each friend left to pay? Assume they split the cost equally.

6 A worker earns £17 per hour overtime.
   a) How much does she earn for 9 hours overtime?
   b) Her basic wage is £418 per week. How much does she earn altogether?

7 A new car has a mileage of 278 at the beginning of June and 548 at the end of June.
   a) How many miles has the car travelled in June?
   b) How many miles is this per day? (There are 30 days in June.)

Sometimes working things out without a calculator is easier, especially with small numbers.

Example

Paint is sold in 5 litre, 2 litre and 1 litre tins.
The bigger the tin, the better the value, i.e. the bigger the tin, the cheaper the price per litre.

A decorator needs 17 litres of red paint.
What tins should she buy?

The 5 litre tins are the best value, so the decorator should buy most of the paint in these.

$17 \div 5 = 3$ remainder 2

Three 5 litre tins

2 more litres are needed

The decorator should buy **three 5 litre tins and one 2 litre tin**.

Working out $17 \div 5$ on a calculator gives 3.4 (5 goes into 17 'three and a bit times'). This means you need 3 whole 5 litre tins and 0.4 (just under half) of the next 5 litre tin. If you decide to buy just three 5 litre tins, you must calculate further to see how much more paint you need:

$3 \times 5 = 15$ $17 - 15 = 2$ i.e. you need another 2 litre tin (as above).

---

8 Find the cheapest combination of tins for:

   a) 27 litres   **b)** 39 litres   **c)** 128 litres    ⟵ This means use a calculator for part c).

9 Paper plates can be bought in packs of 10, 25 or 50.
   What combination of packs gives **exactly**:

   a) 40 plates   **b)** 75 plates   **c)** 275 plates 🔢   **d)** 360 plates? 🔢

> **Example**   A sofa costs £785 in cash. You can also buy it on credit with a deposit of £50 and 9 monthly payments of £90. How much extra is this?
>
> 9 monthly payments of £90 costs $9 \times £90 = £810$
> To find the total cost, add the £50 deposit $£810 + £50 = £860$
> Extra cost = difference between the 2 amounts $£860 - £785 = £\mathbf{75}$
>
> You can do these calculations on a calculator as follows:
>
>  $9 \times 90 = 810$ $+ 50 = 860$ $- 785 = 75$

> On some calculators you must do 9 × 90 first and then + 50. If you enter 50 + 90 × 9, such calculators add 50 and 90, then multiply the total by 9, giving an answer that is much too big! Check how your calculator works.

---

10 Pay £199 cash or £40 deposit and   Pay £995 today or £90 deposit and
6 payments of £35   9 payments of £120

 a) How much more does it cost to buy the armchair on credit?
 b) How much more does it cost to buy the three-piece suite on credit?

11

| 2 AA Batteries **£3** | 4 AA Batteries ONLY **£5** | 10 AA Batteries NEW PRICE **£10** |

3 for the price of 2

The pack of 2 batteries is on special offer.
If you buy 2 packs you get one pack free (so a total of 6 batteries costs £6).

Find the cheapest combination of packs to get:

a) 6 batteries   **b)** 8 batteries   **c)** 10 batteries   **d)** 12 batteries.

12

Theme Park Admission Rates	
Standard Price	£25
Children (16 and under)	£15
Family Ticket (2 adults and 2 children)	£75
Group Ticket (10 people of any age)	£200

Find the cheapest combination of tickets for

a) 2 adults and one child   b) 1 adult and 5 children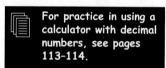
c) 2 adults and 3 children   d) 8 adults
e) 4 adults and 6 children   f) 5 adults and 5 children

> For practice in using a calculator with decimal numbers, see pages 113–114.

Calculate efficiently   N1/L1.10, N2/L1.11

Calculators are useful for working with large or difficult numbers and problems that involve several stages. It is very easy to enter a number incorrectly, so it is important to check that the answer you get is sensible. You can check your answer using estimation or by reversing the process to take you back to the number you started with.

Example   Two jobs are advertised in a newspaper:

Which job pays the higher **annual** salary and by how much?

Administration clerk's annual salary
$= £1580 \times 12 = £18\,960$.
Office assistant's annual salary
$= £369 \times 52 = £19\,188$.
The **office assistant** has the higher annual salary.
Difference between salaries $= £19\,188 - £18\,960$
$\qquad\qquad\qquad\qquad = £\mathbf{228}$.

Checks:
$18\,960 \div 12 = 1580$

$19\,188 \div 52 = 369$

$228 + 18\,960 = 19\,188$

> Job@lert
> **Administration Clerk**
> £1580 per month
>
> **Office Assistant**
> £369 per week

If your calculator has a memory, you can store the answer 18 960 so you don't need to enter it again. On some calculators pressing M+ stores the number. When you want to use it again, press MR (Memory Recall).

However, calculators vary – find out how yours works.

Practice

1   Which job pays the higher **annual** salary and by how much?

a)
| Sales assistant £8 an hour 20 hours a week | or | Part-time receptionist £720 a month |

b)
| Warehouse assistant £380 a week | or | Delivery worker £1590 a month |

> Check your working using division (as in the example).

2 Use the information in the table.

a) Calculate the cost for: 🔢
  i)   2 people spending 3 nights in the Hotel Grand
  ii) 4 people spending 4 nights in the Hotel Splendid
  iii) 4 people spending 5 nights in the Hotel Magnificent.

Hotel	Cost per person for 2 nights (£)	Cost per person for extra night (£)
Grand	199	45
Splendid	227	52
Magnificent	255	63
Supremo	312	71

b) How much more expensive is it for 4 people to spend 4 nights in the Hotel Supremo than the Hotel Magnificent?

c) How much more expensive is it for 4 people to spend 7 nights in the Hotel Supremo than the Hotel Grand?

d) A tourist has a budget of £650 for hotel accommodation. How many nights can she stay at
  i) the Hotel Grand   ii) the Hotel Supremo?

> Check your working by rounding the numbers and estimating.

Calculate efficiently with numbers of any size   N1/L2.5, N2/L2.10

Negative numbers can be entered into most calculators using the [+/−] button. This changes positive numbers into negative numbers. Some calculators have a [(−)] button instead. Consult your calculator booklet if necessary.

Examples

A company has two factories. One makes a profit of £0.8 million, the other a loss of £1.5 million. What is the overall profit?

$$0.8 \text{ million} + -1.5 \text{ million} = -0.7 \text{ million}$$

> Find out how to enter negative numbers on your calculator. You may need to use a [(−)] or [+/−] button.

There should be a minus sign before the 0.7 showing it is a negative number. The company has made an overall profit of −£0.7 million. This means a loss of £0.7 million.

Practice

1 The highest recorded temperature at a weather station was 28.7 °C.
The lowest recorded temperature was −26.9 °C.
What is the difference between these temperatures?

2 The table shows the profit made by a company in three consecutive years.
Calculate the total profit.

	Profit (£)
Year 1	£0.25 million
Year 2	−£0.76 million
Year 3	£1.42 million

3 The table gives the balance in a bank account at the beginning of each day at the end of February.
Calculate the total amount paid into or drawn out of the account each day.

Date	Balance (£)
25th Feb	564.32
26th Feb	−18.49
27th Feb	−835.60
28th Feb	1943.26
1st Mar	−154.42

4 At the beginning of each month Mr Woods pays £38 into his account with a gas company. He receives quarterly bills (at the end of every 3 months) showing the total charge for the gas used. They also show by how much his account is in debit or credit. This amount is then carried forward to the next quarter. The total charge for each quarter in a year is shown below.
Calculate by how much Mr Woods is in debit or credit at the end of each quarter.

1st quarter	£93.95	2nd quarter	£152.70	
3rd quarter	£112.10	4th quarter	£87.15	

5 **a)** Copy the table.
   b) Use the information below to write the profits for each year in your table.
     Write the numbers in full (in figures but not in decimal form) and show whether they are positive or negative.

In its first year of trading in 2004 MT Productions made a profit of £1.2 million. Profits fell by £500 000 in the second year, a further £1 million in the third year and £250 000 in 2007. In 2008 profits rose by £450 000, but 2009 brought about fresh disasters with profits falling again by £0.9 million. Profits then increased steadily by £500 000 per year, reaching £500 000 in 2012.

Year	Profit (£)
2004	
2005	
2006	
2007	
2008	
2009	
2010	
2011	
2012	500 000

   c) Use your table to find:
     i)   the difference between the highest and lowest years' profits
     ii) the total profit made by MT Productions from 2004 to 2012.

Scientific calculators have other useful buttons, e.g. brackets.
If you have a scientific calculator, use the brackets buttons to calculate the answers to questions 7, 9, 12, 13 and 14 on pages 67–68.
Also use the x^2 button to answer questions 1–5 on page 238.
For more practice use the questions on pages 244–246.

Number

8 Fractions

Understand fractions

a Fractions in words, numbers and sketches

Each fraction has a top called the **numerator** and a bottom called the **denominator**.

These numbers are fractions: $\frac{1}{6}, \frac{2}{7}, \frac{7}{10}$ ← **numerator**
← **denominator**

You can show fractions by shading.

Examples

one sixth

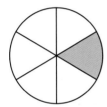

1 out of 6 **equal** parts

$=\frac{1}{6}$

This is a **unit** fraction – it has 1 on the top.

two sevenths

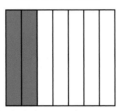

2 out of 7 **equal** parts

$=\frac{2}{7}$

seven tenths

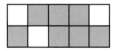

7 out of 10 **equal** parts

$=\frac{7}{10}$

Practice

1 Write these fractions as numbers.

  a) one third (one out of 3) **b)** two fifths **c)** three eighths **d)** five twelfths

2 Write these fractions in words.

  a) $\frac{1}{2}$ **b)** $\frac{2}{3}$ **c)** $\frac{3}{4}$ **d)** $\frac{3}{5}$ **e)** $\frac{5}{8}$ **f)** $\frac{7}{9}$

3 What fraction is shaded? Write your answers in words and numbers.

a)

b)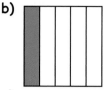

c)

d)

e)

f)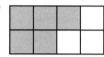

g)

h)

i)

b Shade fractions

Practice

1 In which of these is two thirds shaded?

a)

b)

c)

d)

2 In which of these is $\frac{3}{4}$ shaded?

a)

b)

c)

d)

3 Copy each sketch. Shade the fraction.

a) $\frac{2}{5}$

b) $\frac{3}{10}$

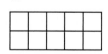

c) $\frac{7}{8}$

d) three sevenths

e) five eighths

f) eleven twelfths

4 Draw a sketch to show each fraction.

a) $\frac{1}{2}$ b) $\frac{1}{3}$ c) $\frac{1}{4}$ d) $\frac{1}{8}$ e) $\frac{1}{10}$ f) $\frac{4}{5}$

g) $\frac{3}{8}$ h) $\frac{2}{9}$ i) $\frac{6}{7}$ j) $\frac{9}{10}$ k) $\frac{5}{12}$ l) $\frac{11}{15}$

c Read about fractions

Common fractions like a half, a third and three-quarters are used in many ways. For example, adverts often use fractions to describe special offers. Government reports and the media sometimes give survey results as fractions. Recipes in cookery books give some amounts in fractions.

Practice

Write down each fraction in words and in numbers.

1

Half price!

One third off!

2

Lentil Patties	
1 cup lentils	1 pint stock
$\frac{1}{4}$ cup peas	1 onion
$\frac{3}{4}$ cup potato	1 clove garlic
$1\frac{1}{2}$ tablespoons oil	$\frac{1}{2}$ tablespoon thyme
1 egg	salt & pepper

3

Christmas Reductions
Shops are reducing prices by as much as two-thirds in post-Christmas sales.
Almost half of the shops started reducing prices before December 25.

4

Social Trends 41
Use of the Internet and other technologies

'Just over a third of adults aged 15 and over in the UK stated that they (or someone else in their household) had watched online television or videos in 2009.'

'Over three-quarters of all those who had accessed the Internet in the three months prior to interview had done so every day or almost every day, while a fifth had accessed it at least once per week but not every day.'

'Just over a quarter of children aged 5 to 15 owning a mobile phone in 2009 in the UK had first acquired one by the time they were 8 years old and just under two-thirds by the time they were 10 years old.'

'In 2009/10, adults in the UK aged 16 and over spent on average three and a half hours a day watching television, two and a half hours using a computer and one hour listening to radio.'

Source: www.statistics.gov.uk

Equivalent fractions

N2/E3.2

a Use sketches

Look at these sketches of pizzas.

 $\frac{1}{3}$

 $\frac{2}{6}$

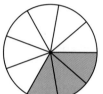 $\frac{3}{9}$

The same amount is shaded in each circle. This shows that $\frac{1}{3} = \frac{2}{6} = \frac{3}{9}$

Using the whole circles gives $\frac{3}{3} = \frac{6}{6} = \frac{9}{9} = 1$

Practice

1   What do these sketches show?

   a)

   b)

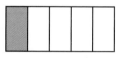

   c)

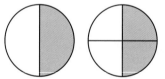

   d)

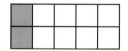

2   Draw sketches to show that:

   a) $\frac{1}{2} = \frac{4}{8}$

   b) $\frac{1}{2} = \frac{5}{10}$

   c) $\frac{1}{4} = \frac{3}{12}$

   d) $\frac{3}{5} = \frac{6}{10}$

   e) $\frac{2}{3} = \frac{6}{9}$

   f) $\frac{2}{5} = \frac{6}{15}$

   g) $\frac{3}{4} = \frac{6}{8}$

   h) $\frac{5}{6} = \frac{10}{12}$

   i) $\frac{4}{5} = \frac{16}{20}$

b Use numerators and denominators

The sketches on page 78 showed that $\frac{1}{3} = \frac{2}{6} = \frac{3}{9}$.

Although these fractions look different, they are the same size.

> The size of a fraction is not altered when you multiply or divide the numerator (top) and denominator (bottom) by the **same number**.

$$\overset{\times 2}{\underset{\times 2}{\frac{1}{3} = \frac{2}{6}}} \qquad \overset{\div 3}{\underset{\div 3}{\frac{3}{9} = \frac{1}{3}}}$$

You can find a missing numerator or denominator.

Example

$$\frac{3}{5} = \frac{?}{10} \qquad\qquad\qquad \frac{12}{20} = \frac{3}{?}$$

The bottom has been multiplied by 2
Do the same to the top:

The top has been divided by 4
Do the same to the bottom:

$$\frac{3}{5} = \frac{6}{10} \qquad\qquad\qquad \frac{12}{20} = \frac{3}{5}$$

Example

In a survey, 40 out of 100 people said they like cats more than dogs.

The report says that this is two-fifths of the people. Is this correct?

$$\overset{\div 20}{\underset{\div 20}{\frac{40}{100} = \frac{2}{5}}}$$

Write 40 out of 100 as a fraction.
Then divide the top and bottom of the fraction by 20.

The report is correct. 40 out of 100 is equivalent to 2 out of every 5 people.

Practice

1 For each part, write 'True' or 'False'.

a) $\frac{1}{2} = \frac{5}{10}$ b) $\frac{3}{7} = \frac{6}{14}$ c) $\frac{3}{4} = \frac{9}{10}$ d) $\frac{2}{5} = \frac{6}{15}$

e) $\frac{14}{21} = \frac{2}{3}$ f) $\frac{12}{20} = \frac{3}{4}$ g) $\frac{8}{10} = \frac{3}{5}$ h) $\frac{16}{24} = \frac{4}{6}$

2 Find the missing numbers.

a) $\dfrac{3}{5} = \dfrac{9}{?}$ b) $\dfrac{1}{4} = \dfrac{6}{?}$ c) $\dfrac{1}{2} = \dfrac{?}{8}$ d) $\dfrac{2}{5} = \dfrac{?}{20}$

e) $\dfrac{6}{20} = \dfrac{3}{?}$ f) $\dfrac{20}{30} = \dfrac{2}{?}$ g) $\dfrac{14}{35} = \dfrac{?}{5}$ h) $\dfrac{20}{25} = \dfrac{?}{5}$

3 a) 50 out of 100 people travel to work on a bus.
   Sam says this is half of the people. Is Sam correct?

   b) 20 out of 100 people walk to work.
   Tara says this is a quarter of the people. Is Tara correct?

4 The fractions $\dfrac{2}{4}$ and $\dfrac{3}{6}$ are both equivalent to $\dfrac{1}{2}$.

Write down six more fractions that are all equivalent to $\dfrac{1}{2}$.

5 Find three other fractions that are equivalent to:

a) $\dfrac{3}{8}$ b) $\dfrac{2}{5}$ c) $\dfrac{4}{11}$ d) $\dfrac{7}{9}$.

6 Twelve out of thirty students in a class are men.
   Ian says that two fifths of the class are men. Is he correct?

7 The following fractions are all equal to 1. What are the missing numbers?

$\dfrac{?}{2} \quad \dfrac{?}{4} \quad \dfrac{5}{?} \quad \dfrac{7}{?} \quad \dfrac{?}{11} \quad \dfrac{15}{?} \quad \dfrac{20}{?} \quad \dfrac{?}{100}$

8 a) List the fractions below that are equivalent to $\dfrac{1}{3}$.

   b) List those that are equivalent to $\dfrac{2}{3}$.

$\dfrac{6}{9}, \quad \dfrac{6}{18}, \quad \dfrac{4}{12}, \quad \dfrac{5}{15}, \quad \dfrac{8}{12}, \quad \dfrac{2}{6}, \quad \dfrac{4}{6}, \quad \dfrac{12}{18}, \quad \dfrac{3}{9}, \quad \dfrac{10}{15}$

9 Fifteen out of twenty people pass a test.
   Lily says three fifths of the people have passed. Is she correct?

10 Copy each diagram. Draw arrows between equivalent fractions.
   One arrow has been drawn on the first diagram.

a)
$\dfrac{1}{5}$	$\dfrac{9}{15}$
$\dfrac{3}{4}$	$\dfrac{15}{18}$
$\dfrac{2}{5}$	$\dfrac{15}{20}$
$\dfrac{5}{6}$	$\dfrac{2}{10}$
$\dfrac{1}{3}$	$\dfrac{5}{15}$
$\dfrac{3}{5}$	$\dfrac{6}{15}$

b)
$\dfrac{3}{4}$	$\dfrac{6}{24}$
$\dfrac{2}{3}$	$\dfrac{4}{24}$
$\dfrac{1}{3}$	$\dfrac{9}{24}$
$\dfrac{1}{4}$	$\dfrac{9}{12}$
$\dfrac{3}{8}$	$\dfrac{8}{12}$
$\dfrac{1}{6}$	$\dfrac{8}{24}$

c Write quantities as fractions

Here are some important facts about length, capacity and weight:

$1\,cm = 10\,mm$ $1\,m = 100\,cm$ $1\,m = 1000\,mm$ $1\,km = 1000\,m$

$1\,c\ell = 10\,m\ell$ $1\,\ell = 100\,c\ell$ $1\,\ell = 1000\,m\ell$ $1\,kg = 1000\,g$

You can use these facts to write some quantities as fractions.

Example

$8\,mm = \dfrac{8}{10}\,cm$ (8 out of 10) $250\,g = \dfrac{250}{1000}\,kg$ (250 out of 1000)

$\qquad\quad = \dfrac{4}{5}\,cm$ $\qquad\quad = \dfrac{25}{100}\,kg$ (dividing by 10)

(by dividing top and bottom by 2) $\qquad\quad = \dfrac{1}{4}\,kg$ (dividing by 25 or 5 then 5 again)

Practice

1 Write these lengths as fractions of a centimetre.

   a) 5 mm   **b)** 4 mm   **c)** 6 mm

2 Write these weights as fractions of a kilogram.

   a) 500 g   **b)** 750 g   **c)** 100 g   **d)** 200 g   **e)** 600 g

3 Write these lengths as fractions of a kilometre.

   a) 5 m   **b)** 50 m   **c)** 500 m   **d)** 100 m   **e)** 150 m

4 Write these volumes as fractions of a litre.

   a) 20 mℓ **b)** 50 mℓ **c)** 200 mℓ **d)** 500 mℓ **e)** 400 mℓ

5 Write these lengths as fractions of a metre.

   a) 20 cm   **b)** 50 cm   **c)** 90 cm   **d)** 80 cm   **e)** 5 cm
   f) 100 mm   **g)** 200 mm   **h)** 400 mm   **i)** 600 mm   **j)** 20 mm

Compare fractions and mixed numbers

N2/L1.1

a Use fraction walls

Example

The fraction wall shows:

$$1 = \frac{2}{2} = \frac{4}{4} = \frac{8}{8} = \frac{16}{16}$$

The shaded part shows:

$$\frac{1}{4} = \frac{2}{8} = \frac{4}{16}$$

The diagram also shows that $\frac{3}{8}$ is bigger than $\frac{1}{4}$ and $\frac{3}{8}$ is less than $\frac{1}{2}$.

Practice

1 Find the missing numbers.

a) $\frac{1}{2} = \frac{?}{4}$ b) $\frac{1}{2} = \frac{?}{8}$ c) $\frac{1}{2} = \frac{?}{16}$ d) $\frac{3}{4} = \frac{?}{8}$

e) $\frac{3}{4} = \frac{?}{16}$ f) $\frac{3}{8} = \frac{?}{16}$ g) $\frac{5}{8} = \frac{?}{16}$ h) $\frac{7}{8} = \frac{?}{16}$

2 Put each set of fractions in order of size. Start with the **smallest**.

a) $\frac{3}{8}, \frac{7}{16}, \frac{5}{16}$ b) $\frac{5}{8}, \frac{9}{16}, \frac{1}{2}$ c) $\frac{13}{16}, \frac{7}{8}, \frac{3}{4}$

3 This fraction wall shows halves, fifths and tenths.
Find the missing numbers.

a) $1 = \frac{?}{5} = \frac{?}{10}$ b) $\frac{1}{2} = \frac{?}{10}$ c) $\frac{1}{5} = \frac{?}{10}$

d) $\frac{2}{5} = \frac{?}{10}$ e) $\frac{3}{5} = \frac{?}{10}$ f) $\frac{5}{5} = \frac{?}{10}$

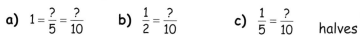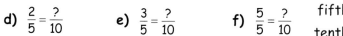

4 Put these fractions in order of size, starting with the **largest**.

$$\frac{2}{5}, \quad \frac{7}{10}, \quad \frac{1}{2}, \quad \frac{1}{10}, \quad \frac{9}{10}, \quad \frac{3}{5}, \quad \frac{3}{10}, \quad \frac{1}{5}, \quad \frac{4}{5}$$

b Compare fractions

These fractions are all **unit fractions**
– they all have 1 in the numerator.

In unit fractions, the larger the
denominator, the smaller the fraction.

For example, $\frac{1}{15}$ is smaller than $\frac{1}{10}$.

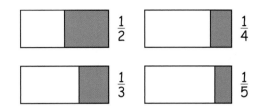

Practice

1 Which is **smaller**?

a) $\frac{1}{7}$ or $\frac{1}{8}$ b) $\frac{1}{10}$ or $\frac{1}{6}$ c) $\frac{1}{20}$ or $\frac{1}{19}$ d) $\frac{1}{12}$ or $\frac{1}{3}$

2 Which is **larger**?

a) $\frac{1}{20}$ or $\frac{1}{25}$ b) $\frac{1}{19}$ or $\frac{1}{17}$ c) $\frac{1}{5}$ or $\frac{1}{50}$ d) $\frac{1}{13}$ or $\frac{1}{11}$

3 Write these fractions in order of size. Start with the **smallest**.

$$\frac{1}{4}, \quad \frac{1}{12}, \quad \frac{1}{2}, \quad \frac{1}{5}, \quad \frac{1}{10}, \quad \frac{1}{15}, \quad \frac{1}{3}, \quad \frac{1}{6}, \quad \frac{1}{20}$$

4 Write these fractions in order of size. Start with the **largest**.

$$\frac{1}{8}, \quad \frac{1}{11}, \quad \frac{1}{21}, \quad \frac{1}{14}, \quad \frac{1}{9}, \quad \frac{1}{16}, \quad \frac{1}{13}, \quad \frac{1}{7}, \quad \frac{1}{17}$$

It is more difficult to compare other fractions.
Shading parts of equal rectangles helps.

The sketch shows that $\frac{4}{5}$ is larger than $\frac{2}{3}$.

Draw sketches to illustrate your answers to questions 5 to 7.

5 Which is **larger**?

a) $\frac{3}{4}$ or $\frac{2}{3}$ b) $\frac{1}{2}$ or $\frac{2}{5}$ c) $\frac{1}{3}$ or $\frac{2}{5}$

6 Which is **smaller**?

a) $\frac{1}{2}$ or $\frac{4}{7}$ b) $\frac{3}{5}$ or $\frac{7}{9}$ c) $\frac{5}{6}$ or $\frac{3}{4}$

7 Meera says that $\frac{3}{7}$ is bigger than a half.
Is she correct?
Explain your answer.

c Improper fractions and mixed numbers

An **improper fraction** has a bigger numerator than denominator, e.g. $\frac{5}{4}$.

A **mixed number** is a mixture of a whole number and a fraction, e.g. $1\frac{1}{4}$.

whole one + 1 quarter = one and a quarter

The diagram shows $1\frac{1}{4} = \frac{5}{4}$.

4 quarters + 1 quarter = 5 quarters

$\frac{3}{4}$ is less than $1\frac{1}{4}$.

3 quarters

Practice

1 What do these diagrams show?
For each part give a mixed number and an improper fraction.

a)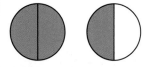

b)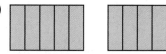

c)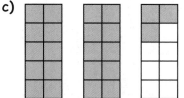

d)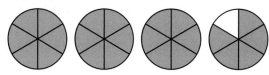

e)

f)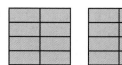

2 Draw diagrams to show that:

a) $1\frac{3}{4} = \frac{7}{4}$ b) $2\frac{1}{2} = \frac{5}{2}$ c) $1\frac{2}{3} = \frac{5}{3}$ d) $3\frac{4}{5} = \frac{19}{5}$

Can you see a quick way to write mixed numbers as improper fractions?

3 a) Draw diagrams to show each of these numbers in eighths.

$$\frac{7}{8}, \quad 1\frac{3}{8}, \quad \frac{5}{8}, \quad 2\frac{1}{8}, \quad 1\frac{1}{4}, \quad \frac{3}{4}, \quad 1\frac{1}{2}$$

b) List the original numbers in order of size, starting with the **smallest**.

d Find or estimate fractions

The first sketch shows a chocolate bar.
The second shows it again after some has been eaten.
What fraction is left?

6 out of 10 squares are left $\frac{6}{10} = \frac{3}{5}$ (dividing by 2)

You can also imagine the bar divided into 5 parts, as in this sketch.
This also shows that $\frac{3}{5}$ is left and $\frac{2}{5}$ was eaten.

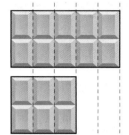

Use your ruler to measure the lengths of the bars in the diagram:
Length of original bar = 30 mm. Length after some was eaten = 18 mm.

The fraction that is left is $\frac{18}{30} = \frac{3}{5}$ (dividing top & bottom by 6).

Practice

1 Here are some more chocolate bars. In each case estimate:
   i) the fraction that is left   **ii)** the fraction that was eaten.

a)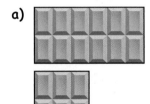

b)

c)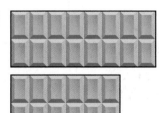

Use your ruler to check your answers.

2 Use a fraction to describe how much is in each jug.

a) 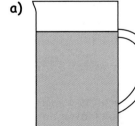   b) 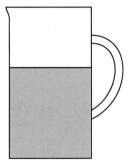   c)

Hint: You can use a ruler to measure the height
of the jug and the height of the liquid.

3 Here is a water tank. Use a fraction to describe how full it is in each sketch.

a) 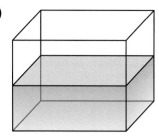   b) 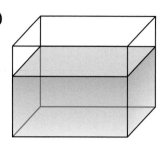   c)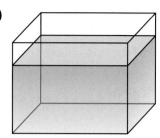

4 Estimate what fraction of the drink is left in each sketch.

a)    b)    c)

5 The sketch shows a pie.
For each part:
i) estimate what fraction of the pie has been eaten
ii) estimate what fraction of the pie is left.

a) 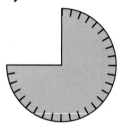   b) 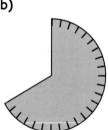   c)   d)

Find a fraction of something

a By dividing when the numerator is 1

To find $\frac{1}{2}$ of something, divide it by 2.

To find $\frac{1}{3}$ of something, divide it by 3.

To find $\frac{1}{4}$ of something, divide it by 4, and so on.

> When the numerator is 1,
> divide by the denominator.

Example

$\frac{1}{3}$ of 84 kg = 84 kg ÷ 3 = **28 kg**

$$\begin{array}{r} 2\ 8 \\ 3\overline{)8\ ^24} \end{array}$$

Example   Five partners in a company each get a fifth of the profits of £64 500.

They each get $\frac{1}{5}$ of £64 500.

£64 500 ÷ 5 = £12 900. They each get **£12 900**.

$$\begin{array}{r} 1\ 2\ 9\ 0\ 0 \\ 5\overline{)6^14^4500} \end{array}$$

Practice

1   Find:

a) $\frac{1}{2}$ of £56   **b)** $\frac{1}{3}$ of 72 cm   **c)** $\frac{1}{5}$ of £780   **d)** $\frac{1}{10}$ of £2500

e) $\frac{1}{4}$ of £816   **f)** $\frac{1}{6}$ of £546   **g)** $\frac{1}{8}$ of 576 m   **h)** $\frac{1}{9}$ of £2574.

Parts f, g and h are at L1.

2   There are 24 students in a class. Half of them are women.
How many are women?

3   A jacket costs £45. This price is reduced by a third in a sale.

a) How much is saved?   **b)** What is the sale price?

4   Seventy people take a driving test. A fifth of them fail.

a) How many people fail?   **b)** How many people pass?

5   A petrol tank holds 60 litres when full. It is a quarter full.
How many litres of petrol are in the tank?

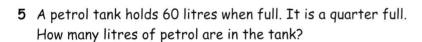

Empty   **FUEL**   Full

6   Four friends share a prize of £340 equally between them.
How much do they get each?

7   A sixth of the adults in a town did not vote. There are altogether 21 000 adults.
How many adults: **a)** did not vote   **b)** voted?

8   A chef uses one eighth of a kilogram bag of flour to thicken a sauce.

a) How many grams of flour does he use?   (1 kg = 1000 g)
b) How many grams are left in the bag?

b Finding more than one part

> ▶ When the numerator is not 1, divide by the
> denominator, then multiply by the numerator.

Example   Find $\frac{3}{4}$ of £540.

To find $\frac{3}{4}$ of £540, divide it by 4, then multiply by 3

£540 ÷ 4 = £135

$$4 \overline{)5^14^20}$$
$$1\ 3\ 5$$

$$\begin{array}{r} 135 \\ 3\times \\ \hline 4\ 0\ 5 \\ \tiny 1\ 1 \end{array}$$

Then £135 × 3 = **£405**

Divide by the denominator.
Then multiply by the numerator.

Practice

1 Find:

a) $\frac{3}{4}$ of £840   b) $\frac{2}{3}$ of £156   c) $\frac{2}{5}$ of 75 m   d) $\frac{3}{5}$ of 250 g

e) $\frac{5}{6}$ of 420 cm   f) $\frac{2}{7}$ of £1610   g) $\frac{5}{8}$ of £2880   h) $\frac{4}{9}$ of 3285 kg

2 A company has 360 employees. Two-thirds of these are men.
How many are men?

3 A box contains 45 chocolates.
Two-fifths of these are dark chocolates.
How many dark chocolates are there?

4 A firm makes £45 000 profit. The firm spends three quarters of the
profit on new equipment. How much does the firm spend on new equipment?

5 There are 34 500 spectators at a football match. Nine tenths of these support the home
team. How many spectators support the home team?

6 Two hundred and seventy five people go to a Christmas pantomime. Three fifths of them
are children. How many children go to the pantomime?

7 A car costs £6900. The buyer gets a loan for five sixths of the price.
How much is the loan?

8 A holiday company sells 480 000 holidays during a year. Three quarters of these are holidays abroad. How many of the holidays are abroad?

c Using other methods

Links between some fractions give other methods. For example, to find a quarter of something you can halve it, then halve again.

Also, the sketch shows $\frac{3}{4} = \frac{2}{4} + \frac{1}{4} = \frac{1}{2} + \frac{1}{4}$.

So, to find $\frac{3}{4}$ of something, you can add $\frac{1}{2}$ and $\frac{1}{4}$

$$\frac{3}{4} \quad = \quad \frac{1}{2} \quad + \quad \frac{1}{4}$$

Example

$\frac{1}{2}$ of £540 = £270

$\frac{1}{4}$ of £540 = £135

So $\frac{3}{4}$ of £540 = £270 + £135

   = £405 ← You can check this by multiplying £135 by 3.

$\frac{1}{2}$ of £540

$\frac{1}{4}$ of £540

$\frac{3}{4}$ of £540

$$\begin{array}{r} 2\,7\,0 \\ 2\,\overline{)5^1 4\,0} \end{array}$$

$$\begin{array}{r} 1\,3\,5 \\ 2\,\overline{)2\,7^1 0} \end{array}$$

$$\begin{array}{r} 2\,7\,0 \\ 1\,3\,5+ \\ \hline 4\,0\,5 \\ {\scriptstyle 1} \end{array}$$

You can also check the answer by adding the amounts for $\frac{1}{4}$ and $\frac{3}{4}$.

You should get back to the whole amount (because $\frac{4}{4}$ = 1).

$$\begin{array}{r} 4\,0\,5 \\ 1\,3\,5+ \\ \hline 5\,4\,0 \\ {\scriptstyle 1} \end{array}$$

Practice

1 Find these and check each answer.

   a) $\frac{3}{4}$ of £1520   **b)** $\frac{3}{4}$ of 56 kg   **c)** $\frac{3}{4}$ of 432 m

2 a) Draw a sketch to show that $\frac{5}{8} = \frac{4}{8} + \frac{1}{8} = \frac{1}{2} + \frac{1}{8}$.

   b) Find: **i)** $\frac{1}{2}$ of £960   **ii)** $\frac{1}{8}$ of £960   **iii)** $\frac{5}{8}$ of £960.

   c) Check **biii)** by multiplying your answer to **bii)** by 5.

3 a) Draw a sketch to show that $\frac{3}{8} = \frac{2}{8} + \frac{1}{8} = \frac{1}{4} + \frac{1}{8}$.

   b) Find: **i)** $\frac{1}{4}$ of 256 m   **ii)** $\frac{1}{8}$ of 256 m   **iii)** $\frac{3}{8}$ of 256 m.

   c) Check **biii)** by multiplying your answer to **bii)** by 3.

4 a) Draw a sketch to show that $\frac{5}{6} = \frac{3}{6} + \frac{2}{6} = \frac{1}{2} + \frac{1}{3}$.

   b) Find: **i)** $\frac{1}{2}$ of £7200   **ii)** $\frac{1}{3}$ of £7200   **iii)** $\frac{5}{6}$ of £7200.

   c) Find $\frac{1}{6}$ of £7200 and use your answer to check **biii)**.

d Scale up and down

Example

A recipe for macaroni cheese for 6 people includes these ingredients:

300g cheese   120g butter   240g macaroni.

Suppose there are only 4 people. You need $\frac{4}{6} = \frac{2}{3}$ of the quantities.

To find what you need, divide each quantity by 3, then multiply by 2:

Cheese: $300g \div 3 = 100g$ $100g \times 2 = \textbf{200g}$
Butter: $120g \div 3 = 40g$ $40g \times 2 = \textbf{80g}$
Macaroni: $240g \div 3 = 80g$ $80g \times 2 = \textbf{160g}$

Note that there is twice as much macaroni as butter in both sets of ingredients.

Note: If there were 9 people, you would need $\frac{9}{6} = \frac{3}{2}$ of the quantities.

In this case you would need to divide each quantity by 2, then multiply by 3.
Check that this gives 450 g of cheese, 180 g of butter and 360 g of macaroni.

Practice

1 The recipe gives the ingredients for leek and
   potato soup for 6 people.
   Find the quantities you need for:

   a) 4 people   **b)** 8 people.

Leek & Potato Soup (for 6 people)	
270 g potatoes	450 g leeks
750 mℓ milk	150 mℓ stock

2 The main ingredients for a nut loaf for 8 people are:
         400g nuts   200g breadcrumbs   240g tomatoes   4 onions.

   a) What fraction of the quantities do you need for 6 people?
   b) Find the quantities for 6 people.

3 These are the ingredients for 20 almond biscuits.
         300g flour   160g sugar   140g butter   60g ground almonds

   a) What fraction of the quantities do you need for 16 biscuits?
   b) Find the quantities for 16 biscuits.

4 For sandwiches for 20 people a caterer uses:
   4 loaves of bread   800g butter   8 tomatoes   2 cucumbers
   400g paté   300g chicken   500g cheese   1 lettuce.
   Find the quantities for sandwiches for
   a) 15 people   **b)** 25 people.

Put fractions in order (using a common denominator)   N2/L2.1

Comparing fractions is easy if they have the same denominator.

For example, $\frac{3}{5}$  is more than $\frac{2}{5}$

Practice

1 Which is **larger?**   **a)** $\frac{1}{3}$ or $\frac{2}{3}$ **b)** $\frac{7}{9}$ or $\frac{5}{9}$ **c)** $\frac{7}{10}$ or $\frac{9}{10}$

2 Put each set of fractions in order. Start with the **smallest.**

a) $\frac{3}{5}$, $\frac{2}{5}$, $\frac{4}{5}$ **b)** $\frac{5}{7}$, $\frac{1}{7}$, $\frac{2}{7}$, $\frac{4}{7}$ **c)** $\frac{4}{9}$, $\frac{5}{9}$, $\frac{2}{9}$, $\frac{8}{9}$, $\frac{1}{9}$

Comparing fractions with different denominators is harder.
You can do this by making the denominators the same.

Or see page 83 for another way.

Example   To compare $\frac{2}{3}$ with $\frac{7}{12}$, write $\frac{2}{3}$ as twelfths: Because 3 divides into 12, you can change the denominator from 3 to 12. 12 is the 'common denominator'.

$$\overset{\times 4}{\underset{\times 4}{\frac{2}{3} = \frac{8}{12}}}$$

To change the bottom to 12, multiply 3 by 4.
Then do the same to the top.

$\frac{8}{12}$ is bigger than $\frac{7}{12}$ so $\frac{2}{3}$ is bigger than $\frac{7}{12}$.

The denominator, 12, is the 'common denominator' for these fractions.

Example   Write $\frac{5}{8}$, $\frac{7}{10}$ and $\frac{3}{5}$ in order of size, starting with the smallest.

Look for a common denominator – a number that 8, 5 and 10 all divide into.
The best common denominator for these numbers is 40.

Now write all the fractions with 40 as the denominator:

$$\overset{\times 5}{\underset{\times 5}{\frac{5}{8} = \frac{25}{40}}} \qquad \overset{\times 4}{\underset{\times 4}{\frac{7}{10} = \frac{28}{40}}} \qquad \overset{\times 8}{\underset{\times 8}{\frac{3}{5} = \frac{24}{40}}}$$

Other common denominators (e.g. 80) could be used.
40 is the **best** because it is the **smallest.**

In order of size the fractions are: $\frac{3}{5}$, $\frac{5}{8}$, $\frac{7}{10}$.

3 Write each pair with the same denominator, then say which is the **larger**.

a) $\frac{1}{4}$ or $\frac{1}{5}$ b) $\frac{3}{5}$ or $\frac{3}{10}$ c) $\frac{2}{3}$ or $\frac{3}{4}$ d) $\frac{1}{2}$ or $\frac{4}{7}$

4 Write each group in order of size, starting with the **smallest**.

a) $\frac{1}{2}, \frac{5}{8}, \frac{3}{8}$ b) $\frac{2}{5}, \frac{3}{10}, \frac{1}{4}$ c) $\frac{5}{16}, \frac{3}{4}, \frac{7}{8}, \frac{1}{2}$

d) $\frac{5}{6}, \frac{3}{4}, \frac{2}{3}, \frac{11}{12}$ e) $\frac{3}{4}, \frac{1}{2}, \frac{5}{7}, \frac{3}{7}$ f) $\frac{1}{3}, \frac{5}{12}, \frac{1}{4}, \frac{1}{2}, \frac{3}{8}$

5 Write each group in order of size, starting with the **largest**.

a) $\frac{1}{2}, \frac{4}{7}, \frac{2}{7}$ b) $\frac{5}{9}, \frac{2}{3}, \frac{5}{6}$ c) $\frac{1}{2}, \frac{4}{9}, \frac{3}{4}, \frac{2}{3}$

d) $\frac{4}{5}, \frac{2}{3}, \frac{9}{10}, \frac{13}{15}$ e) $\frac{5}{9}, \frac{5}{8}, \frac{4}{9}, \frac{7}{12}$ f) $\frac{3}{5}, \frac{2}{3}, \frac{7}{10}, \frac{13}{15}, \frac{2}{9}, \frac{11}{15}$

To put a mixture of fractions and mixed numbers in order, consider them in groups.

> **Example**   To put $\frac{2}{3}, 1\frac{3}{4}, 1\frac{2}{3}, 2\frac{1}{3}$ and $\frac{5}{6}$ in order:
>
> Under 1: $\frac{2}{3}$ and $\frac{5}{6}$ common denominator 6 $\frac{2}{3} = \frac{4}{6}$ is smaller than $\frac{5}{6}$
>
> Between 1 and 2: $1\frac{3}{4}$ and $1\frac{2}{3}$ common denominator 12 $1\frac{3}{4} = 1\frac{9}{12}, 1\frac{2}{3} = 1\frac{8}{12}$
>
> Above 2: $2\frac{1}{3}$ only
>
> In order, starting with the smallest: $\frac{2}{3}, \frac{5}{6}, 1\frac{2}{3}, 1\frac{3}{4}, 2\frac{1}{3}$

---

6 Write each group in order of size, starting with the **smallest**.

a) $1\frac{3}{5}, \frac{3}{4}, 2\frac{2}{5}, 1\frac{1}{2}, 2\frac{1}{3}$ b) $\frac{5}{8}, 1\frac{3}{4}, \frac{4}{7}, 1\frac{7}{12}, 1\frac{5}{6}$

c) $\frac{4}{9}, 2\frac{2}{3}, \frac{5}{7}, 2\frac{3}{7}, 1\frac{3}{4}, 2\frac{1}{2}$ d) $1\frac{2}{3}, 1\frac{5}{8}, \frac{1}{4}, 1\frac{11}{12}, 2\frac{1}{8}, 1\frac{5}{6}$

Write one number as a fraction of another   N2/L1.12, N2/L2.3

a Simplest form

The fractions $\frac{8}{12}, \frac{2}{3}, \frac{4}{6}, \frac{10}{15}$ and $\frac{6}{9}$ are all equal to each other.

$\frac{2}{3}$ is the **simplest form** – its numerator and denominator are the smallest.

> To find the simplest form of a fraction, divide the top and bottom by the same number. Repeat this as far as you can.

Example   Write $\frac{240}{320}$ in its simplest form.

Look for numbers that divide into the top and bottom.

$$\overset{\div 10}{\overset{\frown}{}}\quad\overset{\div 8}{\overset{\frown}{}}$$
$$\frac{240}{320} = \frac{24}{32} = \frac{3}{4}$$
$$\underset{\div 10}{\underset{\smile}{}}\quad\underset{\div 8}{\underset{\smile}{}}$$

There is more than one way of doing this. Here is another way:

$$\overset{\div 2}{\overset{\frown}{}}\quad\overset{\div 2}{\overset{\frown}{}}\quad\overset{\div 20}{\overset{\frown}{}}$$
$$\frac{240}{320} = \frac{120}{160} = \frac{60}{80} = \frac{3}{4}$$
$$\underset{\div 2}{\underset{\smile}{}}\quad\underset{\div 2}{\underset{\smile}{}}\quad\underset{\div 20}{\underset{\smile}{}}$$

Finding the simplest form is sometimes called '**cancelling**'.

If you divide by large numbers it is quicker, but it is sometimes easier to divide by small numbers.
Keep going until you are sure there is nothing else you can divide by.

Practice

Write each fraction in its simplest form.

1 $\frac{20}{50}$ 2 $\frac{12}{16}$ 3 $\frac{9}{27}$ 4 $\frac{12}{30}$ 5 $\frac{80}{100}$

6 $\frac{32}{48}$ 7 $\frac{200}{400}$ 8 $\frac{150}{200}$ 9 $\frac{25}{75}$ 10 $\frac{72}{96}$

11 $\frac{140}{700}$ 12 $\frac{81}{270}$ 13 $\frac{560}{840}$ 14 $\frac{1500}{3500}$ 15 $\frac{1320}{2200}$

b Write one quantity as a fraction of another

Example   Write 60 cm as a fraction of a metre.

$1\,\text{m} = 100\,\text{cm}$, so the fraction is $\frac{60}{100} = \frac{6}{10} = \frac{3}{5}$ (dividing by 10, then 2).

> **Example**   Seventy women and fifty-six men work in a factory.
> What fraction of the workers are men?
>
> The total number of workers $= 70 + 56 = 126$.
>
> The fraction that are men $= \dfrac{56}{126} = \dfrac{28}{63} = \dfrac{4}{9}$ (dividing by 2, then 7).

Practice

1 What fraction of an hour is:

a) 15 minutes   b) 20 minutes   c) 48 minutes?

> Remember metric units:
> 1 kg = 1000 g
> 1 g = 1000 mg
> 1 cℓ = 10 mℓ
> See page 198 and page 217.

2 What fraction of a kilogram is:

a) 100 g   b) 750 g   c) 400 g   d) 640 g?

3 Write each of these volumes as a fraction of a centilitre.

a) 3 mℓ   b) 5 mℓ   c) 2 mℓ   d) 4 mℓ   e) 6 mℓ

4 Find the equivalent weight for each of these weights.

a) 50 g   b) 500 mg   c) 250 g   d) 250 mg   e) 200 g

Choose from $\dfrac{1}{4}$ g, $\dfrac{1}{2}$ kg, $\dfrac{1}{5}$ g, $\dfrac{1}{50}$ kg, $\dfrac{1}{2}$ g, $\dfrac{1}{4}$ kg, $\dfrac{1}{5}$ kg, $\dfrac{1}{20}$ kg.

5 One pound is equal to 16 ounces. What fraction of a pound is:

a) 8 ounces   b) 4 ounces   c) 12 ounces   d) 6 ounces?

6 A caterer buys 36 eggs. She finds that 24 of the eggs are broken.
What fraction of the eggs are broken?

7 Neil earns £2100 per month. His rent costs £420 per month.
What fraction of Neil's wage does he spend on rent?

8 During one day at a driving test centre, eighteen people pass their driving test and twelve people fail.
What fraction of the drivers pass?

9 A car costs £12 800 when it is new. It is sold later for £7200.
What fraction is this of its original value?

10 The table shows the results of a survey at a
factory about how employees travel to work.

a) Copy and complete the table.
b) Find the fraction of i) men ii) women
that travel by each method.
(Hint: Use the total of the Men column for
the denominator in part (i).)
c) What fraction of the total workforce are:
i) men   ii) women?

	Men	Women	Total
Car	240	300	
Bus	120	270	
Train	80	100	
Cycle	15	10	
Walk	25	40	
Total			

c Estimate one quantity as a fraction of another

Example

The table gives the number of
employees in different departments
of a company.

1347 out of the 2054 factory
workers are men.

As a fraction this is $\frac{1347}{2054}$ – not a
very easy or convenient fraction!

Department	Men	Women	Total
Factory	1347	707	2054
Distribution	59	20	79
Administration	48	196	244
Sales	29	13	42
Total	1483	936	2419

There are several ways of giving a simpler estimate. The best way depends on the
numbers in the fraction. In this fraction:

Rounding both numbers to the nearest 1000 gives $\frac{1000}{2000} = \frac{1}{2}$.

This is a much simpler fraction, but not very accurate.

Rounding both numbers to the nearest 100 gives $\frac{1300}{2100} = \frac{13}{21}$.

This is more accurate than $\frac{1}{2}$, but not very simple.

As 1347 is very close to 1350, you could round it up to 1350,
and round 2054 up to 2100.

$\frac{1400}{2100} = \frac{14}{21} = \frac{2}{3}$ About two thirds of the factory workers are men.

This fraction is simpler than $\frac{13}{21}$ and is actually more accurate as well.

When estimating fractions, **try to find near numbers that will cancel.**

Practice

Use the table in the Example on page 95 to answer questions 1–4.

1   By rounding the numbers to the nearest 10, estimate the fraction of the people in Distribution who are:
   a) women   b) men.

2   a) Estimate the fraction of the people in Administration who are women by rounding the numbers:
      i)   to the nearest 10
      ii) to the nearest 50.

   b) What happens if you round the numbers to the nearest 100?

3   a) Use these methods to estimate the fraction of people in Sales who are men.
      i)   Round 29 and 42 to the nearest 10.
      ii) Round 29 to 28, but leave the total as 42.

   b) Which answer do you think gives the most accurate fraction? Why?

4   a) Estimate the fraction of the total work force that work in:
      i) Administration   ii) Sales   iii) Distribution   iv) the Factory.

   b) Estimate the fraction of the total work force that are:
      i) men   ii) women.

5   The number of kilometres travelled by passengers on the railway in one year was 10.1 billion. Of these 7.4 billion kilometres were on ordinary tickets and the rest were on season tickets.

   a) Which of these fractions is the best estimate of the fraction of the distance that was on ordinary tickets?
      i) $\dfrac{7}{10}$ ii) $\dfrac{3}{4}$

   b) Estimate the fraction of kilometres that were on season tickets.

6   The table shows the amount of different types of energy consumed in the UK in a year.

   a) Find the total consumption.
   b) Estimate the fraction of energy that was:
      i)   coal
      ii)   petroleum
      iii) natural gas
      iv)   primary electricity.

Energy type	Amount (tonnes of oil equivalent)
Coal	42 600 000
Petroleum	81 700 000
Natural gas	91 400 000
Primary electricity	21 400 000

Source:
www.dti.gov.uk/energy/inform/energy_stats/
total_energy/1_2inlandenergyconsumption.xls

7   The table shows the amount of waste recycled
    in England and Wales during a year.
    Estimate the fraction of the total that is of each
    type.

Waste type	Amount (thousand tonnes)
Paper	943
Glass	411
Compost	816
Metal	326
Textiles	44
Cans	27
Other	298

Source: Department for Environment,
Food and Rural Affairs/Welsh Assembly
Government

Activity

Visit government or other websites to find some interesting statistics.
Write the information you find in terms of fractions.
Some suggestions are given below.

- Data for employment, industry, population, leisure and tourism etc. from the
  National Statistics website (www.statistics.gov.uk)

- Transport figures from the Department of Transport (www.gov.uk/dft)

- Information about housing from the Land Registry (www.landregistry.gov.uk)

- Information from the Department of Health (www.dh.gov.uk)

- Crime figures from the Home Office website (www.homeoffice.gov.uk)

Add and subtract fractions

N2/L2.4

a Add and subtract fractions with the same denominator

Look at the diagrams.

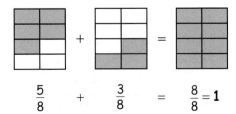

$$\frac{5}{8} + \frac{3}{8} = \frac{8}{8} = 1$$

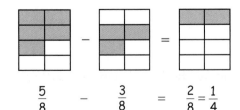

$$\frac{5}{8} - \frac{3}{8} = \frac{2}{8} = \frac{1}{4}$$

When two fractions have the same denominator:

To add them, add the numerators.   To subtract them, subtract the numerators.
The denominator stays the same.   Simplify the answer if possible.

Example

$\frac{7}{12} + \frac{1}{12} = \frac{8}{12} = \frac{2}{3}$ (cancelling by 4) $\frac{9}{10} - \frac{3}{10} = \frac{6}{10} = \frac{3}{5}$ (cancelling by 2)

Practice

Work these out. Write each answer in its simplest form.

1 $\frac{3}{5} + \frac{1}{5}$ **2** $\frac{3}{5} - \frac{1}{5}$ **3** $\frac{5}{6} + \frac{1}{6}$ **4** $\frac{5}{6} - \frac{1}{6}$

5 $\frac{2}{7} + \frac{3}{7}$ **6** $\frac{5}{7} - \frac{2}{7}$ **7** $\frac{5}{8} - \frac{1}{8}$ **8** $\frac{4}{9} + \frac{2}{9}$

9 $\frac{9}{10} - \frac{1}{10}$ **10** $\frac{2}{5} + \frac{3}{5}$ **11** $\frac{4}{15} + \frac{2}{15}$ **12** $\frac{7}{8} - \frac{3}{8}$

13 A café sells drinks, meals and snacks.

It makes $\frac{1}{5}$ of its profits on meals and $\frac{2}{5}$ of its profits on snacks.

a) What is the total fraction of its profits that it makes on meals and snacks?
b) What fraction of its profits does it make on drinks?

14 The audience at a pantomine is made up of children, men and women.
Five eighths of the audience are children.
a) What fraction are adults?

One eighth of the audience are men.
b) What fraction are women?

b Add and subtract fractions with different denominators

If two fractions have different denominators, you cannot add or subtract them immediately.

> **You must change them into equivalent fractions with the same denominator.**

Look for the **smallest** number that all the denominators divide into exactly.
(Here 2 and 3 both divide into 6.)
This is called the **common denominator**.

$\frac{1}{2} + \frac{1}{3} = ?$

$\frac{3}{6} + \frac{2}{6} = \frac{5}{6}$

Example

$$\frac{1}{6}+\frac{3}{8}+\frac{1}{3}$$

6, 8 and 3 all divide into 24
so make 24 the common denominator

$$\frac{1^{\times 4}}{6_{\times 4}}+\frac{3^{\times 3}}{8_{\times 3}}+\frac{1^{\times 8}}{3_{\times 8}}$$

$$=\frac{4}{24}+\frac{9}{24}+\frac{8}{24}$$

$$=\frac{21}{24}=\frac{7}{8}$$

$$\frac{5}{7}-\frac{3}{8}$$

7 and 8 both divide into 56
so make 56 the common denonimator

$$\frac{5^{\times 8}}{7_{\times 8}}-\frac{3^{\times 7}}{8_{\times 7}}$$

$$=\frac{40}{56}-\frac{21}{56}$$

$$=\frac{19}{56}$$

Practice

Work these out. Write each answer in its simplest form.

1 $\frac{3}{5}+\frac{1}{10}$ **2** $\frac{4}{5}-\frac{3}{10}$ **3** $\frac{2}{3}+\frac{1}{6}$ **4** $\frac{1}{2}-\frac{1}{6}$

5 $\frac{2}{3}+\frac{1}{4}$ **6** $\frac{3}{4}-\frac{2}{5}$ **7** $\frac{3}{4}-\frac{5}{8}$ **8** $\frac{5}{9}+\frac{1}{3}$

9 $\frac{2}{3}-\frac{2}{5}$ **10** $\frac{4}{7}+\frac{1}{3}$ **11** $\frac{2}{5}+\frac{1}{3}$ **12** $\frac{9}{10}-\frac{3}{5}$

13 $\frac{1}{4}+\frac{1}{3}+\frac{1}{6}$ **14** $\frac{5}{8}-\frac{2}{5}$ **15** $\frac{1}{2}+\frac{2}{15}+\frac{1}{3}$ **16** $\frac{8}{9}-\frac{5}{6}$

17 $\frac{8}{9}+\frac{1}{3}-\frac{3}{4}$ **18** $\frac{9}{10}-\frac{3}{4}$ **19** $\frac{9}{10}-\frac{2}{5}-\frac{1}{2}$ **20** $\frac{3}{4}+\frac{1}{2}-\frac{3}{8}$

21 A worker spends a quarter of her wages on rent, a third on clothes and a sixth on food.

   a) What is the total fraction that she spends?
   b) What fraction does she have left?

c Improper fractions and mixed numbers

Sometimes adding gives an **improper** fraction, i.e. a 'top-heavy' fraction.

For example, $\frac{9}{10}+\frac{3}{10}=\frac{12}{10}=\frac{6}{5}$ This is $1\frac{1}{5}$

> A **mixed number** has a whole number and a fraction part.

To change improper fractions to mixed numbers:
Make as many whole numbers as possible – the rest is the fraction part.

Example

$$\frac{17}{5} = 3\frac{2}{5} \qquad \frac{19}{7} = 2\frac{5}{7} \qquad \frac{35}{4} = 8\frac{3}{4}$$

How many 4s in 35? (8)
How many are left? (3)

($\frac{15}{5}$ equals 3   ($\frac{14}{7}$ equals 2   ($\frac{32}{4}$ equals 8

$\frac{2}{5}$ left over) $\frac{5}{7}$ left over) $\frac{3}{4}$ left over)

Also see page 84.

Practice

Write these improper fractions as mixed numbers.

1   $\frac{9}{5}$ 2 $\frac{13}{5}$ 3 $\frac{13}{3}$ 4 $\frac{17}{3}$ 5 $\frac{15}{4}$

6   $\frac{9}{2}$ 7 $\frac{13}{2}$ 8 $\frac{10}{7}$ 9 $\frac{21}{4}$ 10 $\frac{17}{6}$

11   $\frac{19}{3}$ 12 $\frac{33}{5}$ 13 $\frac{23}{8}$ 14 $\frac{37}{10}$ 15 $\frac{40}{9}$

If an answer is an improper fraction, change it to a mixed number.

Example

$$\frac{5}{8} + \frac{7}{8} = \frac{12}{8} = 1\frac{4}{8} = 1\frac{1}{2} \qquad\qquad \frac{1}{2} + \frac{3}{8} + \frac{1}{3} = \frac{12}{24} + \frac{9}{24} + \frac{8}{24} = \frac{29}{24} = 1\frac{5}{24}$$

Add these fractions. Give your answers as mixed numbers and cancel where possible.

16   $\frac{7}{9} + \frac{5}{9}$ 17 $\frac{11}{14} + \frac{5}{14}$ 18 $\frac{7}{8} + \frac{3}{4}$ 19 $\frac{1}{2} + \frac{2}{3}$

20   $\frac{4}{5} + \frac{7}{10}$ 21 $\frac{4}{7} + \frac{3}{4}$ 22 $\frac{3}{10} + \frac{2}{5} + \frac{1}{2}$ 23 $\frac{5}{6} + \frac{3}{4} + \frac{5}{8}$

d Add and subtract mixed numbers

It is often useful to change mixed numbers into improper fractions.

> **To change mixed numbers to improper fractions:**
> **Change the whole number, then add the other part.**

Example

$$1\frac{3}{4} = \frac{7}{4}$$

(1 is 4 quarters
+ 3 quarters
= 7 quarters)

$$2\frac{1}{4} = \frac{9}{4}$$

(2 is 8 quarters
+ 1 quarter
= 9 quarters)

$$2\frac{3}{5} = \frac{13}{5}$$

(2 is 10 fifths
+ 3 fifths
= 13 fifths)

$$4\frac{1}{5} = \frac{21}{5}$$

Numerator =
$4 \times 5 + 1 = 21$

Denominator stays
the same

(4 is 20 fifths
+ 1 fifth
= 21 fifths)

Practice

Write these mixed numbers as improper fractions.

1 $1\frac{1}{2}$ **2** $2\frac{1}{2}$ **3** $3\frac{1}{2}$ **4** $1\frac{2}{3}$ **5** $2\frac{2}{3}$

6 $2\frac{1}{5}$ **7** $3\frac{4}{5}$ **8** $2\frac{1}{6}$ **9** $1\frac{5}{6}$ **10** $2\frac{3}{8}$

11 $2\frac{3}{7}$ **12** $5\frac{7}{8}$ **13** $4\frac{9}{10}$ **14** $3\frac{4}{9}$ **15** $6\frac{2}{7}$

To add or subtract mixed numbers:
either
• **work with the whole numbers and fractions separately**
or
• **change them into improper fractions.**   Choose the method you prefer.

Example

$$2\frac{1}{4} + 1\frac{2}{3}$$

$$2\frac{1}{4} + 1\frac{2}{3} = 3\frac{1}{4} + \frac{2}{3}$$ or $$2\frac{1}{4} + 1\frac{2}{3} = \frac{9}{4} + \frac{5}{3}$$ $2 \times 4 + 1 = 9$

The common denominator is 12. (3 and 4 both divide into 12)

$$3\frac{3}{12} + \frac{8}{12} = 3\frac{11}{12}$$ $$\frac{27}{12} + \frac{20}{12} = \frac{47}{12} = 3\frac{11}{12}$$

Example

$$4\frac{2}{5} - 2\frac{7}{10}$$

$$4\frac{2}{5} - 2\frac{7}{10} = 2\frac{2}{5} - \frac{7}{10}$$ or $$4\frac{2}{5} - 2\frac{7}{10} = \frac{22}{5} - \frac{27}{10}$$ $4 \times 5 + 2 = 22$

The common denominator is 10. (5 divides into 10)

$$2\frac{4}{10} - \frac{7}{10} = 1\frac{14}{10} - \frac{7}{10} = 1\frac{7}{10}$$ $$\frac{44}{10} - \frac{27}{10} = \frac{17}{10} = 1\frac{7}{10}$$

Work these out. Write each answer in its simplest form.

16 $1\frac{1}{4} + 1\frac{1}{2}$ **17** $2\frac{3}{5} - 1\frac{3}{4}$ **18** $3\frac{1}{2} + 1\frac{5}{6}$ **19** $3\frac{3}{4} - 2\frac{1}{3}$

20 $5\frac{1}{2} + 2\frac{5}{8}$ **21** $4\frac{1}{3} - 1\frac{1}{2}$ **22** $2\frac{3}{4} - 1\frac{4}{5}$ **23** $3\frac{7}{10} + 2\frac{2}{5}$

24 $5\frac{2}{3} - 1\frac{4}{5}$ **25** $3\frac{4}{7} + 2\frac{1}{2}$ **26** $4\frac{1}{2} + 2\frac{3}{4}$ **27** $2\frac{1}{2} - 1\frac{3}{8}$

28 The preparation of food for a meal takes $1\frac{1}{4}$ hours and it takes $1\frac{3}{4}$ hours to cook it. How long does it take altogether?

29 A $1\frac{1}{2}$ pound piece is cut from a cheese that weighs $5\frac{3}{4}$ pounds. What weight of cheese is left?

30 The table shows the hours worked by a part-time hairdresser.

a) Find the total time she worked.
b) How much longer did she work on Friday than on Saturday?
c) How much longer did she work on Friday than on Thursday?

Day	Time worked
Thurs	$3\frac{3}{4}$ hours
Fri	$6\frac{1}{2}$ hours
Sat	$5\frac{1}{4}$ hours

Multiplying fractions

N2/L2.4

$\frac{1}{2} \times \frac{1}{3}$ means $\frac{1}{2}$ of $\frac{1}{3}$.

> Think of x as meaning 'of'.

The sketch shows a rectangle split into 3 thirds.
Half of one of these thirds equals one sixth.

$\frac{1}{2} \times \frac{1}{3} = \frac{1}{6}$

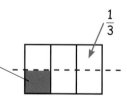

> ▶ **To multiply fractions, multiply the numerators, then multiply the denominators. Change mixed numbers to improper fractions first.**

Example

Find: a) $\frac{3}{4}$ of $\frac{2}{5}$ b) $\frac{2}{3} \times 1\frac{1}{2}$

a) $\frac{3}{4} \times \frac{2}{5} = \frac{6}{20} = \frac{3}{10}$ **Always simplify your answer where possible.**

b) $\frac{2}{3} \times 1\frac{1}{2} = \frac{2}{3} \times \frac{3}{2}$ **Change mixed numbers to improper fractions first.**

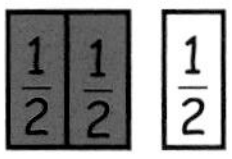

$= \frac{6}{6}$

$= 1$ The sketch shows that $\frac{2}{3}$ of $1\frac{1}{2} = 1$.

Practice

Work these out. Write each answer in its simplest form.

1 $\frac{1}{2} \times \frac{1}{5}$ **2** $\frac{1}{3} \times \frac{1}{7}$ **3** $\frac{1}{2} \times \frac{4}{9}$ **4** $\frac{2}{3} \times \frac{6}{7}$

5 $\frac{3}{4} \times \frac{8}{11}$ **6** $\frac{2}{5} \times \frac{1}{4}$ **7** $\frac{3}{4} \times \frac{2}{3}$ **8** $\frac{7}{8} \times \frac{4}{5}$

9 Kelly says that a half of a quarter is one eighth. Is she correct?

10 Show that a third of a quarter is the same as a half of one sixth.

Work these out.
Cancel and write the answers as mixed numbers where possible.

11 $\frac{1}{2} \times 3\frac{1}{3}$ **12** $\frac{3}{4} \times 2\frac{2}{5}$ **13** $\frac{5}{6} \times 1\frac{1}{2}$ **14** $1\frac{1}{2} \times \frac{4}{9}$

15 $2\frac{1}{4} \times 1\frac{1}{3}$ **16** $3\frac{3}{5} \times 1\frac{1}{4}$ **17** $2\frac{1}{3} \times 4\frac{1}{2}$ **18** $3\frac{4}{7} \times 1\frac{3}{5}$

19 A chef uses two-thirds of a $2\frac{1}{2}$ kilogram bag of potatoes to make chips.
Find the weight of the potatoes used to make the chips.

20 A roll of fabric is $4\frac{4}{5}$ metres long.

A dressmaker uses half of the roll to make a jacket and a third of the roll to make a skirt.

Work out:

a) the length of fabric used for the jacket

b) the length of fabric used for the skirt

c) the length of fabric that is left on the roll.

Show that the length left on the roll is one sixth of the original length.

Dividing fractions

N2/L2.4

To answer $\frac{1}{2} \div \frac{1}{4}$, you need to find how many quarters there are in a half.

The answer is 2.

You can use fractions to find the answer like this:

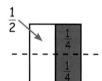

$$\frac{1}{2} \div \frac{1}{4} = \frac{1}{2} \times \frac{4}{1} = \frac{4}{2} = \frac{2}{1} = 2$$

> To divide fractions, turn the fraction you are dividing by upside down,
> then multiply. Change mixed numbers to improper fractions first.

Example

Find: **a)** $\frac{2}{5} \div \frac{3}{10}$ **b)** $\frac{2}{3} \div 1\frac{1}{2}$

a) $\frac{2}{5} \div \frac{3}{10} = \frac{2}{5} \times \frac{10}{3}$ **Turn the second fraction upside down and multiply.**

$$= \frac{20}{15} = \frac{4}{3} = 1\frac{1}{3}$$ **If possible, simplify your answer and change to mixed numbers.**

b) $5\frac{1}{4} \div 1\frac{1}{2} = \frac{21}{4} \div \frac{3}{2}$ **Change mixed numbers to improper fractions first.**

$$= \frac{21}{4} \times \frac{2}{3}$$

$$= \frac{42}{12} = \frac{7}{2} = 3\frac{1}{2}$$ or $= \frac{42}{12} = \frac{21}{6} = \frac{7}{2} = 3\frac{1}{2}$

The answer to b) means that there are $3\frac{1}{2}$ lots of $1\frac{1}{2}$ in $5\frac{1}{4}$.

The sketch below shows that this is true.

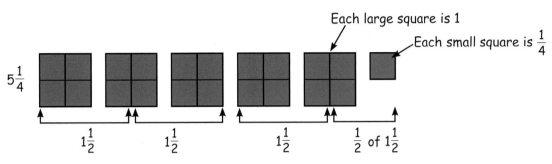

$5\frac{1}{4}$

Each large square is 1

Each small square is $\frac{1}{4}$

$1\frac{1}{2}$ $1\frac{1}{2}$ $1\frac{1}{2}$ $\frac{1}{2}$ of $1\frac{1}{2}$

Practice

Work these out. Write each answer in its simplest form.

1 $\frac{1}{2} \div \frac{1}{6}$ **2** $\frac{1}{6} \div \frac{1}{3}$ **3** $\frac{2}{5} \div \frac{1}{10}$ **4** $\frac{3}{4} \div \frac{3}{8}$

5 $\frac{2}{3} \div \frac{8}{9}$ **6** $\frac{3}{5} \div \frac{6}{7}$ **7** $\frac{5}{9} \div \frac{2}{3}$ **8** $\frac{7}{8} \div \frac{3}{5}$

9 A bag contains $\frac{4}{5}$ kg of icing sugar.

Jan divides this into smaller bags, each containing $\frac{1}{10}$ kg.

How many smaller bags does she use?

10 Show that there are six ninths in two-thirds.
Do this in **two different ways**.

Work these out.
Cancel and write the answers as mixed numbers where possible.

11 $1\frac{2}{3} \div \frac{1}{6}$ **12** $\frac{3}{4} \div 1\frac{1}{2}$ **13** $2\frac{1}{4} \div \frac{3}{5}$ **14** $\frac{5}{6} \div 1\frac{1}{3}$

15 $3\frac{1}{2} \div 1\frac{4}{5}$ **16** $2\frac{1}{7} \div 1\frac{1}{4}$ **17** $2\frac{4}{9} \div 3\frac{2}{3}$ **18** $3\frac{4}{7} \div 1\frac{1}{3}$

19 Katie cuts ribbons to decorate her kite. Each ribbon is $1\frac{1}{4}$ metres long.
How many ribbons does she cut from a roll that is $7\frac{1}{2}$ metres long?

20 It takes three-quarters of an hour to hand-paint a plate.
Use fractions to work out how many plates can be
hand-painted in 6 hours.

Hint: Write 6 as $\frac{6}{1}$.

Fractions on a calculator   N2/L1.11, N2/L2.10

Your calculator may be able to calculate with fractions.

The fraction key often looks like this: [a^b/c] or [⊟] . Try the examples below.

	Example	Press	Display	Answer
Entering a fraction	$\frac{2}{3}$	[2] [a^b/c] [3]	2⌐3 (or 2r3)	
Equivalent fractions – cancelling	$\frac{6}{8}$	[6] [a^b/c] [8] [=]	3⌐4	$\frac{3}{4}$
Improper fractions to mixed numbers	$\frac{8}{5}$	[8] [a^b/c] [5] [=]	1⌐3⌐5	$1\frac{3}{5}$
Mixed numbers to improper fractions	$2\frac{3}{4}$	[2] [a^b/c] [3] [a^b/c] [4] [d/c]	11⌐4	$\frac{11}{4}$
Finding a fraction of a number ('of' is ×)	$\frac{3}{4}$ of 6	[3] [a^b/c] [4] [×] [6] [=] or [6] [÷] [4] [×] [3] [=]	4⌐1⌐2	$4\frac{1}{2}$
Fraction arithmetic use +, −, ×, ÷	$\frac{3}{4} - \frac{2}{3}$	[3] [a^b/c] [4] [−] [2] [a^b/c] [3] [=]	1⌐12	$\frac{1}{12}$
Fraction to decimal – do as a division	$\frac{5}{8}$	[5] [÷] [8] [=]	0.625	0.625
Decimal to fraction – use place value then cancel on the calculator	0.35	Put $\frac{35}{100}$ into the calculator [3] [5] [a^b/c] [1] [0] [0] [=]	7⌐20	$\frac{7}{20}$

Try the earlier fraction questions again using a calculator.

Number

9 Decimals

Understand decimals in context (up to 2 decimal places)   N2/E3.3

Decimal numbers have a whole number part and a fraction part.
The **decimal point** separates the two parts.
Read 7.3 as '**seven point three**.'

We often use decimals to give lengths.
Look at a ruler or tape measure.

Each centimetre (cm) is divided into 10 equal divisions. Each division is $\frac{1}{10}$ cm.

As a decimal this is 0.1 cm.

2 divisions $= \frac{2}{10}$ cm or 0.2 cm, 3 divisions $= \frac{3}{10}$ cm or 0.3 cm and so on.

decimal point

whole number   7 . 3   fraction part
(three tenths)

Example

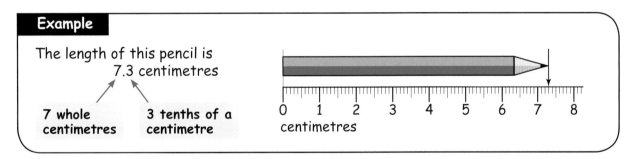

The length of this pencil is
7.3 centimetres

7 whole
centimetres

3 tenths of a
centimetre

Practice

1 Write down the length in centimetres marked by each arrow.

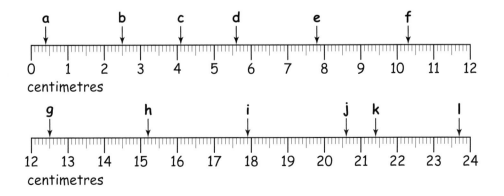

To put lengths in order, compare the whole number part then the fraction part.

| **Example** | The lengths of 5 pencils are: 7.6 cm, 8.1 cm, 6.9 cm, 7.3 cm, 7.9 cm. |

Look at the **whole numbers first**.
8 is the largest so the longest pencil is 8.1 cm.
There are three lengths starting with 7   (7.6 cm, 7.3 cm and 7.9 cm).
To put them in order **compare the tenths**:   7.9 cm then 7.6 cm then 7.3 cm.
The shortest pencil is 6.9 cm long.
The answer is: 8.1 cm, 7.9 cm, 7.6 cm, 7.3 cm, 6.9 cm

Whole nos.	tenths
7 .	6
8 .	1
6 .	9
7 .	3
7 .	9

2 Put each group in order, starting with the **longest**.

   a) Caterpillars: 1.9 cm, 2.4 cm, 1.8 cm, 2.5 cm, 2.1 cm
   b) Leaves: 4.5 cm, 3.8 cm, 3.9 cm, 3.4 cm, 4.1 cm, 4.3 cm
   c) Twigs: 10.6 cm, 9.9 cm, 11.1 cm, 10.5 cm, 9.6 cm, 10.7 cm
   d) Tree heights: 5.3 m, 6.1 m, 5.8 m, 5.5 m, 6.4 m, 5.7 m, 6.2 m, 5.1 m
   e) Road lengths: 1.8 km, 2.9 km, 2.3 km, 1.2 km, 0.9 km, 1.7 km, 1.4 km, 2.1 km

Each length in the last section had **1 decimal place** (1 dp),
i.e. each length had 1 digit after the decimal point.

7.3 cm
↑
1 decimal place

For money we often use **2 decimal places** (2 dp).
There are **100 pence in each pound**
so   £2.31 = 231p (p stands for 'pence').

£2.31 is 2 pounds, 31 pence
↑↑
2 decimal places

Each penny is one hundredth of a pound. $31p = £\frac{31}{100}$ (31 **hundredths** of a £)

3 Write each amount in pence.

   a) £1.00   **b)** £2.18   **c)** £1.09   **d)** £4.07   **e)** £0.62
   f) £8.24   **g)** £5.70   **h)** £9.04   **i)** £10.17   **j)** £16.40

4 Write each amount in £ using decimals.

   a) 213p   **b)** 402p   **c)** 85p   **d)** 350p   **e)** 103p
   f) 600p   **g)** 725p   **h)** 937p   **i)** 1000p   **j)** 2409p

To put prices in order, **compare the £ first then the pence**.

> **Example** Put these prices in order, starting with the **cheapest**.
> £3.45, £2.98, £3.06, £2.80, £2.77
>
> Look at the whole £s first: 2 is less than 3.
> The prices starting with 2 are £2.98, £2.80 and £2.77.
> Now compare the pence. The correct order is £2.77, £2.80, £2.98.
>
> The prices starting with 3 are £3.45 and £3.06.
> Now compare the pence. The correct order is £3.06, £3.45.
>
> The answer is: £2.77, £2.80, £2.98, £3.06, £3.45
>
> | £ | pence |
> |---|---|
> | 3 . | 45 |
> | 2 . | 98 |
> | 3 . | 06 |
> | 2 . | 80 |
> | 2 . | 77 |

5 Put each group in order from cheapest to most expensive.

a) £1.21   £1.37   £1.02   £2.07   £1.46   £1.19   £1.87

b) £74.02   £47.20   £56.78   £66.18   £63.21   £63.02

c) £11.17   £10.08   £11.91   £10.07   £11.01   £10.70

100 centimetres = 1 metre. (See page 178.)
Each centimetre (cm) is **one hundredth** of a metre (m).

15 cm = 0.15 m (just like 15p = £0.15)   and   234 cm = 2.34 m

(Read 0.15 as 'point one five'
and 2.34 as 'two point three four'.)

2 metres   **34 centimetres**

Working with m and cm is like working with £ and pence, but with measurements we usually leave off any zero at the end of the decimal part.

> **Example**
>
> In measurements: 30 cm = 0.3 m   and   750 cm = 7.5 m   **zero not needed**
>
> In money: 30p = £0.30   and   750p = £7.50   **zero needed**

6 How many centimetres are there in each of these measurements?

a) 0.22 m   b) 1.07 m   c) 1.73 m   d) 0.74 m   e) 1.33 m

f) 1.12 m   g) 1.15 m   h) 0.67 m   i) 2.6 m   j) 5.4 m

7 Write each measurement in metres using decimals.

a) 154 cm   b) 315 cm   c) 94 cm   d) 120 cm   e) 762 cm

f) 402 cm   g) 70 cm   h) 250 cm   i) 1134 cm   j) 2230 cm

8 The heights of some groups of children are given below.
Put each group in order starting with the **shortest**.

a) 1.5 m	1.67 m	1.72 m	1.8 m	1.55 m	1.63 m	1.74 m
b) 1.32 m	1.26 m	1.39 m	1.28 m	1.31 m	1.3 m	1.4 m
c) 0.72 m	0.99 m	1.02 m	0.88 m	1.01 m	0.84 m	1.03 m

In fractions, 50 out of 100 = $\frac{50}{100}$ = $\frac{1}{2}$ Half of a metre = 50 cm = 0.5 m

Half of £1 = 50 pence, i.e. £0.50   and $3\frac{1}{2}$ m = 3.5 m

Learn

> 0.5 in decimals = $\frac{1}{2}$ in fractions.

---

9 a) Write these as decimals.   **i)** $1\frac{1}{2}$ m   **ii)** $7\frac{1}{2}$ m   **iii)** $10\frac{1}{2}$ m

b) Write these as mixed numbers.   **i)** 2.5 m   **ii)** 6.5 m   **iii)** 15.5 m

Activities

1 Read prices from adverts, price lists, etc. Select coins and/or notes to match.

2 Measure some items in cm then write the measurements in m using decimals.

Write and compare decimals up to 3 decimal places   N2/L1.4, N2/L2.5

a Place value

Look at the decimal numbers. The headings show the **place value** of each digit.

	Hundreds 100s	Tens 10s	Units 1s	.	Tenths 10ths	Hundredths 100ths	Thousandths 1000ths
14.13		1	4	.	1	3	
7.509			7	.	5	0	9
0.173			0	.	1	7	3

Look at the first decimal 14.13 (fourteen point one three).
It has no hundreds, 1 ten, 4 units, 1 tenth, 3 hundredths and no thousandths.

Note
ten = 10
tenth = $\frac{1}{10}$

In money, £14.13 can be made up from 1 ten-pound note, 4 pound coins, 1 ten-pence coin and

3 one-penny coins. (10 pence is $\frac{1}{10}$ of a £ and 3 pence is $\frac{3}{100}$ of a £.)

There is more than one way of writing the decimal part in fractions.

$$14.13 = 14 + \frac{1}{10} + \frac{3}{100} \qquad \text{or} \qquad 14\frac{13}{100}$$

as separate fractions   **in a single fraction**

This is like having 1 ten-pence coin and 3 one-pence coins or 13 one-pence coins.

Now look at the second decimal in the place value table (on page 110), 7.509.

The **place value** of the **digit** 5 is $\frac{5}{10}$ and the place value of the 9 is $\frac{9}{1000}$.

There are no hundredths, but the zero is important as it keeps the 9 in the thousandths column.

In fractions $7.509 = 7 + \frac{5}{10} + \frac{9}{1000}$ or $7\frac{509}{1000}$ ← decimal digits

heading for last digit

The last decimal, 0.173, has no whole number part.

In fractions $0.173 = \frac{1}{10} + \frac{7}{100} + \frac{3}{1000}$ or $\frac{173}{1000}$

Practice

1 **a)** Draw a place value table. Write the following numbers in it.

  i) 116.74   **ii)** 37.54   **iii)** 12.06   **iv)** 4.52
  v) 536.09   **vi)** 506.926   **vii)** 67.354   **viii)** 127.063
  ix) 107.401   **x)** 231.079   **xi)** 619.102   **xii)** 423.073

  b) Write down the place value of each digit that is coloured orange.

2 Draw a place value table. Write the following numbers in it.

  a) Fifty-seven point three two   **b)** Six point nine
  c) Twelve point four one   **d)** One hundred point three five
  e) Ninety-six point nought seven   **f)** Twelve point five nought six
  g) Four hundred and ninety-seven point six four three

3 Write these as fractions or mixed numbers using tenths, hundredths or thousandths (see page 84 for mixed numbers).

  a) 0.7   **b)** 0.06   **c)** 0.003   **d)** 0.04   **e)** 3.05
  f) 6.4   **g)** 15.007   **h)** 8.02   **i)** 10.6   **j)** 240.009

4 Write each of these numbers in terms of fractions in **two different ways** (i.e. as separate fractions and as a single fraction).

  a) 0.17   **b)** 0.49   **c)** 0.031   **d)** 0.357   **e)** 6.23
  f) 8.043   **g)** 42.107   **h)** 1.241   **i)** 502.69   **j)** 17.653

b Write decimals as common fractions

You can simplify fractions by dividing the top and bottom by the same number.

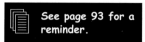 See page 93 for a reminder.

Example   Write 0.75 as a fraction.

$$0.75 = \frac{75}{100} = \frac{15}{20} = \frac{3}{4}$$

As a fraction $0.75 = \frac{3}{4}$

Dividing top and bottom by 5 then 5 again

Practice

Write these decimals as simple fractions.

1 0.5	**2** 0.2	**3** 0.4	**4** 0.6	**5** 0.8
6 0.15	**7** 0.35	**8** 0.55	**9** 0.25	**10** 0.45

c Put decimals in order of size

> Compare the whole number part, then the decimal digits in order of size.

Example

Put 2.97, 3.081, 2.963, 2.79, 2.805, 2.965 in order, starting with the largest.

3.081
2.97
2.965
2.963
2.805
2.79

3rd dp
2nd dp
1st dp
whole

Compare the digits

First look at the whole numbers.
3 is bigger than 2 so 3.081 is the largest.
All the other numbers start with 2.

Look at the 1st decimal place.
9 is the largest, but there are 3 numbers.
To compare 2.97, 2.963 and 2.965:

Look at the 2nd decimal place.
7 is larger than 6 so 2.97 is the largest.
Now to compare 2.963 and 2.965:

Look at the 3rd decimal place.
5 is larger so 2.965 is next, then 2.963
The remaining numbers are 2.79 and 2.805. To compare these:

Go back to the 2nd decimal place.
8 is larger than 7, so 2.805 is next.

The answer is: 3.081   2.97   2.965   2.963   2.805   2.79

Alternative method

You may find it easier to write all the original numbers to 3 decimal places (by adding zeros to the end where necessary): 2.970, 3.081, 2.963, 2.790, 2.805, 2.965
Then compare the whole numbers and decimal parts in two stages.
Try this and see which method you prefer.

Practice

1 Put each group in order of size, starting with the **largest**.

a) 5.02	7.21	6.76	4.32	6.12	5.98	5.03	4.68	7.11
b) 13.1	14.21	12.97	14.26	12.5	13.03	12.43	12.07	
c) 26.42	21.5	25.06	26.11	21.8	25.17	21.43	26.47	
d) 3.19	7.21	3.5	6.48	3.21	7.84	6.96	4.06	5.74

2 Put each group in order of size, starting with the **smallest**.

a) 1.075	0.307	0.003	1.11	1.05	1.761	0.23
b) 12.05	12.502	12.429	12.5	12.121	12.005	
c) 20.095	19.16	21.35	19.006	21.101	19.375	
d) 604.002	483.562	581.879	581.098	428.003	412.438	

3 Put each group of weights in order, **largest** (heaviest) first.

a) 2.009 kg	1.045 kg	1.762 kg	2.032 kg	1.104 kg
b) 1.7 kg	1.07 kg	1.707 kg	1.007 kg	1.077 kg
c) 0.041 kg	0.405 kg	0.318 kg	0.058 kg	0.201 kg

4 The following times (in seconds) are recorded at a race day.
   Write each group in order, starting with the **shortest** (quickest) time.

a) 12.056	12.009	11.909	11.099	12.201	12.037
b) 18.567	19.002	18.411	18.591	18.302	19.041
c) 26.057	25.503	25.601	25.072	26.001	25.021

Use a calculator to solve problems   N2/E3.4

A calculator is useful for working with decimals, but it is easy to press the wrong key.
It is important that you always check the answer.

When using a calculator:

- take care to press the numbers and the decimal point correctly
- include a zero before the decimal point in the answer if there is no whole number
- write down a sensible number of decimal places. For money, use 2 decimal places to give the nearest whole pence and add a zero if necessary (e.g. 3.5 means £3.50)
- write down the units if there are any (e.g. £, m, kg)
- think about whether the answer is sensible
- use estimation, inverse calculations or a different method to check the answer.

Example

Three presents cost £5.99, £2.45 and £3.95.
Using a calculator gives the total: $5.99 + 2.45 + 3.95 = 12.39$ Answer = **£12.39**

Check by estimating (to the nearest £): £6 + £2 + £4 = £12

or

Check using inverse calculations: 12.39 – 3.95 – 2.45 = 5.99

or

Check by adding in a different order: 3.95 + 2.45 + 5.99 = 12.39

There is a choice
of ways to check
– just use one way.

Practice

Use a calculator to work these out. Check each answer.

1 a) $13 + 1.05$
   b) £1.02 + £3.74
   c) $1.5\,m + 0.5\,m$
   d) $5.75 + 3.05 + 1.9$
   e) £4.57 – £0.86
   f) £11.00 – £2.19
   g) $9\,m - 4.5\,m$
   h) $23 - 1.14$
   i) 5.03×3
   j) £3.07 × 0.5
   k) $5 \times 0.2\,kg$
   l) £7.15 × 4
   m) $6.45 \div 3$
   n) £3.13 ÷ 4
   o) $6\,m \div 0.3\,m$
   p) $1.35 \div 0.2$

2 Four friends share the cost of a meal, £46.80. How much do they pay each?

3 A girl spends £1.17 on sweets and her brother spends £4.20.
   How much do they spend altogether?

4 A boy saves £3.50 each week. How much does he save in 8 weeks?

5 a) A dressmaker buys 14.75 metres of material. She only uses 9.5 metres.
       How much material does she have left?
   b) The dressmaker cuts 1.5 metres of ribbon into 5 equal lengths.
       How long is each length?

6 A teenager spends £4.85 on magazines each week.
   How much does she spend during 6 weeks?

7 A shopper spends £4.07, £1.99 and £2.70.
   How much is this altogether?

8 Oliver wins £20. He spends £15.74 of his winnings.
   How much does he have left?

Multiply and divide decimals by 10, 100 and 1000

To multiply by 10, 100 or 1000, move the numbers to the **left** so they become **larger**.

Move the numbers as many times as there are zeros in 10, 100 or 1000:

one column for 10 (as it has 1 zero); two columns for 100 (2 zeros);

three columns for 1000 (3 zeros).

To divide by 10, 100 or 1000, move the numbers to the **right** so they become **smaller**.

Example   4.1 × 100

4.1 × 100 = **410**

The 4 moves 2 places – it is now 4 hundreds instead of 4 units. The 1 is now worth ten instead of a tenth. Put zero in the units column to keep the numbers in place.

Hundreds 100	Tens 10s	Units 1s	.	Tenths 10ths
		4	.	1
	4	1	0	.	

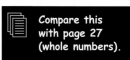 Compare this with page 27 (whole numbers).

Example   32 ÷ 10

32 ÷ 10 = **3.2**

The 3 and 2 both move 1 place to the right.
The 3 is now 3 units.
The 2 is now 2 tenths.

Tens 10s	Units 1s	.	Tenths 10ths
3	2	.	0
	3	.	2

Example   970 ÷ 1000

970 ÷ 1000 = **0.97**

Hundreds 100	Tens 10s	Units 1s	.	Tenths 10ths	Hundreds 100ths	Thousands 1000ths
9	7	0	.	0		
		0	.	9	7	0

This is needed to emphasise the position of the decimal point.

Not needed as there are no thousandths.

Practice

Use a place value table to help you work these out.

1 **a)** 17.4 × 10   **b)** 5.63 × 100   **c)** 17.2 × 1000
    d) 11.743 × 10   **e)** 16.01 × 1000   **f)** 112.2 × 100

2 **a)** 56.22 ÷ 10   **b)** 63.04 ÷ 100   **c)** 1.44 ÷ 100
    d) 102.01 ÷ 10   **e)** 57.4 ÷ 1000   **f)** 106 ÷ 1000

3 **a)** 26.11 × 10   **b)** 5.77 ÷ 10   **c)** 1.04 × 100
    d) 121 ÷ 100   **e)** 47.2 × 1000   **f)** 566.02 ÷ 1000

The metric system is based on tens, hundreds and thousands.

£1 = 100p   1 cm = 10 mm   1 m = 100 cm = 1000 mm   1 km = 1000 m
1 kg = 1000 g   1 cℓ = 10 mℓ   1 ℓ = 100 cℓ = 1000 mℓ

To convert from one unit to another, you need to multiply or divide by 10, 100 or 1000.
To convert from **large** units **to small** units **multiply**.
To convert from **small** units **to large** units **divide**.

Examples

4.6 m to cm
4.6 m × 100 = 460 cm

▶ **There are 100 cm in a metre –
to convert metres to centimetres
multiply by 100.**

1679 g to kg
1679 g ÷ 1000 = 1.679 kg

▶ **There are 1000 g in a kilogram
– to convert grams to kilograms
divide by 1000.**

---

4 Convert these.

a) 30 cm to mm   **b)** 3.7 m to cm   **c)** 145 cm to m   **d)** 45 mm to cm
e) £3.45 to pence   **f)** 789p to £   **g)** 0.66 m to cm   **h)** 65 cℓ to litres
i) 1567 g to kg   **j)** 18 mm to cm   **k)** 0.6 km to m   **l)** 7580 m to km

5 Copy the amounts below. Draw lines between the equivalent amounts.

a) 7500 g   2.3 m
     230 cm   0.75 ℓ
     2300 g   7.5 kg
     75 cℓ   1.25 ℓ
     125 cℓ   2.3 kg

b) 0.5 m   750 g
     1.2 ℓ   5 mℓ
     0.75 kg   50 cm
     1.2 cm   120 cℓ
     0.5 cℓ   12 mm

For more conversions see pages 182, 204 and 216.

Round decimal numbers and approximate   N2/L1.7, N2/L2.5

To round decimal numbers to whole numbers:

● Look at the **first decimal place** (1st dp).

● If it is **below 5, leave the whole number as it is.**

● If it is **5 or above, round the whole number up by 1.**

> Rounding is used to estimate or check calculations.

Example   Round 2.47 to the nearest whole number.

Units 1s	.	Tenths 10ths	Hundreds 100ths
2	.	4	7

4 is **below 5**
so 2.47 = **2** to the nearest whole number.

↑
1st dp   This is the 'deciding digit'

Example Round 9.8 to the nearest whole number.

Whole number   Deciding digit

9.8

1st dp

8 is above 5
so 9.8 = **10**
to the nearest whole number.

9.8

9   9.5   10

9.8 is more than 9.5
– nearer 10 than 9

Practice

1 Round each of these to the nearest whole number.

a) 14.7   b) 13.1   c) 7.4   d) 3.8   e) 9.7   f) 6.21

g) 16.31   h) 11.81   i) 9.84   j) 11.71   k) 15.07   l) 0.71

m) 4.315   n) 5.572   o) 0.469   p) 26.98   q) 249.5   r) 99.67

In these questions round values to whole numbers then estimate the answers.

2 Three friends spend £4.96 each. Estimate how much they spend altogether.

3 A dressmaker buys 1.8 m of blue silk, 1.2 m of red silk and 2.9 m of green silk.
Estimate how much material is bought altogether.

4 A shopper buys items costing £4.21, £0.91 and £1.21.
Estimate the total cost.

5 Estimate the total weight of these parcels.

6 Rita says she has £15 in her purse, to the nearest £.
What is the least amount of money she can have?

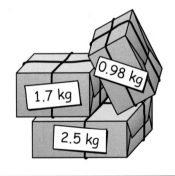

0.98 kg

1.7 kg

2.5 kg

To round a decimal to **one decimal place, look at the** second decimal place.
If this is **5 or above, round the first decimal place up**.

Example Round 7.26 to 1 decimal place.

Deciding digit

7.26

1st dp   2nd dp

6 is above 5
so 7.26 = **7.3** to 1 decimal place.

To round to **two** decimal places, look at the **third** decimal place. If this **is 5 or above, round the second decimal place up** (otherwise leave it as it is).

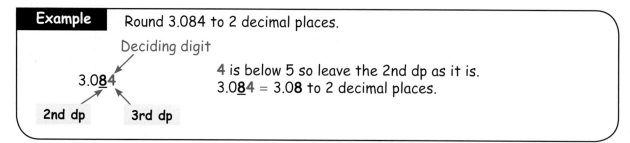

Example   Round 3.084 to 2 decimal places.

Deciding digit

3.0**8**4

4 is below 5 so leave the 2nd dp as it is.
3.0**8**4 = 3.08 to 2 decimal places.

2nd dp   3rd dp

Sometimes you may need to choose an appropriate level of accuracy.
For money written in £, this is usually 2 decimal places (nearest pence).

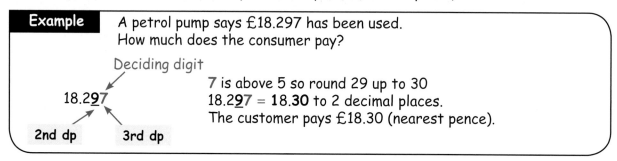

Example   A petrol pump says £18.297 has been used.
How much does the consumer pay?

Deciding digit

18.2**9**7

7 is above 5 so round 29 up to 30
18.2**9**7 = **18.30** to 2 decimal places.
The customer pays £18.30 (nearest pence).

2nd dp   3rd dp

To round to **three** decimal places, look at the **fourth** decimal place and so on.

7 Round these numbers to 2 decimal places.

a) 5.678   b) 16.591   c) 11.325   d) 21.898   e) 155.793
f) 16.224   g) 131.076   h) 6.749   i) 121.903   j) 1.891
k) 2.396   l) 16.7081   m) 15.9152   n) 13.5618   o) 1.0065

8 Round these amounts to 2 decimal places.

a) £1.273   b) 2.726 kg   c) £12.158   d) 17.415 m   e) 22.567 km
f) 14.725 g   g) 16.194 cm   h) £9.6439   i) 0.473 ℓ   j) 15.098 km

9 Round these.

a) £15.39 to the nearest pound   b) 13.606 kg to the nearest kg
c) 5.1128 litres to 2 dp   d) £31.566666 to 2 dp
e) £7.165431 to the nearest pence   f) 6.597 m to 2 dp

10 A student is asked to give answers correct to 1 dp. Round these to 1 dp.

a) 9.533333   b) 12.3567   c) 13.17098   d) 6.6974   e) 121.79311
f) 5.7217   g) 91.7508   h) 151.90854   i) 65.0911   j) 73.64842

11 Round these amounts to the appropriate accuracy suggested.

	Quantity	Amount
a)	Interest on savings to the nearest pence	£7.362
b)	Weight of raspberries to the nearest tenth of a kilogram	6.184 kg
c)	Width of a kitchen to the nearest hundredth of a metre	4.753 m
d)	Cost of electricity to the nearest pence	£123.685
e)	Capacity of a watering can to the nearest tenth of a litre	9.17 litres

L2 **12** Round the following numbers to three decimal places.

a) 5.79241   **b)** 16.9376   **c)** 3.74912   **d)** 0.97386   **e)** 4.03189

f) 16.7701   **g)** 12.08051   **h)** 27.90109   **i)** 1.54973   **j)** 17.6999

> **Activity**
>
> Find out how to round decimals on your calculator or on a spreadsheet.
> Use this method to check your answers to the questions above.

Add and subtract decimals

N2/E3.5, N2/L1.5, N2/L2.6

> ▶ **To add or subtract decimals, line up the decimal points.**

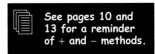

See pages 10 and 13 for a reminder of + and − methods.

You can check your answer by:

- rounding to the nearest whole numbers and estimating the answer
- using inverse (opposite) calculations (see page 14)
- using a calculator.

> **Example**   One package weighs 15.07 kilograms. Another weighs 3.1 kilograms.
> What is the total weight?
>
> **15.07 kg + 3.1 kg**
>
> ```
>   15.07
> +  3.10
>   18.17
> ```
>
> | Check (estimating) | Check (inverse) |
> |:---:|:---:|
> | 15 | 18.17 |
> | + 3 | − 3.10 |
> | 18 | 15.07 |
>
> The total weight is 18.17 kg.

> **Example**   A dressmaker has 24.3 metres of fabric. She uses 3.75 metres to make a dress. How much fabric is left?
>
> **24.3 m – 3.75 m**
>
> 24.30 ⟵——————— Fill gaps with 0s
> − 3.75
> 20.55
>
> Check (estimating)
> 24
> − 4
> 20
>
> Check (inverse)
> 20.55
> + 3.75
> 24.30
>
> There is 20.55 m of fabric left.

Practice

Calculate and check these without a calculator.

1. a) 5.2 kg + 3.9 kg
   b) 4.9 m − 2.1 m
   c) 10.4 cm + 12.96 cm
   d) £72.34 − £49.67
   e) £204.07 − £196.30
   f) £137.60 − £107.06
   g) £18.99 + £108.17
   h) 2.1 m − 0.75 m
   i) 57.06 km + 106.1 km
   j) £13.12 − £12.07
   k) 105.17 kg + 22.4 kg
   l) £49.35 + £107.19

Work these out, then check with a calculator.

2. A saver has £354.60 in the bank. She withdraws £217.74.
   What is her new balance?

3. Two friends book a holiday costing £980.89. They pay a deposit of £98.08.
   How much is left to pay?

4. A shopper receives £13.49 change from a twenty pound note.
   How much has he spent?

5. A student buys three books costing £55.44, £13.62 and £17.29.
   What is the total cost?

6. A factory has 476.5 m of material in stock. It uses 326.25 m. How much is left?

7. A householder receives bills for £372.15, £32.07 and £11.19.
   How much is this altogether?

8. A farmer wants to fence some land. The lengths of the sides are 65.8 m, 67.75 m, 78.95 m and 96.5 m. Add these lengths to find the perimeter.

9. The table shows the time in seconds taken by each swimmer in two relay teams.

   a) Find the total time taken by each team.
   b) Which team won and by how many seconds?

	Team A	Team B
1st swimmer	21.42	22.56
2nd swimmer	25.58	23.10
3rd swimmer	24.17	24.78
4th swimmer	19.96	20.34

L1 10 A worker earns £1120.09. He spends £875.42. How much does he have left?

11 The interest on a loan of £1500 is £427.61. What is the total amount owed?

L2 Work these out without a calculator, then use rounding to check.

12 a) $2.396 + 5.274 + 2.821$ **b)** $9.08 - 2.765$ **c)** $3.24 + 6.382 - 5.493$

13 In a race the fastest time is 43.456 seconds and the slowest time is 59.572 seconds. What is the difference between these times?

Multiply decimals

N2/L1.5, N2/L2.6

To multiply decimals:

See pages 22 and 33 for a reminder.

• Ignore the decimal points and use your usual method to multiply the figures.

• Count the total number of decimal places (dp) in **both** numbers being multiplied. Starting at the end of the answer, count the same number of decimal places and insert the decimal point.
(Add zeros if needed, e.g. $0.3 \times 0.2 = 0.06$)

• Round the answer to a sensible number of figures (e.g. money to the nearest £ or pence).

• Check the answer.

Example 5×1.99

$5 \times 1.99 = 9.95$

0 dp + 2 dp = 2 dp in answer

$$\begin{array}{r} 199 \times \\ 5 \\ \hline 995 \\ {\scriptstyle 4\,4} \end{array}$$

Check by estimating:
5 × 2 = 10

6.24×5.3

2 dp + 1 dp = 3 dp in answer

$6.24 \times 5.3 = 33.072 = 33.07$ (to 2 dp)

Check by estimating:
6 × 5 = 30

$$\begin{array}{r} 624 \times \\ 53 \\ \hline 1872 \\ 31200 \\ \hline 33072 \\ {\scriptstyle 1} \end{array}$$

Practice

1 Calculate these without a calculator. Round answers to 2 dp where necessary. Use estimation to check each answer.

a) 0.6×4	**b)** 0.2×0.1	**c)** 7×0.05	**d)** 0.3×0.4
e) 3.65×4	**f)** 19.7×3	**g)** 11.22×0.5	**h)** 165×0.09
i) 40.7×0.8	**j)** 85.7×0.06	**k)** 5.01×2.1	**l)** 1.8×0.69
m) 5.19×7.3	**n)** 7.43×1.08	**o)** 18.1×3.66	**p)** 4.25×3.07

Work these out, then check with a calculator.

2 A dressmaker wants to make 4 dresses. Each dress uses 3.75 m of material.
How much material is needed altogether?

3 James buys 6 bottles of wine containing 0.75 litres each.
How many litres of wine is this altogether?

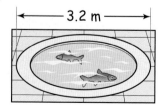

4 A shop assistant's hourly wage is £8.95.
How much does he earn for 37.5 hours?

5 An office puts in an order for stationery.
Copy and complete the table by working out the prices and total cost.

Order No	Item description	Price of item	Quantity	Price
PHF/1357	A4 paper	£4.99	5	
PHF/1359	A3 paper	£8.75	3	
ZL/129	Box black pens	£3.59	8	
FL/376	White board pens	£3.02	6	
GH/76P	Chrome stapler	£6.50	3	
GH/76L	Staples	£1.49	12	
		Total cost of all items		

6 A rectangular lawn is 43.6 m long and 28.4 m wide.
What is its area to the nearest square metre?

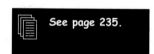
See page 235.

7 Calculate these without a calculator. Use estimation to check each answer.

a) 3.725×8 **b)** 1.864×0.9 **c)** 5.209×4.3

8 A tourist is going to Japan. She changes £800 to yen.
The exchange rate is £1 = 124.894 yen. How many yen does she get?

9 A cuboid is $3.6\,m \times 1.7\,m \times 5.8\,m$. Multiply these to work out its volume to 1 dp.

10 A circular fish pond has a diameter of 3.2 m.
Find its circumference and area to 1 dp. Use $\pi = 3.142$.

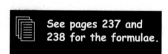
See pages 237 and 238 for the formulae.

Divide decimals

To divide a decimal by a whole number:

See page 36 for
a reminder.

- Divide the figures as usual.
- Keep the decimal points in line.
- Round the answer to a sensible number of figures (where necessary).
- Check the answer.

Example $172.6 \div 4$

Keep the decimal points in line.

$$\begin{array}{r} 4\,3.1\,5 \\ 4\,\overline{)17^12.6^20} \end{array}$$ ◄──── Add 0s to the end if necessary.

Check using inverse:
$$\begin{array}{r} 43.15 \\ \times\ 4 \\ \hline 172.60 \\ {\scriptstyle 1\quad 2} \end{array}$$

Practice

1 Do these without a calculator. Check each answer.

a) $49.2 \div 4$ b) $33.78 \div 3$ c) $79.9 \div 5$ d) $232.47 \div 7$

e) $137.88 \div 9$ f) $67.08 \div 6$ g) $423.06 \div 3$ h) $301.14 \div 8$

To divide a decimal by a decimal:

- Change the number you are dividing by into a whole number.
  Do this by multiplying by 10 or 100 or 1000.

- Do exactly the same to the number you are dividing.

- Divide to find the answer.

Example $7.4 \div 0.05$

$7.4 \div 0.05 = 740 \div 5$

Multiply 0.05 by 100 to make it into the whole number 5.
Do the same to 7.4

$$\begin{array}{r} 1\ \ 4\ \ 8 \\ 5\,\overline{)7^2\,4^4\,0} \end{array}$$ ◄────── Answer = 148

Check by doing the inverse on a calculator:
$148 \times 0.05 = 7.4$

> ▶ Thinking of money sometimes helps
> to make sense of the answer.
> You can think of $7.4 \div 0.05$ as
> 'How many 5p coins make £7.40.
> The answer, 148, then makes sense.

Example   Find $31.52 \div 2.4$ to 2 decimal places.

$31.52 \div 2.4 = 315.2 \div 24$ Multiply 2.4 by 10 to make it into the whole number 24.
Do the same to 31.52

$$24 \overline{)31^{7}5.^{3}2^{8}0} \quad \text{= 1 3. 1 3}$$ ← Answer = **13.13**

This is very difficult!

- Try out some values in rough when you need to or write down
  the multiples of 24 that you need:
  24, 48, 72, 96, ...

- For the last figure, you need to decide which is nearer to 80:
  $24 \times 3 = 72$ or $24 \times 4 = 96$ 3 is nearer
  (Alternatively you could work it out to
  3 decimal places then round your answer.)

e.g.
$$\begin{array}{r} 2\ 4 \\ \times\ 3 \\ \hline 7\ 2 \\ {}^{1} \end{array}$$

Check by estimating:
$30 \div 2 = 15$

---

2   Calculate these without a calculator. Give the answer to 2 dp when not exact.
    Check each answer.

a) $0.8 \div 0.2$ b) $1.5 \div 0.5$ c) $0.4 \div 0.02$ d) $1.8 \div 0.9$
e) $3.9 \div 0.03$ f) $6.8 \div 0.2$ g) $13.25 \div 0.5$ h) $10.8 \div 0.03$
i) $2.5 \div 0.04$ j) $34 \div 0.7$ k) $74.16 \div 0.06$ l) $43.2 \div 1.2$
m) $56.9 \div 3.1$ n) $90.1 \div 5.3$ o) $115.2 \div 0.12$ p) $35 \div 0.25$
q) $5.75 \div 1.1$ r) $120 \div 7.9$ s) $14.3 \div 4.6$ t) $101.52 \div 2.16$

3   A total of 8.4 mm of rain fell in a week. What was the average rainfall per day?

4   Three friends share the cost of a taxi between them as equally as they can.
    The taxi cost £16.50. How much do they each pay?

5   A roll of ribbon is 20 metres long.
    How many pieces of length 0.4 metres can be cut from the roll?

6   A drinks company sells cola in 0.33 litre cans.
    How many cans can be filled from 2000 litres of cola?

 7   Calculate these without a calculator (Note: You will need to $\times 1000$ this time).
    Check each answer.

a) $9.7 \div 0.005$ b) $0.378 \div 0.012$ c) $0.48 \div 0.025$ d) $14.2 \div 0.125$

8   A holidaymaker buys a present costing 18 euros.
    The exchange rate is 1.273 euros to the pound.
    How much does the present cost to the nearest pound?

9 A tourist returns from America with $200. The exchange rate is $1.601 to the pound. What will she receive? Give your answer to the nearest pound.

Here is a mixture of calculations to try.

10 Estimate the answers, then work them out accurately to 2 decimal places.

    a) 15.435 + 16.586   **b)** 3.791 × 1.9   **c)** 21.452 − 7.693

    d) 11.924 ÷ 2.1   **e)** 5.798 × 7.1   **f)** 19.687 − 7.132

    g) 9.687 + 2.763   **h)** 9.782 ÷ 2.5   **i)** 5.7 × 3.793

11 The table shows a caterer's vegetable order.

Item	Weight	Price per kg
Potatoes	50 kg	£0.64
Carrots	25 kg	£0.86
Sprouts	12.5 kg	£1.32
Mushrooms	6.4 kg	£4.35

    a) Find the total cost.

    b) He is given a discount of a third of the price.
        How much does he pay?

12 It takes 25 lengths of wallpaper, each 3.2 m long, to cover a room.
The decorator allows an extra 0.25 m per length for matching the pattern.
Rolls of wallpaper are 10.05 m long.
How many rolls does the decorator need to cover the room?

Number

10 Percentages

Understand percentages

N2/L1.8

Per cent means out of 100. It is also written using the sign %.

100% of something = the whole amount

This grid has been split into 100 squares:

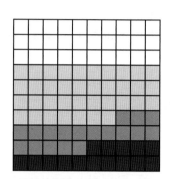

30 out of 100 are white.	30% are white.
37 out of 100 are grey.	37% are grey.
18 out 100 are orange.	18% are orange.
15 out of 100 are black.	15% are black.
The total % = 100%	

Practice

1 Look at the coloured squares in this grid.

a) Write down the percentage for each colour.
b) Check that the total % is 100%.

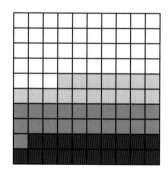

2 Draw a grid of 100 squares.

a) Colour in these percentages:
34% red, 16% green, 21% blue, 19% black.
b) What % of the squares are not coloured?

There are 100 pence in a pound.
10 pence out of a pound = 10% of £1   90 pence out of a pound = 90% of £1

---

3 Write these amounts as percentages of £1.
a) 20 pence   b) 45 pence   c) 16 pence   d) 85 pence

4 The following are percentages of £1. Write each amount in pence.
a) 17%   b) 99%   c) 22%   d) 51%

Equivalent fractions, decimals and percentages

N2/L1.3, N2/L2.2

Fractions, decimals and percentages all represent parts of something.
(Fractions and decimals are parts of 1, percentages are parts of 100.)
You can convert each form to the others.
For example, 9% means 9 out of 100.

As a fraction this is $\frac{9}{100}$ and as a decimal it is 0.09.

a Convert between percentages and decimals

3% = 0.03 and 67% = 0.67
because the value of the second
decimal place is hundredths.

> To change a percentage to a decimal, divide the percentage by 100.
>
> To change a decimal to a percentage, multiply the decimal by 100.

Examples

75% = 0.75	70% = 0.70 = 0.7	7% = 0.07
23% = 0.23	12.5% = 0.125	7.5% = 0.075
0.2 = 20%	0.45 = 45%	0.02 = 2%
0.87 = 87%	1.2 = 120%	0.175 = 17.5%

See page 115 for a reminder of how to x and ÷ by 100.

Practice

1 Change these percentages to decimals.

a) 25%   b) 35%   c) 40%   d) 30%
e) 80%   f) 5%   g) 10%   h) 15%
i) 4%   j) 8%   k) 55%   l) 85%

2 Write these percentages as decimals.

a) 16%   b) 32%   c) 49%   d) 95%
e) 64%   f) 52%   g) 17.5% h) 62.5%
i) 37.8% j) 3.6%   k) 2.5%   l) 1.25%

3 Change these decimals to percentages.

a) 0.75   b) 0.25   c) 0.65   d) 0.6
e) 0.06   f) 0.1   g) 0.01   h) 0.5
i) 0.05   j) 0.9   k) 0.09   l) 0.99

4 Write these decimal numbers as percentages.

a) 0.29   b) 0.57   c) 0.43   d) 1.5
e) 2.5   f) 0.875   g) 0.325   h) 0.065
i) 0.005   j) 0.205   k) 0.168   l) 2.45

b Write percentages as fractions

$53\% = 53$ out of $100 = \dfrac{53}{100}$. This fraction does not cancel, but some others do.

> **To write a percentage as a fraction,
> write it over 100, then cancel if possible.**

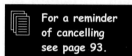

For a reminder
of cancelling
see page 93.

Examples

$$30\% = \frac{30}{100} = \frac{3}{10} \quad (\div 10, \div 10)$$

$$75\% = \frac{75}{100} = \frac{3}{4} \quad (\div 25, \div 25)$$

$$7\% = \frac{7}{100}$$

(doesn't cancel)

Practice

Write each percentage as a fraction in its simplest form.

1 a) 10%   b) 20%   c) 40%   d) 80%
       e) 70%   f) 90%   g) 50%   h) 25%
       i) 5%   j) 15%   k) 45%   l) 3%

2 a) 19%   b) 64%   c) 95%   d) 8%
       e) 86%   f) 72%   g) 41%   h) 96%

L2 Sometimes you need to multiply the top and bottom to give whole numbers.

Examples

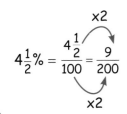

$$4\tfrac{1}{2}\% = \frac{4\tfrac{1}{2}}{100} = \frac{9}{200}$$

×2

×2

$$33\tfrac{1}{3}\% = \frac{33\tfrac{1}{3}}{100} = \frac{100}{300} = \frac{1}{3}$$

×3

×3

Remember this one –
it is used a lot
e.g. in sales.

Examples

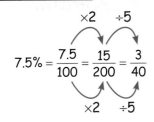

$$7.5\% = \frac{7.5}{100} = \frac{15}{200} = \frac{3}{40}$$

×2 ÷5

×2 ÷5

$$1.25\% = \frac{1.25}{100} = \frac{5}{400} = \frac{1}{80}$$

×4 ÷5

×4 ÷5

or $$7.5\% = \frac{7.5}{100} = \frac{75}{1000} = \frac{3}{40}$$

×10 ÷25

×10 ÷25

or $$1.25\% = \frac{1.25}{100} = \frac{125}{10\,000} = \frac{5}{400} = \frac{1}{80}$$

×100 ÷25 ÷5

×100 ÷25 ÷5

There is often more
than one way to do
this.

3   Write each percentage as a fraction in its simplest form.

a) $1\tfrac{1}{2}\%$

b) 2.5%

c) 6.4%

d) $12\tfrac{1}{2}\%$

e) 8.5%

f) 62.5%

g) 37.5%

h) 10.5%

i) 8.75%

j) $66\tfrac{2}{3}\%$

k) 12.8%

l) $3\tfrac{3}{4}\%$

c Write fractions as decimals and percentages

There are quick ways to write some fractions as decimals and percentages.

For example, $\frac{9}{10} = 0.9$ because the value of the first decimal place is in tenths.

Also $\frac{9}{10} = \frac{90}{100} = 90\%$

×10

×10

Changing the denominator to 100 is a quick
way to find the percentage.

To write a fraction as a decimal, divide the top by the bottom.
Then change the decimal to a percentage by multiplying by 100.

When it is not easy to see how to change the denominator to 100, there is another method
that **always** works:

Example

$\frac{3}{4} = 3 \div 4 = \mathbf{0.75}$ as a decimal

$= \mathbf{75\%}$ as a %

$\begin{array}{r} 0.75 \\ 4\overline{)3.^30^20} \end{array}$ ← Add 0s if needed.

For difficult divisions, use a calculator.

The quick method is $\overset{\times 25}{\underset{\times 25}{\frac{3}{4} = \frac{75}{100}}}$

To change decimals to fractions see page 112.

Practice

1 Write each fraction as a percentage by changing the denominator to 100.

a) $\frac{1}{2}$ b) $\frac{3}{10}$ c) $\frac{1}{4}$ d) $\frac{9}{20}$

e) $\frac{4}{5}$ f) $\frac{7}{25}$ g) $\frac{17}{50}$ h) $\frac{18}{25}$

Use a different method to check each answer.

2 Write these fractions as decimals and then as percentages.

a) $\frac{1}{10}$ b) $\frac{1}{5}$ c) $\frac{2}{5}$ d) $\frac{1}{20}$

e) $\frac{3}{50}$ f) $\frac{1}{25}$ g) $\frac{7}{20}$ h) $\frac{4}{25}$

Use a calculator to check your answers.

3 Copy and complete these tables.

Fraction	Decimal	%
$\frac{1}{2}$		
$\frac{1}{4}$		
$\frac{3}{4}$		
$\frac{1}{10}$		
$\frac{1}{5}$		
$\frac{1}{3}$		$33\frac{1}{3}\%$

Fraction	Decimal	%
	0.4	
	0.7	
	0.35	
		60%
		90%
		15%

Learn those in the first table.

4 Write these fractions as decimals and then as percentages.

a) $\frac{11}{20}$ b) $\frac{5}{8}$ c) $\frac{7}{8}$ d) $\frac{7}{25}$

e) $\frac{27}{40}$ f) $\frac{2}{3}$ (to 3 dp) g) $\frac{1}{16}$ h) $\frac{2}{15}$ (to 4 dp)

Use a calculator to check your answers.

5 Put each group in order of size, starting with the smallest.

a) 0.76 78% $\frac{3}{4}$ b) $\frac{1}{2}$ 0.49 51% **Hint:** Change them to the same type.

c) 24% $\frac{1}{4}$ 0.22 d) 0.21 19% $\frac{1}{5}$

e) $\frac{5}{9}$ 50% 0.58 f) $\frac{8}{9}$ 88% 0.9

g) 59% 0.56 $\frac{7}{12}$ h) $\frac{7}{16}$ 0.49 45%

Find percentages of quantities
N2/L1.9, N2/L2.8

a Finding 1% first

1 square out of the 100 in this grid is black.

$1\% = 1$ out of $100 = \frac{1}{100}$

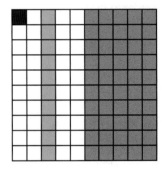

> To find 1% of something, divide it by 100.

| **Example** | 1% of £750 = £7.50 |

Note also that 1% is 1p in each £.
So 1% of £750 = 750 pence = £7.50

| **Example** | 1% of 20 m = 0.20 m = 0.2 m |

See page 115 for help with dividing by 100.

> To find other percentages, find 1%, then multiply.

> **Example**   Find 3% of 250 kg
>
> 1% of 250 kg = 2.5 kg   (dividing by 100)
>
> $\frac{\times\ 3}{}$
>
> 3% of 250 kg = 7.5 kg   (multiplying by 3)

Practice

1   Find 1% of these.
   a) 500   b) 630 km   c) 894 g
   d) £30   e) £480   f) £26

2   Find 3% of these.
   a) 400   b) £540   c) 80 km
   d) £50   e) 35 kg   f) £17

3   Find 4% of these.
   a) 7000   b) 450 m   c) 750 g
   d) 95 km   e) £360   f) £32

4   Find 7% of these.
   a) 500   b) 1600 kg   c) 90 litres
   d) £25   e) 405 cm   f) £89

5   Find 9% of these.
   a) 2000   b) £730   c) 120 m
   d) 70 kg   e) 75 litres   f) £62

b Finding 50%, 25% and 75%

For other percentages the multiplying is often more difficult.
There are quicker ways for some percentages.
Look at the grid on page 131 again. 50 out of the 100 squares are orange.
This is half of the grid.

$50\% = 50$ out of $100 = \frac{50}{100} = \frac{1}{2}$.   Learn $50\% = \frac{1}{2}$.

> To find 50% of something, divide it by 2 (halve it).

Example   Find 50% of 8 m.

$8 \div 2 = 4$ so   50% of 8 m = 4 m

Example   Find 50% of £15.

$15 \div 2 = 7.5$ so   50% of £15 = £7.50

$$2 \overline{)15.\!^{1}0}^{\,7.5}$$

To find 25% of something, halve it (to find 50%), then halve again.
Another way to find 25% is to divide by 4.

Learn 25% = $\frac{1}{4}$.

Example   Find 25% of £80.   Find 25% of 8.4 m.

50% of £80 = £40   50% of 8.4 m = 4.2 m
so 25% of £80 = £20   so 25% of 8.4 m = 2.1 m

You can use the other method to check:
25% of £80 = 80 ÷ 4 = £20   25% of 8.4 m = 8.4 ÷ 4 = 2.1 m

$$4 \overline{)8.4}^{\,2.1}$$

> To find 75% of something, halve it to find 50%, halve
> again to find 25%, then add the two values together.
>
> Another way to find 75% is to divide by 4
> (to find 25%) then multiply by 3.

Learn 75% = $\frac{3}{4}$.

Example   Find 75% of 240 m.

50% of 240 m = 240 ÷ 2 = 120 m
25% of 240 m = 120 ÷ 2 = 60 m
(50% + 25%)   **75% of 240 m = 180 m**

75% = 50% + 25%

Using the other method to check:
25% of 240 m = 240 ÷ 4 = 60 m
75% of 240 m = 60 × 3 = 180 m

Practice

Work these out without using a calculator.

Fruit Crumble

250 g plain flour
120 g butter
90 g sugar
280 g blackberries
200 g redcurrants
150 g strawberries

1   This recipe is for fruit crumble for four people.
   a) Find 50% of each ingredient.
   b) How many people would the new amount of crumble feed?

2   Find 50% of these.
   a) £160   **b)** 50 m   **c)** 96 km
   d) £45   **e)** 780 kg   **f)** 900 g

3   A coat priced at £120 is reduced by 50% in a sale.
   How much is the reduction?

4   A carpenter has 90 metres of skirting board.
   He needs 50% more.
   What extra length does he need?

5   Find 25% of these by halving ($\div 2$), then halving ($\div 2$) again.
   a) £60   **b)** 92 m   **c)** 12.8 m   **d)** £24.80
   e) 768 m   **f)** 196 cm   **g)** £150   **h)** £27.40
   i) 15.4 kg   **j)** 99 m   **k)** 45 km   **l)** £6.60

   Check your answers by dividing each quantity by 4.

6   Find 75% of these using the two different ways shown in the example on page 133.
   a) £180   **b)** 96 m   **c)** 360°   **d)** 520 m
   e) 176 kg   **f)** £36.40   **g)** £12.80   **h)** 124 cm
   i) 4.8 m   **j)** 18.48 kg **k)** 20.6 g   **l)** £74.20

7   Simon says that you can find 75% by finding 25% first, then taking the answer away from the original amount. So to find 75% of 240 m he writes:

   25% of 240 m = 240 $\div$ 4 = 60 m
   75% of 240 m = 240 – 60 = 180 m

            75% = 100% – 25%

Use this method to check three of your answers in question 6.

Use two different methods to answer questions 8–14.

8   Last year 300 people came to the school play.
   This year they have sold 25% more tickets.
   How many more tickets have they sold?

9 There are 60 seats on a coach. 75% of these are taken. How many seats are taken?

10 A social club has 320 members. 25% of the members are men.
   a) How many members are men?   **b)** How many members are women?

11 James buys 480 tiles. 75% of these are blue. How many blue tiles are there?

12 A holiday costs £1870. The travel agent asks for a 25% deposit.
   a) How much is the deposit?   **b)** How much is left to pay?

13 The items in a shopper's basket cost £4.99, £17.98 and £19.99.
   The shop gives a 25% discount on the total bill. What is the discount?

14 A company has a budget of £3.2 million to spend on a new factory.
   The company spends 75% of this on the building and the rest on equipment.
   How much does the building cost?

c Using 10%

Look at the grid on page 131 again. 10 of the 100 squares are grey.
This is a tenth of the grid.

Learn 10% = $\frac{1}{10}$.

10% = 10 out of 100 = $\frac{10}{100}$ = $\frac{1}{10}$

> To find 10% of something, divide it by 10.

Example   10% of 150 km = 15 km

Example   10% of £3.50 = £0.35 or 35p

See page 115 for help
with dividing by 10.

Often you can use links between 10% and other percentages.

> To find 20%, find 10% then double it (multiply by 2).
> 30% = 10% × 3
> To find 30%, find 10% then multiply by 3.
> To find 40%, find 10% then multiply by 4.
> You can use a similar method for other multiples of 10% (e.g. 60%, 70%, etc.)
> To find 5% of something, find 10% then halve it (divide by 2).

Example   Find 30% of 80 kg.

10% of 80 kg = 8 kg   (because 80 ÷ 10 = 8)
30% of 80 kg = 8 × 3 = 24 kg

30% = 10% × 3

Example

A television set costs £470 without Value Added Tax (VAT).
VAT is charged at a rate of 20%. Find the cost of the VAT.
10% of £470 = £47   (because £470 ÷ 10 = £47)

$$\frac{\times 2}{}$$

20% of £470 = **£94**

20% = 10% × 2

Example

VAT is charged at a rate of 5% on electricity bills.
Find the VAT charged on an electricity bill of £275.

10% of £275 = £27.50
5% of £275 = £13.75

$$\begin{array}{r} 13.\,7\,5 \\ 2\,)\overline{27.\!^{1}5\!^{1}0} \end{array}$$

5% = 10% ÷ 2

Practice

Work these out without using a calculator.

1 Find 10% of these.
   a) £450   b) 90 m   c) £2400   d) 54 cm
   e) 62 kg   f) 325 m   g) £4.20   h) £8
   i) £23.70   j) £105   k) 2.8 tonnes   l) 0.7 km

2 Work these out.
   a) 20% of £120   b) 30% of 170 kg c) 60% of 560 g   d) 40% of 960 m
   e) 20% of 54 kg f) 40% of £15   g) 30% of £12   h) 20% of £574
   i) 70% of £63.20 j) 80% of 91 m   k) 90% of 14.2 kg l) 60% of 18.5 km

3 Find 5% of these.
   a) 80 kg   b) £12 800   c) £45   d) £62.40
   e) 436 m   f) 72.8 litres   g) £129.60   h) 9.5 km

4 A couple pay a 10% deposit on a house costing £329 900. How much is the deposit?

5 30% of the proceeds from ticket sales for a concert go to charity.
   Ticket sales total £4300. How much of this goes to the charity?

6   Work out the VAT on these at a rate of 20%.
   a) £900   **b)** £2400   **c)** £180   **d)** £360
   e) £98   **f)** £46   **g)** £25.80   **h)** £9.60
   i) £22.40   **j)** £96.80   **k)** £120.80   **l)** £27.50

7   A plumber charges £480 plus VAT at 20%.
   Work out the cost of the VAT.

8   An electricity bill is £274.80 plus 5% VAT.
   Find the cost of the VAT.

9   Work out the VAT at 20% on these.
   Round answers to the nearest pence where necessary.
   a) sofa costing £980   **b)** camera costing £249
   c) coat costing £59.90   **d)** shoes for £69.50
   e) toaster at £14.95   **f)** electronic game £59.99

10 A nurse earns £1575 per month. He gets a 5% pay rise.
   How much extra does he earn per month?

11 A saver has £965.72 in a savings account. The interest rate is 5% per year.
   How much interest does she earn in a year?

12 An insurance agent sells a policy for £449. His commission is 15%.
   How much commission does he make?   **Hint: 15% = 10% + 5%**

13 A salesman gets 35% commission on any furniture he sells.
   How much does he get for selling a table costing £590.

14 A teacher asks how you can find $12\frac{1}{2}\%$ of £640.

   Tanya says 'Find 25% of £640, then halve it.'

   Jack says 'Find 10%, then 5%, halve that to find $2\frac{1}{2}\%$ then add the 10% and $2\frac{1}{2}\%$ together.'

   Show that both of these methods give the correct answer of £80.

Percentage increase and decrease   N2/L1.10, N2/L2.7
Wage and price rises are often given as percentages.
In a sale, shops often advertise price reductions as percentages.

> ▶ To find the new amount after a % change, **add the increase** or **subtract the reduction.**

Example   An employee who earns £9 per hour gets a 6% increase.

What is his new rate of pay?

6% is 6p for each £,
so the increase is 9 × 6p = 54p.

New rate of pay = £9 + 54p = **£9.54**

Check (by a different method):
1% of £9 = 900p ÷ 100 = 9p
6% of £9 = 9p × 6 = 54p

Example   A coat that costs £89.90 is reduced by 30% in a sale.

Find the sale price.

Find the reduction: then take it away:

10% of £89.90   = £8.99   £89.90 −
                              × 3   £26.97
30% of £89.90 = £26.97   Sale price = £62.93

Check (by rounding): 9 × 3 = 27   Check: 90 − 30 = 60

Practice

1   A hotel charges the following prices:
Breakfast £8   Lunch £16   Evening meal £24
Single room £70 per night   Double room £120 per night
The manager increases these prices by 5%. Find the new prices.

2   The table shows the weekly wage of each
assistant in a shop before they all get a pay
rise of 4%.
Find each worker's new weekly wage.

Worker	Weekly wage
Jan	£250
Kate	£275
Luke	£320
Wes	£295

3   A stationery shop gives a 9% discount to employees.
How much does an employee pay for goods priced at £20?

4   For each price, find the sale price after a reduction of: **a)** 20%   **b)** 25%.
i) £160   ii) £500   iii) £1500
iv) £1840   v) £10.60   vi) £892

5 For each price, find the new price after an increase of: **a)** 10%   **b)** 30%.
   i) £80   **ii)** £120   **iii)** £780
   iv) £64   **v)** £10.70   **vi)** £64.40

6 A saver puts £600 into a bank account. The interest rate is 5% per year.
How much will the saver have in the bank after one year?

7 A dressmaker needs 25 m of material. She decides to buy 10% extra in case of mistakes.
How much material does she buy altogether?

8 A shop is having a sale.
   a) A dress priced at £65 is reduced by 20%.
      What is the sale price?
   d) A blouse priced at £39 is reduced by 30%.
      What is the sale price?

9 The number of tickets available for a play is 900. Only 75% of these tickets are sold.
   a) How many tickets are sold?   **b)** How many tickets are left?

10 There were 120 sweets in a bag. 40% have been eaten.
How many sweets are left?

11 A television is priced at £950.
The assistant offers a 10% discount for cash.
What is the discounted price?

12 A deposit of 25% has been paid on a holiday costing £1040.
How much is there left to pay?

13 For each price, find the sale price after a reduction of 7%.
   a) £12   **b)** £89   **c)** £592
   d) £1460   **e)** £15 300

14 For each price, find the new price after an increase of 8%.
    a) £780   **b)** £1920   **c)** £84.50
    d) £21.25   **e)** £16.75

15 The prices of some items before VAT are shown below.
    Find the prices including VAT at 20%.

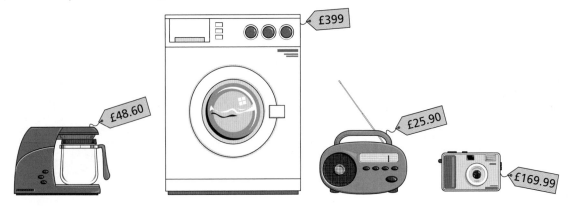

16 The old design of a greenhouse costs £920. A new design costs 35% more.
    How much is the new design?

Another way to increase or decrease by a percentage   N2/L2.7

You can add or subtract the % from 100% first.

Example   A worker who earns £9 an hour gets a 4% wage rise.

The new wage is 100% + 4% = 104% of the old wage.
        1% of the old wage = £9 ÷ 100 = 9p
        New wage = 104%   = 104 × 9p = 936p = £9.36

> Add the %
> increase to 100%.

Example   A coat costing £85 is reduced by 20% in a sale.

        The sale price is 100% − 20% = 80% of the original price.
        10% of the original price = £85 ÷ 100   = £8.50
                                            × 8
        Sale price = 80% of the original price = £68.00

> Subtract the %
> increase from 100%.

Practice

1   A worker who earns £10 an hour gets a 3% wage rise.
    What is his new wage rate?

2   A shop makes 10% profit on confectionery. The cost price of some items are shown below.
    Find how much the shop sells these goods for.

3   A jacket costing £56 is reduced by 30% in a sale.
    What is the sale price?

4   A car bought for £13 500 loses 20% of its value in a year.
    How much is it worth at the end of the year?

    Check your answers to some of the Practice questions on pages 138–140 using this method.

Use a calculator to find percentages   N2/L1.11, N2/L2.10

Two methods of using a calculator to work out percentages are given below.

Method A
- Divide the amount by 100 to find 1%.
- Multiply by the required percentage.
- Add or subtract the result if you need to give the final amount.

Method B
- Divide the % by 100 to write it as a decimal.
- Multiply the decimal by the amount.
- Add or subtract the result if you need to give the final amount.

> Your calculator may have a % key.
> The way this works varies from one calculator to another. Consult your instruction booklet to find out how yours works and use it if you wish.

Example   VAT is charged at 5% on a gas bill for £249.32.

Find: **a)** the VAT charged   **b)** the total amount charged.

a)

Method A
1% of £249.32 = £249.32 ÷ 100
            = £2.4932
5% of £249.32 = £2.4932 × 5
            = £12.466 = £12.47 (to 2 dp)

Method B
5% = 5 ÷ 100   = 0.05
0.05 × £249.32 = £12.466
            = £12.47 (to 2 dp)

For part b), carry
on the working on
your calculator.

b) Total amount = VAT + original amount = £12.466 + £249.32
            = £261.786 = £261.79 (to 2 dp)

See pages 116–117 for
help with rounding.

Whichever method you use:

● Think about whether the answer is sensible.
● Carry out a check – you could use the other method or find an estimate.

Practice

Use a calculator to work these out. Round answers to 2 dp where necessary.
Use one method to work out the answer, then another method to check.

1   Calculate these.
   a) 22% of 65   **b)** 37% of £170   **c)** 41% of 12 m   **d)** 82% of £520
   e) 19% of 25 kg   **f)** 36% of 920 g   **g)** 28% of 94   **h)** 11% of 46 cm
   i) 14% of 16 litres **j)** 17% of 154 m   **k)** 72% of £672   **l)** 9% of £178
   m) 63% of 74 km   **n)** 15% of £45   **o)** 1.4% of £85   **p)** 7.5% of 120 cm
   q) 8.2% of £165   **r)** 68% of 13 m   **s)** 28% of 6.25 m   **t)** 0.5% of £6350

2   The prices shown below do not include VAT. Find each price including VAT at 20%.

3   Two shops sell the same digital camera.
Work out the prices to find out which is
the best buy.

Shop A
Original price £259
now
reduced by **30%**

Shop B
Original price £249
and now ¼ off

4   A worker who earns £19 000 a year gets a 3.5% wage rise. What is her new salary?

5   A car costs £14 800 when new. It is resold later for 65% of this price.
How much is it sold for?

L2 6   This table shows the annual
percentage rate (APR)
for loans from four
different lenders.

Lender	Bayleys	Anchor	Rock Solid	Direct
APR	6.6	6.5	6.3	6.7

Work out the interest charges from each lender for one year on loans of:
a) £1000   **b)** £2500   **c)** £1750
d) £3225   **e)** £12 500   **f)** £0.8 million.

7   A photocopier can enlarge or reduce the
dimensions of anything it copies. Find the new
length and width when the photocopier is set at:
a) 75%   **b)** 125%
c) 133%   **d)** 141%.

6 inches

4 inches

Remember you can increase or decrease the %
first to find the final amount.

> **Example**   VAT is charged at 5% on a gas bill for £249.32.
>
> Find the total amount charged.
>
> 100% plus 5% VAT = 105%
> Total amount = 105% of £249.32 = £2.4932 × 105 = £261.79 (method A)
>   or   1.05 × £249.32 = £261.79 (method B)

---

8   Find the final amounts.
a) A 22% discount on £42   **b)** 105 g plus an extra 8%
c) 600 m plus an extra 15%   **d)** 42 cm minus 19%
e) 24 km increased by 32%   **f)** 106 kg minus 12%
g) 57 m plus 15%   **h)** 25 litres minus 35%
i) £14 increased by 60%   **j)** A 25% wage rise on £7 an hour

9   A shopper buys the following goods:

    a radio for £21.99   3 DVDs at £14.99 each   a DVD rack for £14.95.

    These prices do not include VAT.

    What is the total price of the goods including VAT at 20%?

Write one number as a percentage of another   N2/L2.9

- Write the information as a **fraction**.
- If possible, **make the denominator 100** – the numerator then gives the percentage.
- If this is not possible, **simplify** the fraction then **divide the top by the bottom** to write the fraction as a **decimal**. Finally, **multiply by 100** to change into a percentage (see page 129).

Example   A student gets 32 out of 40 in a test.

What percentage is this?

$$32 \text{ out of } 40 = \frac{32}{40} = \frac{16}{20} = \frac{80}{100}.$$ The percentage is 80%.

(÷2 and ×5 shown between the fractions)

Example   9 students out of a class of 24 have blonde hair.

The fraction having blonde hair is $\frac{9}{24} = \frac{3}{8}$.

$$8\overline{)3.\,{}^{3}0\,{}^{6}0\,{}^{4}0} = 0.375$$

As a decimal this is 0.375

The % who have blonde hair = $0.375 \times 100 = $ **37.5%**

Check on a calculator:

$9 \div 24 = 0.375$

$0.375 \times 100 = 37.5$

Practice

1   In a class of 20 students, 12 are men.

    What % of the students are:

    a) men

    b) women?

2   Ten chocolates in a box of twenty-five are soft centred.

    What % is this?

3   The table shows the marks a student gets in his course assignments.
Find his % mark for each assignment.

Asst.	1	2	3	4
Mark	18	56	52	66
out of	30	70	80	120

▶   Quantities must be in the **same units**.

▶   For a % change, write the change as a % of the **original amount**.

Example   A fare costing £2.50 goes up by 75p to £3.25.

What is the percentage increase?

Write 75p as a fraction of 250p $\dfrac{75}{250} = \dfrac{3}{10} = \dfrac{30}{100}$

The original price £2.50 = 250 pence

The percentage increase is **30%**.

---

4   **a)**   What is 450g as a % of 900g?
   b)   What % of £60 is £20?
   c)   What % of £6 is 30p?
   d)   What % of 1kg is 250g?
   e)   What is 50mm as a % of 1m?
   f)   What is 33cℓ as a % of 1 litre?

1kg = 1000g
1m = 1000mm
1 litre = 100cℓ
1km = 1000m

5   A water company is laying 2 km of pipe. In the first week it lays 160m of pipe.
What % of the total length is this?

6   A bank loan of £5000 is repaid over 3 years at £150 a month.
Write the interest as a % of the original amount.

7   £1000 is invested in a savings account and receives interest of £3.75 per month.
   a) How much is the money worth after a year?
   b) What is the % interest rate?

8   Out of 75 people who take a driving test, 45 pass. What percentage is this?

9   A new car is bought for £16 000 and sold one year later for £14 000.
What is the % loss?

10 There is £600 in a bank account before a withdrawal of £168 is made.
What percentage is withdrawn?

Use a calculator for the rest. Give your answers to 1 dp when they don't work out exactly.
11 There are 150 houses on Sherbourne Close.
    a) 102 houses are lived in by families. What % are not families?
    b) A survey carried out in Sherbourne Close shows that 21 of the households
        are vegetarian. What percentage is this?
    c) 86 of the houses on Sherbourne Close have satellite TV. What % is this?

12 In one month a travel company sells 780 holidays. 520 of these are in Europe.
    a) What % of the holidays are: **i)** in Europe   **ii)** outside Europe?
    b) Check your answers to part a) by adding the percentages. The total should be 100%.

13 A school has 378 pupils. 203 of these are blond. What % are not blond?

14 In a college with 1254 students, 358 students are part-time.
    a) What percentage are part-time?
    b) 25 of the 358 part-time students study maths. What % is this?

15 Look at the menu.
If you order this meal between 6pm and 7pm, it costs £19.
What % do you save?

Set meal	
Onion Bhajee	£2.20
Chicken Chat	£2.45
Balti Chicken	£6.45
Chicken Tikka Masala	£5.99
Aloo Gobi	£1.95
Pilau Rice	£1.80
Keema Nan	£1.30

> **Example**   1546 out of 6012 students in a college are full-time.
>
> This gives the fraction $\frac{1546}{6012}$.
>
> Rounding to the nearest 100 gives $\frac{1500}{6000} = \frac{1}{4}$.
>
> The % of students who are full-time is approximately **25%**.
>
> Note: Rounding to the nearest 1000 gives $33\frac{1}{3}\%$, where $\frac{1}{3}$ is a proper fraction – this is not as accurate.

Sometimes you may just want a rough %, or you may want to estimate a % to check an answer from a calculator.

16 Estimate these.
  a) What percentage of 587 is 294?
  b) What percentage of 1960 is 978?
  c) What percentage of 987 is 254?
  d) What percentage of £124 is £27?

17 In the 20 162 crowd who watch a football match, there are 4213 women and 4936 children. The rest of the crowd are men.
  a) Estimate the % of the crowd that are:
     (i) women   (ii) children   (iii) men.
  b) Check your answers by adding these percentages together.

18 The table shows a householder's gas bills during a year.
  a) What is the total for the year?
  b) Approximately what % of this total was for:
     i) spring   ii) summer
     iii) autumn   iv) winter?
  c) Check your answers to part b) by adding the percentages together.

Season	Bill
Spring	£122.88
Summer	£57.26
Autumn	£176.60
Winter	£236.52

Measures, Shape and Space

1 Money

Add and subtract sums of money

MSS1/E3.1

a Line up the £s and pence

To add or subtract money, line up the decimal points.

Examples

£2 + £1.20 + 5p

```
  £     p
  2  .  00  +
  1  .  20
  0  .  05
 £3  .  25
```

£5 – 75p

```
  £     p
  5  .  00  –
  0  .  75
 £4  .  25
```

If less than 10p, put 0 here.

> For writing money as decimals, see page 108.

Practice

1 Write these in columns, lining up the decimal points. Work out the answers.

   a) £5.80 + 45p + 6p **b)** £16.50 + 25p + 5p **c)** £2 + 60p + 7p
   d) £5 + 5p + 35p **e)** 75p + £2 + 50p + £1.10 **f)** £3.99 – 54p
   g) £5.50 – 45p **h)** £5 – £2.40 **i)** £5 – £3.99
   j) £5 – £1.49 **k)** £2.76 + £5.80 + £17 **l)** £80.05 – £15.99

2 A boy has £5 in his pocket. He finds 50p, then meets a friend who gives him £2.50. What is the total amount of money he has now?

3 A girl has £5. She spends £2.35 on a bus ticket. How much does she have left?

4 **a)** A workman buys a beefburger, chips and a cup of coffee at the burger bar. The prices are in the list. What is the total cost?
   b) He pays with a £20 note. How much change should he get?

Burger Bar Prices

Beefburger £1.90
Cheeseburger £2.25
Chips 99p

Tea 95p
Coffee £1.35
Fruit juice £1.20

b Use a calculator to add and subtract money

Parts of a £1	
Before the point	After the point
£	● 10p 1p

£3 . 4 2

3 pounds

40 pence

2 pence

Either work in £s using decimals

£1.20 + 60p − 5p   enter 1.2   [+]   0.6   [−]   0.05   [=]   **answer 1.75**

£1   2 × 10p   £0   6 × 10p   £0   0 × 10p   5 × 1p

Remember to include the £ sign.   **£1.75**

or work in pence (using £1 = 100p).

£1.20 + 60p − 5p   enter 120   [+]   60   [−]   5   [=]   **answer 175**

Remember to include the pence sign.   **175p**

Do not use both the £ and p signs. Your answer is in £s or p. It cannot be in both.

£1.75 ✓   175p ✓   £1.75p ✗

For more money calculations, see page 114.

Practice

1   Add these amounts together using a calculator.
    Give the answer in £ if it is 100p or more.
    Check your answers using another method.

Note:
Add a zero to the end if the calculator gives just 1 decimal place when you work in £.
For example, 4.5 means £4.50.

a) £1.50 + 50p + 5p   b) £2.25 + 35p − 10p   c) £2.54 + 42p + 2p
d) 60p + 42p + £2   e) £3.13 + 35p − 9p   f) 50p + 3p + 4p + 40p
g) £3.30 + 3p − 50p   h) 5p + 90p + £2 + £1.50   i) £1.10 − 8p − 40p − 3p
j) £3 + 65p + £2.50 + 10p k) £18.37 − 90p − £4.49   l) £7.48 − £3.96 + £9

Round money to the nearest 10p or £   MSS1/E3.2

a Round to the nearest 10p

The rounding rule is:

> ‣ If the **amount ends in 5p or more, round up** to the next 10p above.
> ‣ If the **amount ends in less than 5p, round down** to the 10p below.

Examples

Look at the grid below.
54p ends in **4**, which is **less than 5**. **Round down** to **50p**.
68p ends in **8** which is **more than 5**. **Round up** to **70p**.
Round amounts in the **unshaded** area **down**. Round amounts in the **shaded** area **up**.

	1p	2p	3p	4p	5p	6p	7p	8p	9p	
0										10p
10p										20p
20p										30p
30p										40p
40p										50p
50p	←	Round down		54						60p
60p								68	Round	70p
70p									up	80p
80p										90p
90p										£1.00

Practice

1 Find where each amount is on the grid, then round it to the nearest 10p.

a) 12p   b) 27p   c) 34p   d) 41p   e) 85p   f) 59p
g) 25p   h) 62p   i) 46p   j) 77p   k) 90p   l) 3p

2 Round these to the nearest 10p.

a) £2.13   b) £4.43   c) £5.71   d) £7.24   e) £6.93   f) £9.11
g) £5.69   h) £4.18   i) £3.33   j) £8.85   k) £18.04   l) £27.97

3 Round the cost of each item to the nearest 10p.
Then add to find the approximate total cost for each person.

	Val			Yui			Carl	
a)	pen	43p	**b)**	pen	29p	**c)**	pen	39p
	ruler	27p		ruler	33p		ruler	28p
	pencil	38p		pencil	54p		pencil	45p

b Round to the nearest £

The rounding rule is:

> ▸ If the **amount ends in 50p or more, round up** to the £ above.
>
> ▸ If the **amount ends in less than 50p, round down** to the £ below.

Examples

£5.42 = **£5** to the nearest £   (because **42p** is less than 50p)
£62.59 = **£63** to the nearest £   (because **59p** is more than 50p)

Practice

1   Round each amount to the nearest £.

a) £2.60   b) £3.10   c) £4.80   d) £5.20   e) £6.90   f) £7.30
g) £8.20   h) £9.40   i) £1.70   j) £0.50   k) £10.00   l) £19.70

2   Round these figures to the nearest £.

a) £12.13   b) £34.43   c) £15.71   d) £27.24   e) £336.93   f) £9.11
g) £25.69   h) £24.18   i) £13.33   j) £228.85   k) £76.50   l) £76.49

3   a)   Round the cost of each item to the nearest pound, then add to estimate the total cost.

i)	Veg	£1.99	ii)	Cake	£2.49	iii)	Files	£4.10
	Cheese	£2.27		Coffee	£3.05		Paper	£3.89
	Meat	£5.90		Milk	99p		Calculator	£5.10

b)   Use a calculator to find the exact totals.

Activity

Round each amount on a shopping receipt (or bank statement, payslip, etc.) to the nearest £ and find the total. Use a calculator to add the exact amounts and compare the results.

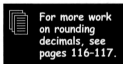

For more work on rounding decimals, see pages 116–117.

Add, subtract, multiply and divide sums of money

MSS1/L1.1

a Understand place value of £s and pence

Two hundred and forty-six pounds, thirty-five pence is written in numbers as £246.35.

The 2 represents £200
The 4 represents £40
The 6 represents £6
The 3 represents 30p
The 5 represents 5p

WHOLE POUNDS				PARTS OF A £	
£			•	PENCE	
Hundreds of £s	Tens of £s	Single £s	DECIMAL POINT	Tenths of a £ 10p	Hundredths of a £ 1p
2	4	6	.	3	5

Practice

1 What does the 3 represent in each amount?

  a) £531.42   **b)** £26.43   **c)** £143.15   **d)** £84.32   **e)** £306

2 What does the 9 represent in each amount?

  a) £92.12   **b)** £20.96   **c)** £119.20   **d)** £45.59   **e)** £931.45

3 Write these amounts in numbers.

  a) one hundred and ten pounds   **b)** twenty-five pounds, twenty pence
  c) twelve pounds and fifty pence   **d)** thirty pounds, five pence
  e) two hundred and thirty pounds, fifteen pence
  f) four hundred and thirteen pounds, twenty-five pence
  g) three hundred and three pounds, three pence
  h) three hundred and twenty-five pounds

4 List these amounts in numbers, starting with the **smallest**.
  seven hundred and thirty-two pounds, sixty-one pence
  seven hundred and forty-one pounds, twelve pence
  seven hundred and thirty-six pounds, two pence
  seven hundred and thirty-one pounds, thirty-four pence

5 Write these amounts in words.

  a) £10.50   **b)** £5.99   **c)** £12.50   **d)** £505.00
  e) £2.15   **f)** £410.04   **g)** £45.49   **h)** £209.09

6 Put these amounts in order, starting with the **largest**.
  £324.14   £234.41   £324.41   £342.14   £234.14   £342.41

b Add, subtract, multiply and divide money

Use the decimal methods given in Section 1, Chapter 9 (pages 119–125).

> ▶ To **add** or **subtract**, **line up the decimal points**.

Examples

£125 + £1.25
  125.00 +
    1.25
  £126.25

£125 − £1.25
  125.00 ←—— **Fill the gaps with 0s.**
    1.25
  £123.75

Line up the decimal points.

> To multiply decimals, multiply the numbers, then count the decimal places.

> To divide a decimal by a whole number, divide as usual, keeping the decimal points in line.

Examples

£4.15 × 3

$$£4.15 \longleftarrow \text{2 decimal places in the question.}$$
$$\underline{\times 3}$$
$$£12.45 \longleftarrow \text{2 decimal places in the answer.}$$

£14.46 ÷ 2

$$\begin{array}{r} 7.23 \\ 2\overline{)£14.46} \end{array} = \mathbf{£7.23}$$

Divide as usual, keeping the decimal points in line.

If the answer has one decimal place (e.g. £4.2) add a zero to show the pence (£4.20).

Practice

1 Work these out without a calculator, then use a calculator to check.

a) £14.59 + £4.15 + 69p   b) £6.95 × 4   c) £14.99 × 6

d) £2.99 + £30 + 99p   e) £20 − £6.99   f) £18 ÷ 4

g) £25.80 ÷ 5   h) £15 − £2.79   i) £50 − £3.76 − £14.99

j) 7 × £15.40 + £8.75   k) £7.80 ÷ 3 − 50p   l) 57p × 9 − £4.30

2 a) A student's course costs £84.50 for one term.
How much does it cost for three terms?

b) The exam fee is £28.60.
What is the total cost of the three-term course and the exam?

3 A student pays £411.60 for a course for the full year. There are three equal terms.
How much is the course per term?

4 A student pays £42.97 for a shirt and tie.
He returns the tie and gets a refund of £12.99.
How much did the shirt cost?

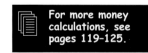 For more money calculations, see pages 119–125.

Activities

Use a catalogue to find the total cost for a list of goods, or work out what you can buy for £100.
Fill in an expenses claim form.
Check bills or wage slips.

Convert between currencies

To find exchange rates, look on the Internet, or in the daily papers.

Country/area	Currency name	Bank SELLING rate £1 =	Bank BUYING rate £1 =
Europe	Euro (€)	€1.24	€1.27
India	Rupee	87.52 rupees	89.61 rupees
Japan	Yen	118.48 yen	121.32 yen
Poland	Zloty	5.06 zloty	5.18 zloty
South Africa	Rand	12.77 rand	13.08 rand
USA	Dollar ($)	$1.56	$1.61

If you want foreign currency, the bank SELLS it to you at the SELLING rate.

Example   For each £ you get €1.24 (euros) or 118.48 Japanese yen.

$$£ \xrightarrow[\text{(using a calculator)}]{\text{MULTIPLY by selling rate}} \text{Foreign Currency}$$

If you exchange £500:   for euros you get $500 \times 1.24 = €620$

for Japanese yen you get $500 \times 118.48 = 59\,240$ yen

You can check these by dividing (the inverse) or rounding

Practice

1   The table shows a bank's customers for foreign currency one day.
Copy and complete the table using the rates given above.

	£s to exchange	Currency type	Currency amount
Mr Blank	£900	US dollars	
Mrs Smith	£750	Polish zloty	
Miss Patten	£1200	South African rand	
Ms Chang	£2000	Japanese yen	
Mr Davies	£150	Indian rupees	
Miss Bailey	£87 500	Euro	

If you want to change foreign currency back to £s, the bank BUYS it from you at the BUYING rate.

Examples

$$\text{Foreign Currency} \xrightarrow[\text{(using a calculator)}]{\text{DIVIDE by buying rate}} £$$

For: $300 (USA)   you get $300 \div 1.61 = £186.34 = £186$ (nearest £)

10 000 rupees   you get $10\,000 \div 89.61 = £111.59 = £112$ (nearest £)

2 A bank has customers returning from holiday who want to convert their foreign currency back into £s. Copy and complete the following table to show to the nearest £ what they should receive.

	Amount	Currency type	£
Mr Wragg	$720	US dollars	
Miss Yen	950	Polish zloty	
Mr Ennis	3400	Japanese yen	
Ms Masters	5600	Indian rupees	
Miss Dank	€1200	Euro	
Mr Caine	256 000	Japanese yen	
Ms Wright	350	South African rand	

3 A family used 500 litres of petrol at 1.59 euros per litre whilst touring France.
The exchange rate they received was 1.23 euros to the pound.
How much did the petrol cost them to the nearest pound?

Travel agents and banks often charge a commission for changing currency.

For the following questions, assume that the commission charged is £10 for exchanges up to £1000 and $1\frac{1}{2}$% on amounts over £1000. Assume that the customers only receive the whole units of currency (e.g. they would only get €642 of €642.73).

---

4 a) A businessman changes £950 into US dollars for a trip to America.
   How many dollars does he receive? (Use the rates given on page 154)
 b) The businessman is ill, and has to cancel his trip. He changes the dollars back into £s.
   How much does he get?
 c) Altogether how much does he lose because of the cancellation?

5 a) A couple changed £1200 into Polish zloty at 4.97 zloty to the £ when they booked a flight to Poland. How many zloty did they get?
 b) The exchange rate later rose to 5.06 zloty to the pound.
   How much extra, in zloty, would they have gained if they had waited to exchange their money?
 c) The couple returns from Poland with 750 zloty.
   How much do they receive after paying commission when they exchange these zloty for pounds at a rate of 5.24 zloty to the £?

Activity

Use information from the Internet or holiday brochures to cost a holiday.
Decide how much money you would take and estimate what foreign currency you would receive.

Measures, Shape and Space

2 Time

Read and record dates

	Month	Abbreviated form	Number of days
1	January	Jan	31
2	February	Feb	28 or 29
3	March	Mar	31
4	April	Apr	30
5	May	May	31
6	June	Jun	30
7	July	Jul	31
8	August	Aug	31
9	September	Sep	30
10	October	Oct	31
11	November	Nov	30
12	December	Dec	31

There are different ways to write the date.

Full date   16th November 2013

Medium date   16/Nov/2013

Use the abbreviated month

Short date   16/11/13

November is the 11th month   **13 is 2013 (83 is 1983)**

Practice

1   Fill in the missing words.

   a) January is the _____ month of the year. It has _____ days.

   b) _____ is the second month of the year. It has _____ or _____ days.

   c) March is the _____ month of the year. It has _____ days.

   d) April is the _____ month of the year. It has _____ days.

   e) _____ is the 5th month of the year. It has _____ days.

   f) _____ is the sixth month of the year. It has _____ days.

   g) _____ is the 7th month of the year. It has _____ days.

   h) The eighth month of the year is _____. It has _____ days.

   i) The 9th month of the year is _____. It has _____ days.

   j) October is the _____ month of the year. It has _____ days.

   k) The 11th month of the year is _____. It has _____ days.

   l) The 12th month of the year is _____. It has _____ days.

2 What is today's date? Write it in all three ways.

3 a) A student was born on 02/04/98. Write her birth date in full.
   b) Her friend was born on 22nd February 2001. Write this is in the short date format.

4 Alexander Bell made the first ever telephone call on 10th March 1876.
   Write this in the medium date format.

5 The first test-tube baby was born on 25/07/78. Write this date in full.

Activities   Essential

1 You need a list of the months of the year and a list of college term dates.
   a) Write each college term date next to the correct month on the list.
       Use the short date format.
   b) i) Write in your family's birthdays in the medium date format.
       ii) Ask members of the class for their birthdays and add them to the list.
           Use the medium date format.
   c) Add any other special dates to your list.

2 You need a blank calendar with the bank holidays marked.
   a) Which year is your calendar for?
   b) Find which days of the week these are on.
       i) Christmas Day, i.e. 25th December
       ii) New Year's Day, i.e. 1st January
       iii) May Day, i.e. 1st May
       iv) May 17th
       v) October 25th
   c) Find the dates of:
       i) Good Friday
       ii) Easter Monday
       iii) the August bank holiday.
   d) Write these on your calendar.
       i) your birthday
       ii) your name by today's date
       iii) the days your college terms begin and end
       iv) any other dates that are important to you

Read, measure and record time   MSS1/E3.3

a am and pm

8:30 could be when you get up in the morning, or when you watch TV in the evening.
To tell us which it is, the day is split into two halves: **am** and **pm**.

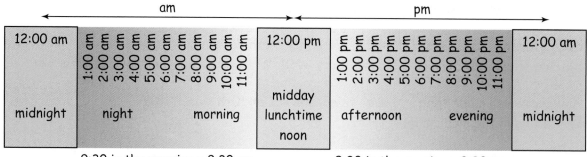

9:30 in the morning = 9:30 am   9:30 in the evening = 9:30 pm

Practice

1 Write these times in numbers using am or pm.

    a) 7:00 in the morning   **b)** 4:00 in the afternoon

    c) 11:30 before lunch   **d)** 5:00 when the shops are closing

    e) 1:00 when most people are asleep   **f)** 8:45 the morning rush hour

2 Match the times to the events.

    a) Waking up with a bad dream   **i)** 7:45 pm

    b) Football kick off time   **ii)** 11:00 am

    c) Coffee break   **iii)** 1:00 am

3 Suggest activities for the following times.

    a) 9:00 am   **b)** 12:00 pm   **c)** 5:30 pm   **d)** 7:45 am

4 How many hours are there from:

    a) 3:00 am to 3:00 pm   **b)** 5:30 am to 6:30 pm

    c) 4:15 am to 9:15 pm   **d)** 10:00 am to 7:30 pm

    e) 8:00 pm Mon to 7:00 am Tue   **f)** 7:00 pm Tue to 6:30 am Wed?

b Read and record time to the nearest 5 minutes

The **short** hand is the **hour** hand.

The **long** hand is the **minute** hand.

There are **60 minutes** in an **hour**.

There are **30 minutes** in **half an hour**.

There are **15 minutes** in a **quarter of an hour**.

There are **5 minutes** between each number and the next.

On the clock, the hour hand has gone past 5.

The minute hand has gone round from 12 to 3. This means it is 15 minutes past.

The time is five fifteen. This is written as 5:15. (We also say 'quarter past five'.)

Look at these clocks. The times are given underneath.

Examples

10 minutes past 7
or

7:10

30 minutes past 9
or

9:30

or half past nine

40 minutes past 3
or

3:40

Practice

1 Write each time in two different ways.

a)

b)

c)

d)

When the minute hand has gone past 30, we can count minutes **to the next hour.**

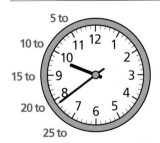

We often read the time to the nearest 5 minute mark.
To the nearest 5 minutes, this clock says **20 minutes to 10**

> **To the nearest 5 minutes means to the nearest number marked on the clock.**

Look at these clocks.

They both say 50 minutes past 1
or
10 minutes to 2.

1:50

2 Write the time to the nearest 5 minutes in three ways.

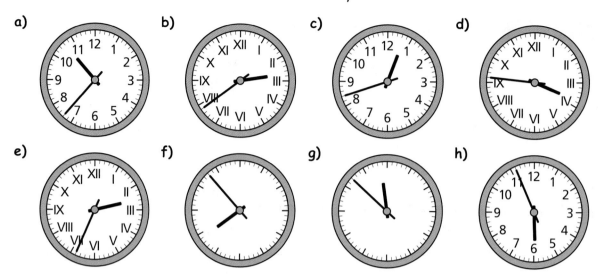

a)　　　　b)　　　　c)　　　　d)

e)　　　　f)　　　　g)　　　　h)

Use the 12 and 24 hour clock

MSS1/E3.3, MSS1/L1.2, MSS1/L2.2

a The 12 hour and 24 hour clock

The 24 hour clock is used for timetables, TV recorders, digital clocks and other timers.

The hours are numbered up to 24 instead of using am and pm.

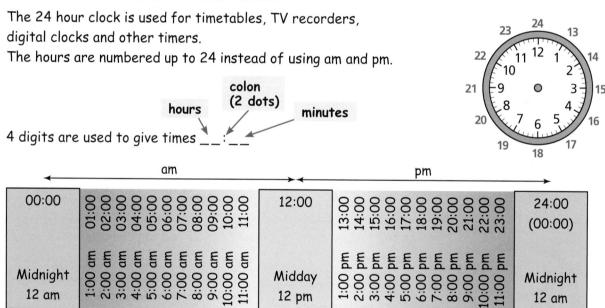

hours

colon
(2 dots)

minutes

4 digits are used to give times __ : __

am		pm	
00:00	01:00 02:00 03:00 04:00 05:00 06:00 07:00 08:00 09:00 10:00 11:00	12:00	13:00 14:00 15:00 16:00 17:00 18:00 19:00 20:00 21:00 22:00 23:00	24:00 (00:00)
Midnight 12 am	1:00 am 2:00 am 3:00 am 4:00 am 5:00 am 6:00 am 7:00 am 8:00 am 9:00 am 10:00 am 11:00 am	Midday 12 pm	1:00 pm 2:00 pm 3:00 pm 4:00 pm 5:00 pm 6:00 pm 7:00 pm 8:00 pm 9:00 pm 10:00 pm 11:00 pm	Midnight 12 am

It is easy to write am times as 24 hour clock times:

Example	6:20 am = **06:20**	10:35 am = **10:35**	
			Timetables often omit the colon.

It is more difficult with pm times – usually you need to add 12 hours:

| Example | Write 4:36 pm in 24 hour time. |

hours:minutes
4:36 (remove the pm)
add 12 hours   +12:00
4:36 pm   is   **16:36**

▶ Add 12 hours to pm times to find the 24 hour clock time.

| Example | Write quarter to seven in the morning as a 24 hour clock time. |

15 minutes before 7 = 45 minutes past 6 = **06:45**
(For quarter to seven in the evening add 12 hours to give 18:45.)

Take care near midnight and midday: 12:15 am = 00:15
12:15 pm = 12:15

Practice

1 Write these as 24 hour clock times.

a) 3:50 pm   b) 7:00 am   c) 9:15 am   d) 10:26 pm   e) 3:42 pm   f) 1:30 pm
g) 6:45 am   h) 12:05 am   i) 7:40 pm   j) 6:59 pm   k) 6:42 am   l) 8.25 pm

2 Write these as 24 hour clock times.

a) quarter past 3 in the morning   b) quarter to 5 in the evening
c) half past 7 in the morning   d) 10 to 9 in the morning
e) 10 past 11 at night   f) midday
g) 5 to 4 in the afternoon   h) quarter to 8 in the morning
i) 25 to 6 in the evening   j) 10 to 3 in the afternoon
k) 5 past 4 in the afternoon   l) 10 past 11 in the morning

3 Write these times in figures as 24 hour clock times.

a) six minutes past four in the afternoon
b) fifteen minutes past three in the afternoon
c) eleven minutes past eight in the morning   d) fifteen minutes to ten in the morning
e) ten minutes to seven in the evening   f) twenty five minutes to midnight
g) five minutes to eleven in the morning   h) twelve minutes past midday
i) twenty minutes past nine in the evening   j) half past six in the evening

> **Example** Write 07:15 as a 12 hour clock time.
>
> Remove the 0 and write am (morning time) **7:15 am**

> **Example** Write 23:50 in 12 hour time.
>
> | | Subtract 12 hours to find the pm time. |
>
> 23:50
> subtract 12 hours − 12:00
> 2350 = **11:50 pm** (evening time, so write pm)

4 Write these 24 hour clock times using am or pm.

a) 23:30	**b)** 06:45	**c)** 07:12	**d)** 16:45	**e)** 12:10	**f)** 01:10
g) 12:16	**h)** 14:10	**i)** 00:00	**j)** 06:05	**k)** 13:46	**l)** 09:15

5 Match the times.

a) 15:52	14 minutes to 4 in the morning	5:05 pm
b) 03:46	9 minutes past 7 in the evening	11:54 pm
c) 17:05	28 minutes to 5 in the afternoon	3:46 am
d) 23:54	8 minutes to 4 in the afternoon	4:32 pm
e) 19:09	6 minutes to midnight	3:52 pm
f) 16:32	5 minutes past 5 in the afternoon	7:09 pm

b Use timetables

Most timetables are in 24 hour clock time. The colon is often omitted.

> **Example** Find the first train leaving Bournville for Birmingham.
>
> **1** Look down the first column – find Bournville.
>
> **Extract from the Redditch – Birmingham timetable**
>
> | Redditch | d | | | | 0628 | | | 0658 |
> |---|---|---|---|---|---|---|---|---|
> | Longbridge | d | 0615 | 0623 | 0633 | 0644 | 0654 | 0702 | 0713 |
> | Kings Norton | d | 0620 | 0627 | 0637 | 0649 | 0658 | 0707 | 0719 |
> | Bournville 2—d | | 0622 | 0630 | 0640 | 0651 | 0701 | 0710 | 0721 |
> | Five Ways | d | 0629 | 0638 | 0648 | 0659 | 0708 | 0718 | 0729 |
> | Birmingham | a | 0633 | 0642 | 0652 | 0703 | 0713 | 0722 | 0733 |
>
> **2** Read along the row. The first train leaves Bournville at 0622.
>
> **3** Check for any further information,
> e.g. a = arrival (the train arrives at 0633), d = departure.
>
> With some timetables, trains only run on certain days.

Practice

1   Use the timetable above to answer these questions. Give answers as 24 hour clock times.

   a) i)   When is the first train from Redditch to Birmingham?
       ii) When does this train arrive in Birmingham?
   b) When is the first train after 7 o'clock in the morning to leave from Longbridge?
   c) i)   When does the first train leave from Kings Norton?
       ii) At what time does this train arrive in Birmingham?
   d) Which is the latest train from Bournville that you can catch if you want to be in Birmingham by 7 o'clock in the morning?
   e) Which is the latest train you can catch from Kings Norton if you want to be in Birmingham by half past 7 in the morning?
   f) A commuter arrives at Kings Norton at half past six in the morning.
       i)   Which is the next train she can catch?
       ii) When does this train depart from Five Ways?

2   This timetable gives the bus times going in both directions. It uses 12 hour clock time. Give your answers as am or pm times.

EASY RIDER 629		Woodgate Valley North – Selly Oak			
From **Woodgate**	Woodgate Valley North	8:45	10:10	1:55	3:15
	Moat Meadow	8:53	10:18	2:03	3:23
	Quinbourne Centre	9:00	10:25	2:10	3:30
	Onneley House	9:15	10:40	2:25	3:45
to	Harborne, High Street	9:25	10:50	2:35	3:55
	Old Tokengate	9:39	10:55	2:40	4:00
	Queen Elizabeth Hospital	9:40	—	2:45	4:05
Selly Oak	Selly Oak	9:50	—	3:00	4:10
From **Selly Oak**	Selly Oak	—	12:20	4:10	
	Queen Elizabeth Hospital	—	12:30	4:20	
	Old Tokengate	—	12:40	4:25	
	Harborne, High Street	11:25	12:45	4:30	
to	Onneley House	11:35	12:55	4:40	
	Quinbourne Centre	11:50	1:10	4:50	
	Moat Meadow	11:57	1:17	4:55	
Woodgate	Woodgate Valley North	12:05	1:25	5:00	

   a) When does the first bus leave Woodgate Valley North?
   b) When does the first bus return from Selly Oak to Woodgate Valley North?
   c) i)   Which is the first bus from Quinbourne Centre after 10 am to go to Selly Oak?
       ii) When does this bus arrive at Selly Oak?
   d) Which is the first bus from Moat Meadow to go to Queen Elizabeth Hospital?
   e) You are at Queen Elizabeth Hospital at midday.
       What is the earliest time you could get to Harbourne High Street by bus?
   f) i)   You are at Onneley House. Which bus should you catch to be at Queen Elizabeth Hospital by ten thirty in the morning?
       ii) If you leave the hospital at quarter to twelve, when will you catch your bus back to Onneley House?
       iii) When would you get back to Onneley House?
   g) You are at Moat Meadow. You are due to meet friends at the Old Tokengate at 3 pm. What is the latest bus you could catch?

Units of time
MSS1/E3.3, MSS1/L1.3, MSS1/L2.2

60 seconds = 1 minute	52 weeks = 1 year
60 minutes = 1 hour	365 days = 1 year
24 hours = 1 day	366 days = 1 leap year (a leap year has an extra day)
7 days = 1 week	100 years = 1 century
12 months = 1 year	1000 years = 1 millennium

Practice

1 Copy and complete these.

a) $\frac{1}{2}$ minute = _____ seconds

b) $\frac{1}{4}$ year = _____ months

c) $\frac{1}{2}$ hour = _____ minutes

d) 2 centuries = _____ years

e) $\frac{1}{4}$ hour = _____ minutes

f) 1 millennium = _____ years

g) $\frac{1}{2}$ year = _____ months

h) 2 years = _____ weeks

i) 2 weeks = _____ days

j) $\frac{1}{2}$ year = _____ weeks

Do you know this poem?   See page 156.

30 days have September, April, June and November

All the rest have 31

Except February alone, which has 28 days clear and 29 each leap year.

---

2 a) How many days are there in   i) June   ii) November   iii) February in a leap year?

b) How many days are there altogether in   i) July and August   ii) March and April?

3 Which unit of time would you use to measure the following events?

a) Boil an egg

b) 100 metre sprint

c) The time taken for 2nd class mail delivery

d) A holiday abroad

e) Flying to Europe

f) Travelling by boat to Australia

g) The natural age of an oak tree

seconds?	minutes?
hours?	days?
weeks?	months?
years?	centuries?

4 Copy and complete these.

a) $3\frac{1}{2}$ minutes = _____ seconds

b) 180 seconds = _____ minutes

c) $4\frac{1}{2}$ hours = _____ minutes

d) 300 minutes = _____ hours

e) 3 years = _____ months

f) 30 months = _____ years

g) 7 weeks = _____ days

h) 35 days = _____ weeks

i) $4\frac{1}{2}$ centuries = _____ years

j) 500 years = _____ centuries

k) 3 millennia = _____ years

l) 10 000 years = _____ millennia

5 How many millennia have there been since the birth of Christ?

> **per annum** means **per year** > **a quarter** means $\frac{1}{4}$ **of a year**

6 A workman's wage is £15 000 per annum. How much is this each month?

7 A hospital patient has to wait 28 days to hear the results of a test.
How many weeks is this?

8 The instructions say, 'Cook the food in a microwave for $2\frac{1}{2}$ minutes', but the microwave works in seconds. Write the cooking time in seconds.

9 It takes $1\frac{3}{4}$ hours to cook a chicken in the oven. What is this in minutes?

10 How many weeks are there in 1 quarter?

11 A student's phone bill is £26 per month. What is the phone bill for
a) a quarter of the year　　**b)** a full year?

12 a) A family go on a 7 night holiday on July 28th. On what date do they return?
　b) A couple go on holiday on April 19th and return on May 3rd.
　　How many nights are they away?

Calculate time

MSS1/L1.3, MSS1/L2.2

a Add and subtract time in hours, minutes and seconds

> **Remember 60 seconds = 1 minute　 60 minutes = 1 hour**

Example　　1 minute + 40 seconds + $1\frac{1}{2}$ minutes

```
min   sec
 1    00  +   1 minute
      40      40 seconds
 1    30      1 minute 30 seconds
 2    70      70 seconds = 1 minute + 10 seconds
```

> Write $\frac{1}{2}$ minute as 30 seconds.

> Change 70 seconds to minutes and seconds.

2 min + 1 min + 10 sec = **3 minutes 10 seconds**

Examples　　4 hours 36 minutes − 49 minutes

```
h   min
4   36  −
    49
```

> can't take 49 from 36 so change an hour to minutes

4 h 36 min = 3h 36 min + 1 hour = 3h 36 min + 60 min

```
h   min
3   96  −
    49
3   47      3 hours 47 minutes
```

Practice

1 Add these times together.
   a) 3 hours 20 minutes $+ 1\frac{3}{4}$ hours $+ 15$ minutes
   b) $6\frac{1}{2}$ minutes $+ 2\frac{1}{4}$ minutes $+ 12$ seconds
   c) $2\frac{1}{2}$ minutes $+ 40$ seconds $+ 25$ seconds
   d) $3\frac{3}{4}$ minutes $+ 30$ seconds $+ 2$ minutes
   e) 7 hours 40 minutes $+ 35$ minutes $+ 1\frac{1}{2}$ hours
   f) $1\frac{1}{2}$ hours $+ 40$ minutes $+ 1$ hour 55 minutes

2 Subtract these times.
   a) $3\frac{1}{4}$ hours $- 55$ minutes
   b) $2\frac{1}{4}$ hours $- 35$ minutes
   c) $2\frac{1}{4}$ minutes $- 55$ seconds
   d) $1\frac{1}{2}$ minutes $- 38$ seconds
   e) $4\frac{1}{4}$ hours $- 40$ minutes
   f) 2 minutes 10 seconds $- 49$ seconds

3 a) Cooking instructions for a curry in the microwave say, 'Cook for $6\frac{1}{2}$ minutes, stand for 3 minutes, then cook for a further $1\frac{1}{2}$ minutes.' What is the total time?
   b) For rice, the cooking time is $4\frac{1}{2}$ minutes, followed by 2 minutes standing, then another 45 seconds cooking. Altogether how long does it take to cook the rice?

4 A decorator takes altogether $1\frac{1}{2}$ hours to paint a window. He spends 35 minutes preparing the wood and the rest of the time painting. How long does he spend painting?

b Find the difference between times by adding on

Example	How long is it from 09:45 to 11:30?

Start time   09:45
Number of minutes to the next hour   09:45 ⟶ 10:00 =   15 minutes
Number of hours   10:00 ⟶ 11:00 = 1 hour   **Add**
Number of minutes after the hour   11:00 ⟶ 11:30 =   30 minutes
                          Time from 09:45 to 11:30 = **1 hour 45 minutes**

To check this, round to the nearest hour: 10:00 to 12:00 is 2 hours

Practice

1 Find the time from: (also check by rounding the times)
   a) 04:55 to 07:36   b) 06:42 to 09:24   c) 09:45 to 12:30
   d) 08:42 to 10:15   e) 07:26 to 11:14   f) 06:35 to 10:25
   g) 9:55 am to 2:10 pm   h) 7:50 am to 4:26 pm   i) 8:35 am to 12:16 pm
   j) midday to 17:50   k) 2:15 pm to midnight   l) 2:15 am to midnight

Example   A part-time worker starts work at 11:15 am and finishes at 3:45 pm.
For how long does he work?

Start time	11:15	
Number of minutes to the next hour	11:15 ⟶ 12:00 =	45 minutes
Number of hours	12:00 ⟶ 3:00 = 3 hours	**Add**
Number of minutes after the hour	3:00 ⟶ 3:45 =	45 minutes
	Time between 11:15 and 15:45 (or 3.45) =	**3 hours 90 minutes**

▶ **Remember 60 minutes = 1 hour**

3 hours + 1 hour 30 minutes
He works 4 hours 30 minutes

2 Find the time worked by each worker.

	Started	Finished		Started	Finished		Started	Finished
a)	9 am	12:30 pm	b)	10:15	16:30	c)	12:10 pm	5:40 pm
d)	14:30	18:45	e)	7:30 pm	1 am	f)	22:45	06:30

3 Find each journey time.

	Departure	Arrival		Departure	Arrival		Departure	Arrival
a)	0615	1136	b)	1305	1723	c)	1910	2316
d)	1024	1436	e)	0816	1251	f)	1513	1726
g)	0822	1054	h)	1212	1436	i)	1806	2131
j)	1825	2134	k)	0628	1042	l)	2102	2314

c Find the difference between times by subtracting

Example   A journey starts at 10:15 am and ends at 1:46 pm. How long does it take?

Put the times into 24 hour clock time (if necessary)	10:15 to 13:46	
Start with the final time	13:46 –	
Subtract the start time	10:15	**Subtract**
	Time for the journey = 3:31	**3 hours 31 minutes**

Example   How long is it from 7:55 am to 2:10 pm?

You can't take 55 from 10, so change an hour into 60 minutes.

In 24 hour clock time this is 07:55 to 14:10		
Start with the final time	14:10 –	13:10 + 1 hour
Subtract the start time	07:55	

13:10 + 60 = 13:70 –
07:55

Time between =   6:15   **6 hours 15 minutes**

Practice

1 Find the time taken for these journeys.

a) 1:15 pm to 3:42 pm   b) 11:26 am to 1:55 pm   c) 2:36 pm to 3:49 pm

d) 1:26 pm to 2:59 pm   e) 7:24 am to 2:50 pm   f) 6:54 am to 1:05 pm

g) 11:43 am to 1:32 pm   h) 9:43 am to 1:28 pm   i) 7:49 pm to 9:16 pm

j) 11:56 am to 4:13 pm   k) 3:33 pm to 5:13 pm   l) 10:44 am to 9:13 pm

2 The timetable below is for Eurostar.

London	Brussels
0609	1001
0653	1037
0822	1210
0827	1210
1022	1405

London	Calais
0614	0856
0619	0856
1148	1431
1153	1431
1518	1756

London	Paris
0922	1329
0927	1329

a) How long do these journeys take?

   i) 0822 from London to Brussels   ii) 1022 from London to Brussels

   iii) 1153 from London to Calais   iv) 0922 from London to Paris

b) What is the difference in journey time between the 0614 and the 0619 from London to Calais?

Activity

Plan a journey, a concert or an evening watching TV.

Write down the times for the start and end of each part/programme.

Work out how long each part takes.

Use timers

Alarm clocks can be set for any time. They use 12 or 24 hour clock time, with an analogue or digital display. **Analogue** means on a clock or scale.

These clocks have all been set to ring at 7 pm (to wake a nightshift worker).

The alarm hand is usually
a different colour

(12 hour)   (24 hour)

Practice

1 For each part sketch three clocks like these.

Show the alarm time on each clock.

a) 10:30 pm   **b)** 6 am

c) 11:15 pm   **d)** 8:30 am

 (12 hour)   (24 hour)

Ovens often have timers – generally in 24 hour clock time. The diagram shows one type.

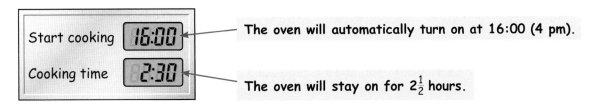

Start cooking **16:00** ← The oven will automatically turn on at 16:00 (4 pm).

Cooking time **2:30** ← The oven will stay on for $2\frac{1}{2}$ hours.

The meal will be cooked by 18:30, i.e. 6:30 pm.

Example   A meal takes 1 hour 20 minutes to cook. It must be ready by 18:00 (6 pm).

To find the time it needs to start cooking:

$$\begin{array}{r} 18:00\ - \\ 1:20 \\ \hline 16:40 \end{array}$$

The cooker times are as shown.

Start cooking **16:40**

Cooking time **1:20**

Meal ready for 18:00 (6 pm)

---

2 Copy the cooker display for each part. Fill in the times needed.

Start cooking [ ]

Cooking time [ ]

a) You want a meal to be ready for 19:00.
It takes $1\frac{1}{2}$ hours to cook.

b) You want a meal to be ready at 17:30.
It takes $2\frac{1}{4}$ hours to cook.

c) A meal takes 1 hour 40 minutes to cook. You want it to be ready at 20:15.

d) A meal takes $1\frac{3}{4}$ hours to cook. You want it to be ready at 7 pm.

Television recorders show the date, channel and the start and end times of programmes to be recorded. Time displays are in 24 hour clock time.

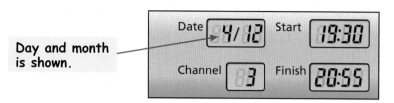

Day and month is shown.

The recorder will start recording Channel 3 at 7:30 pm on 4th December and end at 8:55 pm.

> **Activity**
>
> You need a listing of television programme times. Pick four programmes you would like to record. Write down the information you would need to set a TV recorder.

Some digital timers (such as stopwatches) show minutes and tenths and hundredths of a second.

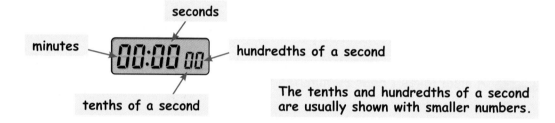

minutes

seconds

hundredths of a second

tenths of a second

The tenths and hundredths of a second are usually shown with smaller numbers.

For example, these show the times for runners in a race:

1st **01:05 20**   1 minute, 5 seconds and 2 tenths of a second (1 min 5.20 s)

2nd **01:15 32**   1 minute, 15 seconds and 32 hundredths of a second (1 min 15.32 s)

3rd **01:16 02**   1 minute, 16 seconds and 2 hundredths of a second (1 min 16.02 s)

3 The times for the rest of the runners are given below. Write them in words and figures.

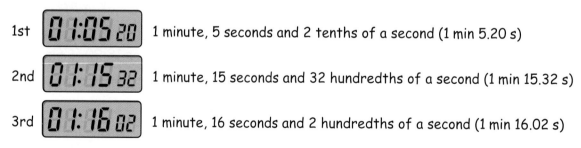

a) **01:16 47**   b) **01:21 05**   c) **01:16 59**

d) **01:22 20**   e) **01:17 08**   f) **01:22 40**

Example

The times of the first and second runners in a race are as shown. How much quicker is the winner?

03:56 23 **04:02 38**

	minutes		seconds		minutes		seconds	
longer				can't take 56.23				
time	4	:	02.38	from 2.38, so	3	:	62.38 –	
first	3	:	56.23	change a minute	3	:	56.23	subtract
				to seconds	0		6.15	

The winner is **6.15 seconds** quicker.

4 Find the difference between each pair of times.

a) **01:25 31** **01:27 52**   b) **02:43 03** **02:42 47**   c) **01:58 23** **02:00 38**

d) **05:10 24** **04:59 56**   e) **04:36 01** **04:22 49**   f) **13:57 30** **14:22 32**

5 The times for the swimmers in the six swimming lanes in a race are given below.

02:34 05   **02:49 30**   **02:16 43**   **03:01 09**   **03:00 24**   **02:44 51**

Put the times in order, then find the difference between each time and the next.

Measures, Shape and Space

3 Length

Understand distance: kilometres and miles

MSS1/E3.4

In Britain we use **miles** to measure long distances on the roads.
In the rest of Europe, long distances are measured in **kilometres**.

km stands for kilometres.   1 km = 1000 m (metres)

1 mile is longer than 1 km.   1 km is about $\frac{2}{3}$ of a mile.

It takes about 30 minutes to walk 1 mile. It takes about 20 minutes to walk 1 km.

Practice

1 A City 10 miles   B 3 km   C City 5 km   D 1 mile

   a) Which of the signs above might you see in Britain?
   b) Which might you see in other parts of Europe?

2 Copy and complete the table below. You will need to:

   - estimate how far in miles you travel to get to each place (starting from home)
   - estimate how far it is in kilometres
   - say how you would get there
   - estimate how long it takes to get there.

Journeys

	Approximate distance in miles	Approximate distance in kilometres	Do you use the car, bus, bike or walk?	How long does the journey take?
Class				
Shops				
Dentist				
Bus stop				
Work				
The nearest hospital				

3 For each journey choose an approximate distance.

   a) 1 hour by car
   b) 40 minutes walking
   c) 15 minutes walking
   d) 30 minutes walking
   e) 1 hour by bike
   f) 20 minutes walking
   g) 1 hour by train

   | 50 miles | 1 km | 80 miles | 1 mile | 2 km | 10 miles | ½ mile |

Find the distance

MSS1/L1.5

a Use distances marked on a map

This is a sketch map of part of Somerset. The towns are shown with dots.
The numbers give the distances in miles between the towns by road.

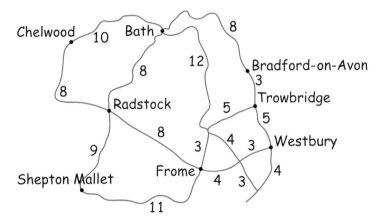

1 Find the shortest distance by road in miles between:

   a) Radstock and Bath
   b) Bradford-on-Avon and Bath
   c) Chelwood and Radstock
   d) Shepton Mallet and Bath
   e) Chelwood and Frome
   f) Westbury and Shepton Mallet
   g) Trowbridge and Bath
   h) Frome and Bath.

2 A taxi driver starts from Shepton Mallet. He travels to Frome, then Westbury and then
   Trowbridge by the most direct route. How far does he travel altogether?

3 A travelling salesman starts from Westbury and travels to Trowbridge.
   He then goes via Bradford-on-Avon to Bath and finally to Chelwood.
   How far does he travel altogether?

b Use a mile/kilometre chart

Distances in miles

Aberdeen						
216	Carlisle					
129	92	Edinburgh				
146	95	43	Glasgow			
261	45	137	140	Kendal		
84	132	45	62	177	Perth	
227	371	284	296	416	239	Thurso

To find the distance from Carlisle to Perth:

Look down from Carlisle.
Look across from Perth.

Distance = **132 miles**

Practice

1 Find the distance between:

a) Aberdeen and Kendal

b) Edinburgh and Thurso

c) Carlisle and Glasgow

d) Thurso and Perth

e) Perth and Aberdeen

f) Thurso and Kendal.

Some charts show both kilometres and miles like the chart below.
Check which side of the chart you need to look at.

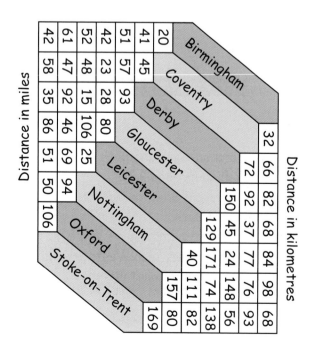

2 Use the chart on page 174 to answer the following questions.

   a) What is the distance in kilometres from Birmingham to Oxford?
   b) What is the mileage between Gloucester and Nottingham?
   c) How many miles are there between Coventry and Derby?
   d) How many kilometres are there between Leicester and Stoke-on-Trent?
   e) What is the mileage from Coventry to Nottingham?
   f) How much longer, in miles, is the journey from Birmingham to Nottingham than the journey from Birmingham to Derby?
   g) How many more kilometres do you travel to get from Oxford to Derby than you travel to get from Oxford to Coventry?
   h) You travel from Birmingham to Oxford and then to Nottingham.
   How many kilometres do you travel altogether?

Activity

Use a road atlas to find the distance in miles from your home town to other towns in England. List the towns in order of distance from your home town.

Measure length in metric units   MSS1/E3.5, MSS1/L1.4, MSS1/L2.3

a Measure in millimetres and centimetres

A **millimetre** is about the width of this fullstop •

It is the smallest measurement most people use.
The width of your little finger is probably about a **centimetre**.

This ruler measures centimetres (cm) and millimetres (mm)   **10 mm = 1 cm**

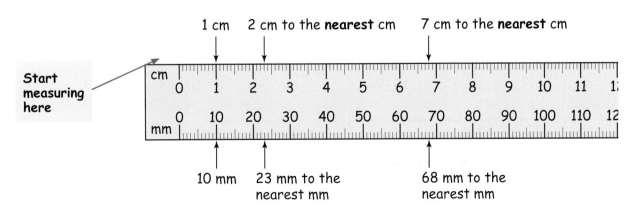

Which is the best way to measure the length of the needle?

A

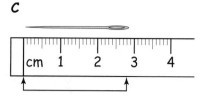

B

C

C is the best way.

The needle is **3 cm** long to the **nearest cm**.
More accurately, it is **28 mm** long.

Practice

On the ruler below, the measurement
marked by **a** is **1 cm** to the nearest cm
and　　　　　**13 mm** to the nearest mm.

> ▶ Use the **numbered marks** for **the nearest cm**.
> ▶ Use the **other marks** for the **nearest mm**.

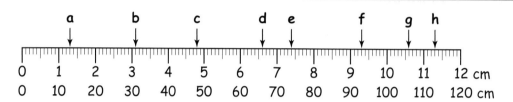

1 Write down the measurements marked by each of the other letters.
　Give each answer　　**i)** to the nearest cm　　**ii)** to the nearest mm.

> ▶ **The ruler shows that 10 mm = 1 cm so**　to change mm to cm, divide by 10.
> 　　　　　　　　　　　　　　　　　　　　　to change cm to mm, multiply by 10.

2 Copy and complete these.

　　a) 20 mm = _____ cm　　　b) 3 cm = _____ mm　　　c) 50 mm = ____ cm
　　d) 7 cm = _____ mm　　　e) 60 mm = _____ cm　　　f) 8 cm = _____ mm
　　g) 40 mm = _____ cm　　　h) 9 cm = _____ mm　　　i) 120 mm = ____ cm

> ▶ You can write lengths using decimals.
> **10 mm = 1 cm** means **1 mm = 0.1 cm**

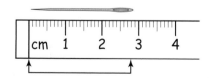

The needle's length = 28 mm or 2 cm 8 mm
This is the same as 2.8 cm

 Also see page 107.

| **Examples** | 42 mm = **4.2 cm**　　1.7 cm = **17 mm** |

3 Copy and complete these.

   a) 15 mm = ____ cm   **b)** 1.2 cm = _____ mm   **c)** 26 mm = ____ cm

   d) 1.3 cm = _____ mm   **e)** 14 mm = ____ cm   **f)** 2.1 cm = _____ mm

   g) 84 mm = ____ cm   **h)** 3.9 cm = _____ mm   **i)** 125 mm = ____ cm

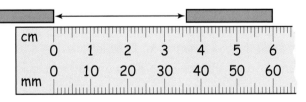

Check that the length of this line is 57 mm

This is the same   as   5 cm 7 mm

                      or   5.7 cm

The first decimal place gives mm.

4 Measure each line. Give each measurement in 3 different ways.

   a) _____

   b) _____

   c) _____

   d) _____

   e) _____

   f) _____

   g) _____

   h) _____

5 List the lines in question 4 in order of size, longest first.

Example

The distance between the rods is 36 mm or 3 cm 6 mm or 3.6 cm.

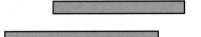

---

6 Measure the distance between each pair of rods. Write each distance in three ways.

   a)

   b)

   c)

7 Measure the length of each item to the nearest mm. Write each answer in three ways.

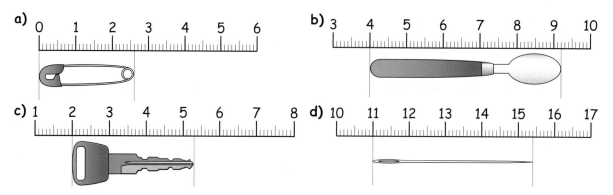

e) Which item was the easiest to measure? Why?

b Measure in metres, centimetres and millimetres

Long tape measures are used to measure distances like the length of a room. They are labelled in a variety of ways. This tape measure is labelled in centimetres and millimetres.

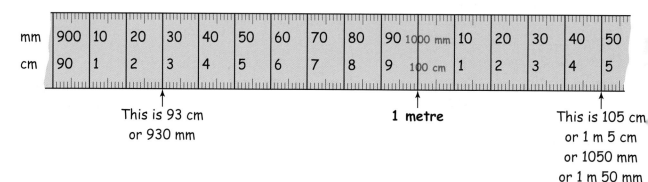

This is 93 cm
or 930 mm

1 metre

This is 105 cm
or 1 m 5 cm
or 1050 mm
or 1 m 50 mm

The tape measure shows that   ▶ 1 m = 100 cm = 1000 mm

Here is another tape measure near 1 m. It is labelled in a different way.
The measurement at **a** on the tape is **98 cm** or **980 mm**. **b** is at **111 cm** or **1110 mm**.

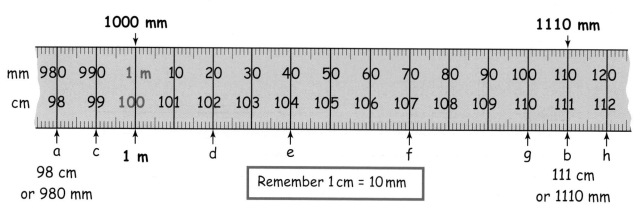

a
98 cm
or 980 mm

Remember 1 cm = 10 mm

111 cm
or 1110 mm

The mark at a is **98 cm** which is the same as **0.98 m**
The measurement at b can also be written as: **1 m 11 cm** or **1.11 m**

The first 2 decimal
places are cm.

Practice

For questions 1 and 2 use the ruler at the bottom of page 178.

1 Write the measurement at each of the other letters in cm then in mm.

2 Write the measurement at each of the other letters in metres using decimals.

3 The diagram below shows part of a tape measure labelled in centimetres.

    To the **nearest cm:**   a is at **365 cm** or **3.65 m**   and   b is at **367 cm** or **3.67 m**.

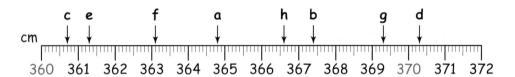

    i) Give the measurement at each of the other letters to the nearest cm.
    ii) Write each measurement in m using decimals.

4 You can use the small divisions on the tape measure to give lengths more accurately.
    a is at **364.8 cm**. In millimetres this is **3648 mm** (multiplying by 10).
    You can write this using metres as **3 m 648 mm** or **3.648 m**.
    b is at **367.4 cm** or **3674 mm** or **3 m 674 mm** or **3.674 m**.

                                                                          The first 3 decimal places are mm.

Write the measurement at each of the other letters in these four different ways.

Activity

1 Copy the table. Use a tape measure, rule and/or trundle wheel to complete it.

Name	Length in m	Length in cm	Length in mm	Width in m	Width in cm	Width in mm
Room						
Window						
Table						

2 Measure your height in metres, then centimetres to the nearest centimetre.
   Copy and extend the table. Fill it in for yourself and other members of the group.

Name	Height in m	Height cm	Height in mm

3 List everyone in your group in height order, starting with the tallest.

Estimate and measure lengths   MSS1/E3.5, MSS1/E3.8, MSS1/L1.4, MSS1/L2.3

Sometimes you need to measure very accurately to make sure something will fit. For example, when checking to see whether a kitchen unit will fit into a space, you need to measure the space accurately.

The table suggests the most appropriate units and equipment to use when measuring something accurately.

Length	Most appropriate units	Equipment
Short e.g. insect	millimetres (mm)	ruler micrometer
Medium e.g. book	centimetres (cm)	ruler tape measure
Long e.g. room	metres (m)	tape measure trundle wheel
Very long e.g. journey	kilometres (km)	trundle wheel pedometer mileometer

People don't always use the suggested units. For example, builders usually use metres or millimetres, but not centimetres.

Sometimes you might round up your measurement to the next full unit. For example, when buying carpet you might round the length of the room up to the next full metre to make sure you have enough.

Sometimes you may just need a rough idea of the length; for example, when estimating the size of a lawn to work out how much lawn food you need.

Practice

1   In the following situations, do you need to measure accurately, or round up?
    Write **accurate** or **round up**.
    a) Measuring the gap for your new sink unit.
    b) Buying worktops that are sold in whole metres for your new kitchen.
    c) Measuring a space to fit a new piece of furniture.
    d) Measuring the height of your windows for new ready-made curtains.
    e) Buying a pane of glass to replace a broken one.

2 Match each item with a likely measurement
   a) Width of a coffee table   **b)** Length of a marathon
   c) Height of a door   **d)** Length of a nail
   e) Length of a book   **f)** Length of a pencil
   g) Length of a computer mouse   **h)** Width of a nail head
   i) Width of a pencil lead

   30 cm   3 cm   14 cm   2 m   1 m   3 mm   40 km   1 mm   11 cm

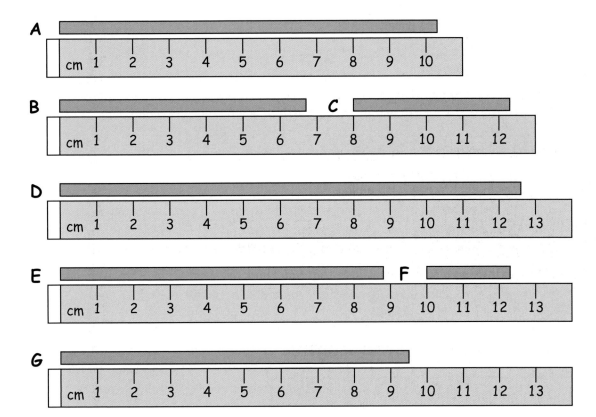

The length of the rod above is about 4 cm.
A more accurate estimate is 3 cm 7 mm or 3.7 cm

Imagine the millimetres between 3 and 4.

3 **a)** Estimate the length of each rod below: **i)** to the nearest cm   **ii)** to the nearest mm.
   b) Measure the lengths accurately.

Activities

1 Copy and complete this table. Estimate, then measure, each length.

Length to be measured	Estimate to nearest cm	Accurate measurement
Width of your little finger		
Your handspan (across open hand)		
Length of your foot		
Length of your arm		
Length of a stride		

2 Copy and complete the following table.
  Estimate, then measure, each object to the nearest centimetre.
  Measure the items more accurately in millimetres.

Item	Estimate to nearest cm	Measurement to the nearest cm	Measurement in mm
Length of this book			
Length of the fire exit notice			
Width of the fire exit notice			
Length of a piece of paper			
Width of a piece of paper			
Length of your pencil			
Length of an eraser			

3 You will need a collection of nails, screws, screw plugs and pieces of wood.
  a) Measure the length and pin head width of each nail, screw and plug.
     Label each nail, screw and plug with its measurements.
  b) Sort the nails and screws in order of length, starting with the shortest.
  c) Sort the nails and screws again, this time in order of pin head width.
  d) Match each screw with a plug.
  e) Sort the screws depending on the width of screwdriver they need.
  f) Measure the thickness of each piece of wood. Label them.
  g) Find the nails and screws that are the best for each piece of wood.

4 Without using a ruler, draw lines that you think are 1 cm, 2 cm, 3 cm, 4 cm and 5 cm long. Measure your lines with a ruler to see how accurate they are.

Convert between metric lengths

MSS1/L1.7, MSS1/L2.5

) **1 m = 100 cm = 1000 mm and 1 km = 1000 m**

To convert lengths between units, you must multiply or divide by 10, 100 or 1000.

from **mm** $\xrightarrow{\div 10}$ **cm** $\xrightarrow{\div 100}$ **m** $\xrightarrow{\div 1000}$ **km**   from **km** $\xrightarrow{\times 1000}$ **m** $\xrightarrow{\times 100}$ **cm** $\xrightarrow{\times 10}$ **mm**

$\div 1000$ $\times 1000$

Examples

Convert:

600 mm to cm	**120 cm to m**	**2 km to m**	**1.3 m to cm**
$600 \div 10 = 60$ cm	$120 \div 100 = 1.2$ m	$2 \times 1000 = 2000$ m	$1.3 \times 100 = 130$ cm
3400 m to km	**250 mm to m**	**0.5 cm to mm**	**0.04 m to mm**
$3400 \div 1000 = 3.4$ m	$250 \div 1000 = 0.25$ m	$0.5 \times 10 = 5$ mm	$0.04 \times 1000 = 40$ mm

Practice

1 How many mm?

  a) 2.3 cm   **b)** 4.3 cm   **c)** 43.5 m   **d)** 4.5 m   **e)** 12.45 m   **f)** 2 km

2 How many cm?

  a) 13 m   **b)** 40 m   **c)** 9.6 m   **d)** 13.7 m   **e)** 65 mm   **f)** 2.5 km

3 How many m?

  a) 4000 mm   **b)** 5 km   **c)** 2400 cm   **d)** 5.84 km   **e)** 0.74 km   **f)** 3534 mm
  g) 530 cm   **h)** 680 mm   **i)** 96 cm   **j)** 13.75 km   **k)** 0.659 km   **l)** 0.05 km

4 How many km?

  a) 7000 m   **b)** 9400 m   **c)** 450 000 cm   **d)** 95 000 m   **e)** 2 500 000 mm

5 Copy and complete these.

  a)   4 km = _____ m   **b)**   5 m = _____ cm   **c)**   45 cm = _____ mm
  d) 5000 m = _____ km   **e)** 250 mm = _____ cm   **f)** 360 cm = _____ m
  g) 275 cm = _____ m   **h)**   275 m = _____ km   **i)**   3.6 km = _____ m
  j)   2.7 m = _____ cm   **k)** 12.2 cm = _____ mm   **l)** 7563 m = _____ km

6 Put these in order of size. Start with the smallest.
  150 cm   650 mm   2 m   0.5 m   6 mm   2.6 cm   200 mm

7 A fence panel is 250 centimetres long. What is this in metres?

8 A race is 10 000 metres long. What is this in kilometres?

9 A worktop is 600 millimetres wide. Convert this to: **a)** centimetres   **b)** metres.

10 A door is 0.75 metres wide. What is this in: **a)** centimetres   **b)** millimetres?

Lengths can also be given using fractions.

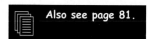 Also see page 81.

Examples

$\frac{1}{5}$ m = 1000 mm ÷ 5 = 200 m   75 cm = $\frac{75}{100}$ = $\frac{3}{4}$ m   (dividing top and bottom by 25)

---

11 Copy and complete these.

a) $\frac{1}{2}$ m = _____ cm   **b)** $\frac{1}{4}$ m = _____ cm   **c)** $\frac{1}{2}$ m = _____ mm

d) $\frac{1}{4}$ m = _____ mm   **e)** 400 mm = _____ m   **f)** 30 cm = _____ m

Calculate with metric lengths

MSS1/L1.6, MSS1/L2.5

a Metres and centimetres

Decimal point

Units	.	Tenths	Hundredths
metres	.	$\frac{1}{10}$ metre	$\frac{1}{100}$ metre
100 cm	.	10 cm	1 cm

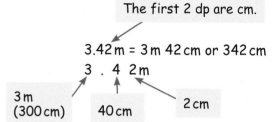

The first 2 dp are cm.

3.42 m = 3 m 42 cm or 342 cm

3 . 4 2 m

3 m
(300 cm)   40 cm   2 cm

Example   1.5 m + 50 cm + 5 cm

 See page 109 to revise decimals.

Either **work in centimetres:**

To add lengths they must be in the same units

1.5 m + 50 cm + 5 cm = 150 + 50 + 5 = **205 cm or 2.05 m**

× 100 = 150 cm

Remember the units.

```
 150 +
  50
   5
 205
   1
```

Use a calculator to check.

or **work in metres using decimals:**

1.5 m + 50 cm + 5 cm = 1.5 + 0.5 + 0.05 = **2.05 m**

50 cm   5 cm   Remember the units.

```
 1 . 5  +
 0 . 5
 0 . 05
 2 . 05
     1
```

Keep the decimal points in line.

Practice

1 Work these out. Check your answers using a calculator or another method.

   a) $1.5\,m + 75\,cm + 5\,cm + 3\,m$ **b)** $3\,m + 5\,cm + 25\,cm$

   c) $2.5\,m + 45\,cm + 5\,cm + 1.5\,m$ **d)** $5\,m - 75\,cm$

   e) $3.25\,m + 90\,cm + 5\,m + 1.5\,m$ **f)** $4\,m + 4\,cm - 40\,cm$

   g) $5\,m - 125\,cm - 1.5\,m$ **h)** $3\,m + 1.25\,m - 150\,cm$

   i) $2.5\,m + 1.1\,m + 15\,cm + 5\,cm$

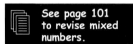

See page 101 to revise mixed numbers.

L2

2 Work these out in metres.

   a) $5\,m + 2\frac{1}{2}\,m$ **b)** $3\frac{1}{2}\,m - 1\frac{1}{4}\,m$ **c)** $2\frac{1}{4}\,m + 1\frac{1}{2}\,m$ **d)** $5\frac{1}{2}\,m - \frac{3}{4}\,m$

   e) $2\frac{1}{2}\,m + 1\frac{3}{4}\,m$ **f)** $4\,m - 2\frac{1}{4}\,m - \frac{1}{2}\,m$ **g)** $5\,m + 3\frac{3}{4}\,m + 2\frac{3}{4}\,m$ **h)** $8\,m - 3\frac{3}{4}\,m - 1\frac{3}{4}\,m$

b Metres and millimetres

Units	.	Tenths	Hundredths	Thousandths
metres	.	$\frac{1}{10}$ metre	$\frac{1}{100}$ metre	$\frac{1}{1000}$ metre
100 cm	.	100 mm	10 mm	1 mm

Decimal point

The first 3 dp are cm.

$4.275\,m = 4\,m\ 275\,mm$ or $4275\,mm$

4 . 2 7 5 m

4 m
(4000 mm)

200 mm   70 mm

5 mm

Example $3.5\,m - 725\,mm$

Either **work in milliimetres:**

$3.5\,m - 725\,mm = 3500 - 725 = \mathbf{2775\,mm}$ or $\mathbf{2.775\,m}$

    $\times 1000 = 3500\,mm$ Remember the units.

$$\begin{array}{r} 3500\ - \\ 725 \\ \hline 2775 \\ \hline \end{array}$$ Use a calculator to check.

or **work in metres using decimals:**

$3.5\,m - 725\,mm = 3.5 - 0.725 = \mathbf{2.775\,m}$

    700 mm   20 mm   5 mm

$$\begin{array}{r} 3.500\ - \\ 0.725 \\ \hline 2.775 \\ \hline \end{array}$$ Fill up with 0s

Practice

1 Work these out. Check your answers using a calculator or another method.

    a) 2.5 m + 125 mm + 50 mm
    b) 3.5 m − 475 mm
    c) 520 mm + 95 mm + 1.5 m
    d) 1.4 m + 825 mm + 25 mm
    e) 1.25 m + 950 mm + 3 m + 2.5 m
    f) 2.6 m + 54 mm − 620 mm

c Mixed calculations

Practice

1 Use metres or centimetres or millimetres. Give your final answer in metres.
Check each answer using a calculator or another method.

    a) 25 cm + 375 mm + 2 m
    b) 4.36 m + 48.5 cm + 625 mm
    c) 35 cm + 195 mm + 2.5 m

2 A kitchen wall is 3 m long.
Will a cooker of width 90 cm and units of width 60 cm, 1.2 m and 0.5 m fit along this wall?
Show your working and explain your answer.

3 A gardener needs 20 m of lawn edging.
He already has some pieces with lengths 2 m 40 cm, $1\frac{1}{2}$ m, $3\frac{1}{4}$ m, 2 m 80 cm and 4 m 60 cm.
What extra length does he need?

4 A householder wants to make two new curtains, each 1.3 m long.
He allows 15 cm extra on each curtain for hems.
What length of fabric does he use?

5 **a)** Mick has 5 m red bunting, 7.5 m green bunting, $4\frac{1}{2}$ m white bunting and 520 cm mixed red and white bunting. What length of bunting has he got altogether?
   b) Mick needs 36 m. How much more bunting does he need to buy?

6 **a)** An interior designer has $3\frac{1}{2}$ m of fabric for cushions.
She makes 3 cushions, each using $\frac{3}{4}$ m.
How much fabric is left?

   b) She also has $7\frac{1}{2}$ m of velvet.
She makes 2 curtains, using $1\frac{3}{4}$ m each and 4 cushions using $\frac{3}{5}$ m each.

    i) How much velvet does she use?   **ii)** How much velvet is left?

Use a scale on a plan or map

Plans and maps use **scales**. The scale of a plan tells you the relationship between lengths on the plan and the real distances.

In this plan of a flat the scale is **1 cm to 2 m**. Each centimetre on the plan represents two metres in the actual flat.

Measure the length of the bedroom.
It is 3 cm on the plan.
The actual length $= 3 \times 2 = 6$ m.

Measure the width of the kitchen.
It is 1.5 cm on the plan.
The actual width $= 1.5 \times 2 = 3$ m.

These measurements have been put into the table in question 1 below.

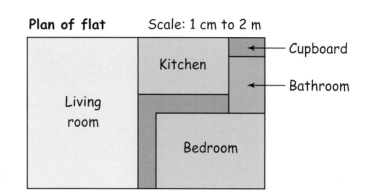

Practice

1 Copy and complete the table.

Room	Length on plan	Actual length	Width on plan	Actual width
Bedroom	3 cm	$3 \times 2 = 6$ m		
Kitchen			1.5 cm	$1.5 \times 2 = 3$ m
Living room				
Bathroom				
Cupboard				

2 Here is a plan of a group of houses.
Each house has its own plot of land.

Scale: 1mm to 1m

Copy and complete the following table.

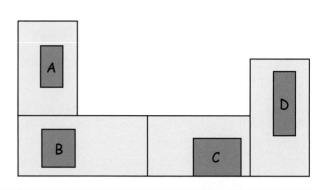

	Measurements on the plan				Actual measurements			
	House length	House width	Plot length	Plot width	House length	House width	Plot length	Plot width
House A								
House B								
House C								
House D								

3 This drawing shows the front of a house.

The scale is 10mm to 1m

a) On the drawing, measure in mm the heights of points A, B, C and D above the ground.

b) Use the scale 10mm to 1m, to find the actual heights of points A, B, C and D.

c) Copy and complete the table below by measuring from the plan in mm and then working out the actual measurements.

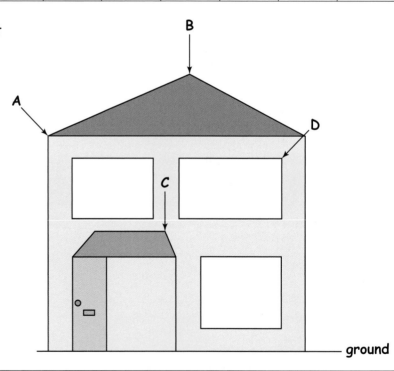

	Measurement on drawing in mm	Actual measurement
i) The length of the house		
ii) The width of the downstairs window		
iii) The width of the smaller upstairs window		
iv) The width of the larger upstairs window		
v) The height of the downstairs window		
vi) The height of the upstairs windows		
vii) The width of the front porch		

Maps often show the scale like this.
Use your ruler to measure the length of the line
representing 10 km.
It is 2 cm long.
On this map the scale is 1 cm to 5 km.
You can use this scale to measure
distances on the map.

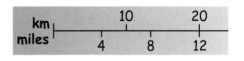

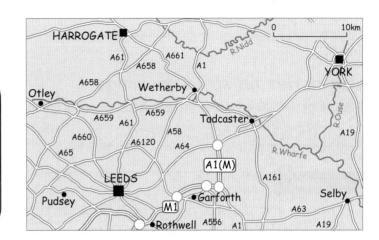

Example

Use a ruler to measure the
distance between Harrogate
and Otley. It is 2.8 cm.

The **direct** distance between
Harrogate and
Otley = 2.8 cm × 5 = 14 km.

Sometimes you might want to measure distances that are not straight.

Example

To estimate the distance **along the road** from
Leeds to Pudsey lie a piece of cotton (or string)
along the road.

Then lie it along the scale.
The actual distance is 11 km or (7 miles).

If the distance is too long to lie along the scale, measure the length of the cotton
with your ruler. Then use the scale 1 cm to 5 km.

4 a) Use a ruler to find the **direct** distance in kilometres between:
- **i)** York and Tadcaster
- **ii)** Leeds and Garforth
- **iii)** Harrogate and York
- **iv)** Garforth and Tadcaster.

b) Find the distance in kilometres **along the road** from:
- **i)** Leeds to Wetherby
- **ii)** Tadcaster to York
- **iii)** Selby to Wetherby
- **iv)** Wetherby to Harrogate.

5 A tutor uses this map when visiting college sites at A, B, C and D.

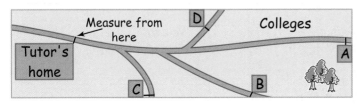

Scale: 1 cm to 500 m

Use the map at the bottom of page 189 to copy and complete the table:

	Length on map (cm)	Actual distance (km)
Tutor's home to A		
Tutor's home to B		
Tutor's home to C		
Tutor's home to D		

Use ratios on scale drawings   MSS1/L2.10

A scale of **1 : 20** on a plan means **actual measurements are 20 times those on the plan**.

So   5 cm on the plan   represents $20 \times 5 = 100$ cm (i.e. 1 m)   in real life.

Another way of giving this scale is **5 cm to 1 m**.

Example

The plan of a shop has a scale of 1 : 500. On the plan the shop is 12 cm long and 9 cm wide. What are its actual dimensions?

Length = $500 \times 12 = 6000$ cm   Width = $500 \times 9 = 4500$ cm
        = **60 m**   = **45 m**

Multiply by 500 to find the real length.

Divide by 100 to change cm to m.

Practice

1   Some of the other measurements from the plan of the shop are given below.
    Use the scale 1 : 500 to find the actual measurements.

   a) Length of car park 10 cm   b) Width of car park 6 cm
   c) Length of stockroom 5 cm   d) Width of stockroom 4 cm
   e) Length of office 2 cm   f) Width of office 1.2 cm
   g) Length of display area 3.2 cm   h) Width of display area 2.3 cm

2   The plan of a flat has a scale of 1 : 200.

   a) Copy and complete the table to give
       the actual measurements of the rooms.

   b) What does 1 cm on the plan represent?
       Give your answer in metres.

Room		on plan	actual
Lounge	Length	30 mm	
	Width	25 mm	
Kitchen	Length	18 mm	
	Width	14 mm	
Bathroom	Length	15 mm	
	Width	12 mm	
Bedroom	Length	24 mm	
	Width	16 mm	

3 Here is a plan of the downstairs layout of a house.

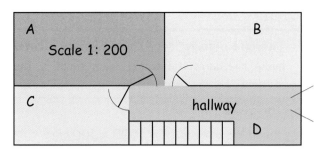

a) Measure the dimensions (i.e. length and width) of Rooms A, B and C and the hallway D on the plan.

b) Find the actual dimensions.

4 This is a plan of a garden.
Find the **actual**:

a) length of the patio
b) width of the patio
c) length of the shed
d) width of the shed
e) length of the greenhouse
f) width of the greenhouse
g) length of the whole garden
h) width of the whole garden.

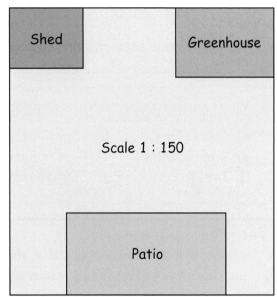

5 This map has a scale of 1 : 25 000.
Copy and complete the table below.
Measure the distances by road on the map as accurately as you can.
(Places are shown by black dots.)

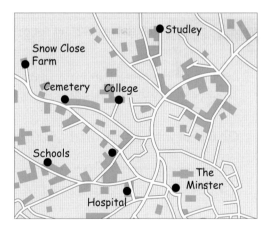

Distance	On map	Actual
a) The Minster – Hospital		
b) The Minster – Cemetery		
c) Snow Close Farm – Hospital		
d) College – The Minster		
e) Snow Close Farm – The Minster		
f) Schools – Hospital		
g) Studley – The Minster		

> **To write a scale** like 4 cm to 1 km **as a ratio scale** you must **make the units the same,** then **simplify the ratio** (if possible).

Example

$4 \text{ cm} : 1 \text{ km} = 4 \text{ cm} : 1 \times 1000 \times 100 \text{ cm}$
$= 4 : 100\,000$
$= \mathbf{1 : 25\,000}$

> $1 \text{ km} = 1000 \text{ m}$
> $1 \text{ m} = 100 \text{ cm}$

This is the scale used on some Ordnance Survey maps.

6   Write each of these scales as a ratio.

a) 1 cm to 1 m   b) 2 cm to 1 m   c) 5 cm to 1 m   d) 1 cm to 20 m
e) 1 cm to 500 m   f) 2 cm to 5 km   g) 5 cm to 8 m   h) 4 cm to 5 km
i) 1 mm to 1 m   j) 5 mm to 1 km   k) 1 mm to 0.5 cm l) 1 mm to 2.5 m

> **Drawing a plan or map**
> If the scale of a plan or map is $1 : n$, then you need to divide actual distances by n to find the corresponding distances on the plan or map.

Example   A drive is 20 m long. How long is it on a plan with a scale of 1 : 500?

Length of drive = 20 m.
In cm 20 × 100 = 2000 cm.

> It is more convenient to **work in cm for lengths on a plan or map.**

On a plan with a scale of 1 : 500, the length of the drive $= \dfrac{2000}{500} = 4 \text{ cm}$.

7   In each part draw lines to represent the given distances to the given scale.

a) Using scale 1 : 10, draw lines to represent   i) 60 cm   ii) 1.3 m
b) Using scale 1 : 500, draw lines to represent   i) 16 m   ii) 24 m
c) Using scale 1 : 200, draw lines to represent   i) 26 m   ii) 17.4 m
d) Using scale 1 : 1000, draw lines to represent   i) 70 m   ii) 43 m
e) Using scale 1 : 10000, draw lines to represent   i) 450 m   ii) 360 m
f) Using scale 1 : 20000, draw lines to represent   i) 840 m   ii) 1.8 km
g) Using scale 1 : 50000, draw lines to represent   i) 3.5 km   ii) 5.7 km

8 The sketch shows the dimensions of an office.
It contains

a) a computer desk 1m by 1.5m
b) a table 1.2m by 0.6m
c) a filing cabinet 60cm by 50cm
d) a shelf 40cm wide.

Draw a scale diagram using a scale of 1 : 25.

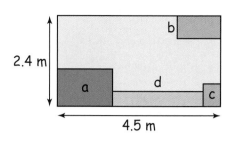

2.4 m

4.5 m

Activity

1 Use Ordnance Survey maps or maps from the internet to work out actual distances.
2 Measure a room and the largest items in it (e.g. a classroom, lounge, kitchen or bedroom). Draw a scale diagram of the room.

Measure lengths in imperial units   MSS1/L1.4, MSS1/L2.3

The imperial units for measuring lengths are **inches**, **feet**, **yards** and **miles**.
This ruler is marked in inches. Each inch is divided into 16 parts, called $\frac{1}{16}$ ths.

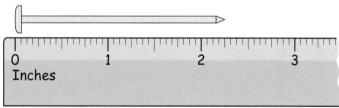

Screws, nails and wood thickness are sometimes measured in inches.
The length of this nail $= 2\frac{4}{16}$ inches $= 2\frac{1}{4}$ inches.

This can be written as $2\frac{1}{4}$ in or $2\frac{1}{4}''$.

See page 78 for equivalent fractions.

Practice

1 Measure these screws and nails in inches.

a)  b)  c)

d) 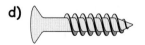 e)  f)

g)  h)  i)

j)  k)  l)

Longer lengths, such as the length of a room, are measured in feet or yards.

1 foot = **12 inches**   3 feet 6 inches can be written as 3 ft 6 in or 3', 6"
1 yard = **3 feet**   4 yards can be written as 4 yd

Activity

1   Estimate your height in feet and inches then measure it to the nearest inch.
Estimate, then measure the heights of other students in your group.
Put your results in a table.

2   a) Measure the length and width of the room in feet and inches.
   b) Write each measurement in yards, feet and inches.

Convert imperial lengths

MSS1/L2.5

> 12 inches = 1 foot   3 feet = 1 yard   1760 yards = 1 mile

$\div 12$ $\div 3$ $\div 1760$ $\times 1760$ $\times 3$ $\times 12$

inches $\longrightarrow$ feet $\longrightarrow$ yards $\longrightarrow$ miles   miles $\longrightarrow$ yards $\longrightarrow$ feet $\longrightarrow$ inches

---

Examples

Convert:

72 inches to yards
72 inches ÷ 12 = 6 feet
6 feet ÷ 3 = 2 yards
so **72 inches = 2 yards**

1 mile to inches
1 mile × 1760 = 1760 yards
1760 yards × 3 = 5280 feet
5280 feet × 12 = 63 360 inches
so **1 mile = 63 360 inches**

Use a calculator when the
arithmetic is very difficult.

4' 6" to inches
4' × 12 = 48 inches
48 ÷ 6 = 54 inches
so **4' 6" = 54 inches**

Practice

1   How many inches?

a) 5 yards   b) 6'   c) 28 feet   d) 16 feet
e) 2 yards   f) 12 yards   g) 9 feet 5 inches   h) $4\frac{1}{2}$ feet

2   How many feet?

a) 2 yards   b) 80 yards   c) 30 inches   d) 144"
e) 5 yards   f) 60"   g) $1\frac{1}{3}$ yards   h) 28 inches

3 How many yards?

  a) 72 inches **b)** 15' **c)** 4 miles **d)** 144 inches

4 How many miles?

  a) 5280 yards **b)** 15 840 yards **c)** 17 600 yards **d)** 10 560 yards

5 A room measures 9' 9" by 12" 9".
   What is the perimeter of the room

  a) in feet **b)** in yards?

> **See page 232 for perimeter.**

6 Three children measure their heights in inches:

   Jan 58 inches   Imran 65 inches   Carl 73 inches

   Write each height in feet and inches.

Convert between metric and imperial lengths

MSS1/L2.6

a Approximate metric/imperial conversions

> 1 inch ≈ 2.5 cm   1 foot ≈ 30 cm   1 m is slightly more than 1 yard

cm $\xrightarrow{\div 25}$ inches   feet $\xrightarrow{\times 30}$ cm

inches $\xrightarrow{\times 25}$ cm   cm $\xrightarrow{\div 30}$ feet   > ≈ means approximately equal to.

> **Example**   This pencil is 4 inches long. How many cm?

1 inch ≈ 2.5 cm   so   4 inches ≈ 4 × 2.5 = 10 cm.   The pencil is about 10 cm long.

Practice

1 Copy the tables below, then use approximate conversions to fill the gaps.
   Remember to write the measurement carefully, e.g. 3yd 2ft, not 3.2yd

inches	centimetres
2 in	
6 in	
8 in	
	7.5 cm
7 in	
	25 cm
	80 cm
18 in	
	100 cm

yards/feet	centimetres	metres
2 ft		
$2\frac{1}{2}$ ft		
	90 cm	
	105 cm	
$4\frac{1}{2}$ ft		
	150 cm	
	210 cm	
2 yd 2 ft		
3 yd		

This is $\frac{2}{10}$ of a yard.

2 feet = $\frac{2}{3}$ of a yard.

> **5 miles ≈ 8 km**

1 km ≈ $\frac{5}{8}$ mile or 0.6 miles   1 mile ≈ $\frac{8}{5}$ km or 1.6 km

km ――――――→ miles   miles ――――――→ km
　÷ 8, then × 5   ÷ 5, then × 8

Examples

Convert:
40 km to miles   **30 miles to km**
40 km ÷ 8 = 5 then 5 × 5 = **25 miles**   30 miles ÷ 5 = 6 then 6 × 8 = **48 km**

Alternative method
1 km ≈ 0.6 miles   **1 mile ≈ 1.6 km**
40 km ≈ 40 × 0.6 = **24 miles**   30 miles ≈ 30 × 1.6 = **48 km**

Not quite the same as before, but both 24 miles and 25 miles are good **estimates**.

2 **a)** Convert these distances to miles.   **i)** 80 km   **ii)** 208 km   **iii)** 30 km
   b) Convert these distances to kilometres. **i)** 60 miles **ii)** 240 miles **iii)** 30 miles

3 The table below gives the distances between pairs of European cities.
   Copy the table, then use approximate conversions to fill the gaps.
   Use a calculator if you wish.

	miles	km
a) Paris to Boulogne	159	
b) Brussels to Calais		216
c) Berlin to Cherbourg	847	
d) Brussels to Dieppe		310
e) Paris to Dunkerque	174	
f) Frankfurt to Amsterdam	283	
g) Munich to Boulogne		944
h) Turin to Ostend		967
i) Zurich to Calais	476	
j) Warsaw to Dieppe		1675

Give answers to the
nearest whole number.

b More accurate conversions

Some more accurate conversion factors are given below.

> **1 inch = 2.54 cm   1 yard = 0.914 m   1 mile = 1.61 km**

To convert, either multiply or divide by the conversion factor.

Look carefully at the examples below. Use your calculator to check the answers.

> ### Examples
>
> Convert:
>
> | **6 inches to cm** | **5 yards to metres** | **4 miles to km** |
> |---|---|---|
> | 6 inches × 2.54 = 15.24 cm | 5 yards × 0.914 = 4.57 m | 4 × 1.61 = 6.44 km |
>
> | **15 cm to inches** | **5 m to yards** | **6 km to miles** |
> |---|---|---|
> | 15 cm ÷ 2.54 = 5.91 inches | 5 m ÷ 0.914 = 5.47 yards | 6 km ÷ 1.61 = 3.73 miles |
> | | | (rounded to 2 decimal places) |

Practice

Use a calculator where necessary.

1 Change these into cm. (First change into inches if necessary.)

   a) 4 inches   **b)** 9 inches   **c)** 10.5 inches   **d)** 1 ft   **e)** 1 foot 6 inches

2 Change these into metres. Give the answers to 2 decimal places.

   a) 10 yards   **b)** 6 yards   **c)** $4\frac{1}{2}$ yards   **d)** 7 ft   **e)** 180 inches

3 Change these into inches. Give the answers to 2 decimal places.

   a) 15 cm   **b)** 45 cm   **c)** 24.8 cm   **d)** 2.5 m   **e)** 3 m

4 Change these distances to miles. Give the answers to 1 decimal place.

   a) 48 km   **b)** 16 km   **c)** 40 km   **d)** 30 km   **e)** 23.5 km

5 Change these distances to km. Give the answers to 1 decimal place.

   a) 15 miles   **b)** 25 miles   **c)** 42 miles   **d)** 30 miles   **e)** $9\frac{1}{2}$ miles

6 **a)** A map says it is 640 miles from Berlin to Calais. How many kilometres is this?
   b) Rome to Cherbourg is 1090 miles. How many kilometres is this (to the nearest km)?

7 How many miles is this (to the nearest mile)?
Cologne
160 km

8 The maximum speed limits on a road in Europe are 120 km per hour and 100 km per hour. What are these speeds in miles per hour? Give answers to the nearest mile per hour.

Measures, Shape and Space

4 Weight

Measure weight in metric units

MSS1/E3.6, MSS1/L1.4, MSS1/L2.3

a Know metric units of weight

The common metric units for weight are **grams** (g) and **kilograms** (kg).

1 g is about the weight of 10 matchsticks.

A bag of sugar weighs 1 kg.

> **1 kg = 1000 g**

The weight of very small items like tablets is measured in **milligrams** (mg).

> **1 g = 1000 mg**

Practice

1 Which unit would you use to measure the weight of:
   a) a packet of tea
   b) a bag of flour
   c) a small tin of tuna
   d) a packet of crisps
   e) a frozen chicken
   f) a pill?

b Weigh on balance scales

Here are some weights that can be used with balance scales:

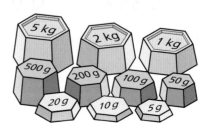

You can use more than one of each.

Example

To measure 750g of flour you could use these weights:
500g + 200g + 50g

Pour in flour until the scales are level.

Put a 500g, a 200g and a 50g weight in this pan.

There are other ways, e.g. you could use 200g + 200g + 200g + 100g + 50g.

Practice

1 What weights could you use to measure each amount?

a) 300g   b) 450g   c) 800g   d) 430g
e) 150g   f) 25g   g) 290g   h) 40g

2 Write down three different ways to balance 1kg.

> **Remember**   1kg = 1000g

You can write weights in g, kg or a mixture of these units.

L1

Examples

5000g = 5kg   5200g = 5kg 200g

1000 g = 1 kg so
5000 g = 5 kg

3 Which weights could balance each parcel?

a) 4 kg

b)  3½ kg

c) 2½ kg

d)  1 kg 550 g

e) 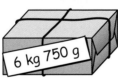 6 kg 750 g

f) 7 kg 400 g

4 Write each of these weights in kilograms or a mixture of kilograms and grams.

a) 7000g   b) 4000g   c) 2000g   d) 8000g
e) 4200g   f) 7300g   g) 2600g   h) 2870g
i) 5080g   j) 5389g   k) 5020g   l) 1005g

5 Write each of these weights in grams.

a) 4kg   b) 2kg   c) 3kg   d) 1kg
e) 1½ kg   f) 1¼ kg   g) 2½ kg   h) 3¾ kg

6 Each bag of sweets should weigh 1 kg.
Some bags are only partly filled. Their weights are given below.
What weight of sweets must be added to each bag to make 1 kg?

a) 500 g	**b)** 300 g	**c)** 470 g	**d)** 780 g
e) 250 g	**f)** 970 g	**g)** 840 g	**h)** 580 g
i) 360 g	**j)** $\frac{1}{4}$ kg	**k)** $\frac{1}{2}$ kg	**l)** $\frac{3}{4}$ kg

c Measure in grams and kilograms

Practice

This spring balance weighs up to 10 kg.

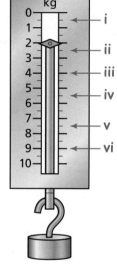

1 a) How many parts is each kilogram divided into?

b) How many grams does each small division represent?

c) Write down in kg and g the weight shown by each arrow:
i, ii, iii, iv, v, vi.

2 a) What is the maximum number of
kilograms that you can weigh on
these scales?

b) How many parts is each kilogram
divided into?

c) How many grams are represented by
each small division?

d) Write down in kg and g the weight
shown by each arrow: i, ii, iii, iv, v.

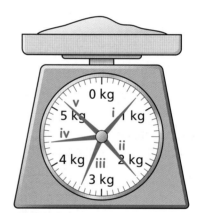

> Each kg on the scale is divided into 10 parts.
> Each part represents $\frac{1}{10}$ kg = 0.1 kg (= 100 g).
>
> So the weight shown by arrow (v) can be written as 5.2 kg.

e) Write the weight shown by each of the other arrows in kg using decimals.

3 Some scales weigh light objects like letters at a post office.

a) What is the maximum weight these scales can measure?

b) What weight is shown by each arrow?

Look at the scale below. The first arrow is nearer 1 kg than 2 kg.
This means the first reading is 1 kg **to the nearest kilogram**.

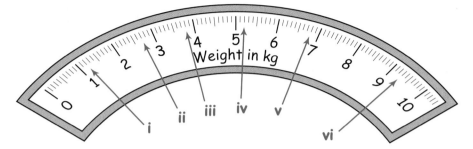

The second arrow is nearer 3 kg than 2 kg.
The second reading is 3 kg **to the nearest kilogram**.

4   Write down the rest of the readings, giving each reading to the nearest kilogram.

> Each small division on the scale represents $100\,g = \frac{1}{10}\,kg = 0.1\,kg$.
>
> So the weight shown by arrow (vi) can be written as 9 kg 300 g or 9.3 kg.

5   Write the weight shown by each of the other arrows:

   a) in kg and g   **b)** in kg using decimals.

6   Look carefully at this scale.

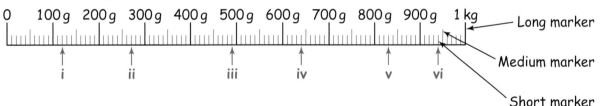

There is 100 g between each long marker and the next.

   a) What weight is between a long marker and the next medium marker?
   b) What weight is between each short marker and the next short marker?
   c) What is the weight shown by each arrow?

7   **a)** Write down, in kilograms, the weights shown by each arrow on this scale.

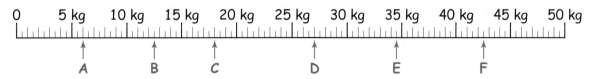

   b) What is the difference in weight between:
      i) A and B   **ii)** C and D   **iii)** E and F?

8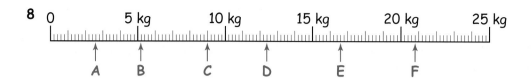

a) What do the medium markers show on this scale?

b) How many grams do the smallest divisions represent?

c) Write down the weights at A, B, C, D, E and F.
Give your answers in kilograms and grams.

d) What is the difference in weight between:
   i) A and B   ii) C and D   iii) E and F?

Estimate weights between marked divisions   MSS1/L2.3

Some scales look like this.
You have to estimate between the
labelled divisions.

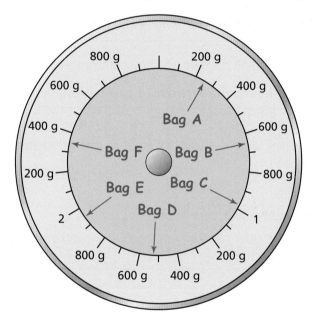

Practice

1 Look at the scale shown above.

a) Estimate the weight of each bag as accurately as you can.

b) Estimate the difference in weight between:

   i)   bags A and B   ii) bags C and D   iii) bags E and F
   iv) bags A and F   v)   bags C and E   vi) bags B and D.

c) How much needs to be added to bags A, B and C if they should each weigh $1\frac{1}{4}$ kg?

d) How much needs to be added to bags D, E and F if they should each weigh $2\frac{1}{2}$ kg?

e) There is also a bag G. The total weight of the seven bags is 10 kg.
   What is the weight of bag G?

Activities

1 Write down an estimate of the weight of 10 items in grams or kilograms.
Use a table like this.
Use scales or a spring balance to weigh each item accurately.

Item	Estimated weight	Measured weight

2 Estimate your own weight in kg. Weigh yourself. How accurate was your estimate?
Ask other people you know if you can weigh them.
(Remember that people are often sensitive about their weight.)
Put your results in a table.

3 Use a set of kitchen scales to weigh 100g of sugar. Put it in a bag.
Weigh 100g of paper.
Find a book that you think weighs about 100g. Weigh the book as accurately as you can. How close was it to 100g?

4 Choose 10 objects that you have not weighed before.
(For example, you could use things like a bunch of 5 pens.)
Estimate the weight of each object to the nearest 50g.
Weigh the objects. Record your results in a table.
Now list the objects in order of their weights, starting with the lightest.

Choose units and instruments   MSS1/E3.8, MSS1/L1.4, MSS1/L2.3

Practice

1 Copy this list of items. For each item choose a likely weight.

   a) Medium tin of beans
   b) Packet of crisps
   c) Bag of potatoes
   d) Yoghurt
   e) Bag of flour
   f) Egg
   g) Packet of butter
   h) Box of 160 tea bags
   i) Bag of sugar
   j) Adult's weight

   25 g, 250 g, 410 g, 5 kg, 50 g, 70 kg, 125 g, 500 g, 1.5 kg, 1 kg

2 Items are often weighed to the nearest **gram**, **10 grams**, **100 grams** or **kilogram**.
How accurately would you weigh these items?

   a) A bag of concrete mix
   b) Airmail letters
   c) Sugar for a recipe
   d) Your weekly weight loss during a diet
   e) Mushrooms

Scale A   **Scale B**

3 a) Which of these scales would be the best to
weigh each parcel?

P — 9 kg   Q — 4 kg 400 g   R — 8.5 kg   S — 5.4 kg

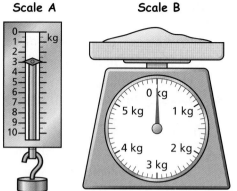

b) Put the parcels in order of weight, lightest first.

c) A postman weighs a parcel on scale A and says it
weighs 3 kg 500 g.
He then measures it on scale B and says it weighs 3 kg 600 g.
Which answer do you think is likely to be more accurate? Why?

Convert between metric weights

MSS1/L1.7, MSS1/L2.5

> 1000 mg = 1 g   1000 g = 1 kg   1000 kg = 1 tonne

To convert between these units you need to multiply or divide by 1000.

$$\text{from mg} \xrightarrow{\div 1000} \text{g} \xrightarrow{\div 1000} \text{kg} \xrightarrow{\div 1000} \text{tonnes} \qquad \text{from tonnes} \xrightarrow{\times 1000} \text{kg} \xrightarrow{\times 1000} \text{g} \xrightarrow{\times 1000} \text{mg}$$

Examples   Convert:

5000 mg to g
$5000 \div 1000 = 5$ g

6700 g to kg
$6700 \div 1000 = 6.7$ kg

200 kg to tonnes
$200 \div 1000 = 0.2$ tonnes

1.5 g to mg
$1.5 \times 1000 = 1500$ mg

0.625 kg to g
$0.625 \times 1000 = 625$ g

0.08 tonnes to kg
$0.08 \times 1000 = 80$ kg

The first 3 figures after the decimal point are grams.

Practice

1 How many kilograms?

a) 4000 g	**b)** 6500 g	**c)** 5485 g	**d)** 34 500 g
e) 800 g	**f)** 250 g	**g)** 300 000 g	**h)** 750 000 mg
i) 4 tonnes	**j)** 3.54 tonnes	**k)** 0.5 tonnes	**l)** 0.25 tonnes

2 How many grams?

a) 7 kg	**b)** 4.5 kg	**c)** 8.268 kg	**d)** 16.4 kg
e) 3.05 kg	**f)** 0.3 kg	**g)** 0.06 kg	**h)** 1.025 kg
i) 2000 mg	**j)** 7500 mg	**k)** 500 mg	**l)** 0.001 tonne

3 How many milligrams?

a) 44g   b) 30.5g   c) 3.3g   d) 0.83g

e) 12.3g   f) 0.07g   g) 0.05kg   h) 3.75kg

4 How many tonnes?

a) 503000kg   b) 42000kg   c) 4500kg   d) 589000g

5 Copy and complete these.

a) 3 tonnes = _____ kg   b) 0.5g = _____ mg   c) 1.5kg = _____ g

d) 0.5 tonnes = _____ kg   e) 7 tonnes = _____ kg   f) 0.25g = _____ mg

g) 750kg = _____ tonnes   h) 5000kg = _____ tonnes   i) 10kg = _____ g

6 Put each group in order of size, starting with the **smallest**.

a) 450g   0.4kg   4000000mg   0.4 tonnes   450kg

b) $\frac{1}{2}$kg   0.25kg   330g   25600mg   0.45kg

Calculate using metric weights

a Kilograms and grams

Parts of a kilogram

Units	.	Tenths	Hundredths	Thousandths
kg	.	$\frac{1}{10}$ kg	$\frac{1}{100}$ kg	$\frac{1}{1000}$ kg
1000 g	.	100 g	10 g	1 g

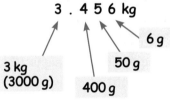

The first 3 dp are grams.

3 . 4 5 6 kg

3 kg (3000 g)   400 g   50 g   6 g

= 3 kg 456 g or 3456 g

Example   1.5kg + 400g + 30g + 5g

Either **work in grams**:

1.5kg + 400g + 30g + 5g = 1500 + 400 + 30 + 5 = **1935g** or **1.935kg**

     × **1000 = 1500 g**

Remember the units.

$$\begin{array}{r} 1500\ + \\ 400 \\ 30 \\ \underline{5} \\ 1935 \end{array}$$

or **work in kilograms using decimals**:

1.5 kg + 400 g + 30 g + 5 g = 1.5 kg + 0.4 kg + 0.03 kg + 0.005 kg

       = **1.935 kg**

Remember the units.   4 × 100g   3 × 10g   5g

Use a calculator to check.

$$\begin{array}{r} 1.5\ \ \ + \\ 0.4 \\ 0.03 \\ \underline{0.005} \\ 1.935 \end{array}$$

Practice

1   Work these out. Give the answer in kilograms if it is more than 1000 g.
Check your answers using a calculator or another method.

a) $3 \, kg + 500 \, g + 2 \, kg + 200 \, g$

b) $2.5 \, kg + 500 \, g + 75 \, g$

c) $700 \, g + 2 \, kg + 400 \, g + 50 \, g$

d) $1.5 \, kg + 500 \, g + 0.25 \, kg$

e) $1 \, kg + 750 \, g + 20 \, g + 2 \, kg$

f) $4 \, kg + 20 \, g - 1.75 \, kg$

g) $5\frac{1}{2} \, kg - 800 \, g$

h) $2.5 \, kg - 900 \, g$

i) $4 \, kg - 60 \, g$

j) $2 \, kg - 400 \, g + 25 \, g$

k) $2 \, kg - 100 \, g - \frac{1}{4} \, kg$

l) $4 \, kg + 40 \, g + 4 \, g$

m) $2 \, kg + 300 \, g + 40 \, g + 5 \, g$

n) $1 \, kg - 4 \, g$

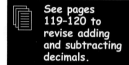

See pages
119-120 to
revise adding
and subtracting
decimals.

2   Find the difference in weight between:

a) $1.75 \, kg$ and $5 \, kg$ b) $3.5 \, kg$ and $1.75 \, kg$ c) $475 \, g$ and $6.5 \, kg$.

3   Three parcels weigh $2.75 \, kg$, $2.5 \, kg$ and $800 \, g$.
What is their total weight?

4   How much flour must be added to $285 \, g$ to make $570 \, g$?

5   A bag of frozen peas holds $2\frac{1}{2} \, kg$. A chef uses $700 \, g$.
What weight is left?

6   A piece of cheese weighs $1.2 \, kg$. Two pieces, each weighing $120 \, g$ are used for lunch packs.
What weight of cheese is left? Give the answer in grams.

7   A cook makes a cake. She uses $150 \, g$ flour, $150 \, g$ butter, $150 \, g$ sugar, $250 \, g$ mixed dried
fruit, $100 \, g$ fruit peel and $50 \, g$ chopped cherries.
What is the total weight of these ingredients?

8   A fruit grower picks $2.4 \, kg$ of strawberries and $1.75 \, kg$
of raspberries.

a) How many more grams of strawberries than
raspberries does she pick?

b) She uses $500 \, g$ of strawberries and $400 \, g$ of
raspberries in a fruit pudding.
What total weight of strawberries and raspberries
are left?
Give your answer in kg.

b Mixed calculations

Practice

1 A caterer buys 10 kg of potatoes, $1\frac{1}{2}$ kg of tomatoes, $1\frac{1}{4}$ kg of carrots and $1\frac{3}{4}$ kg of apples. What is the total weight of the shopping?

2 Small pots hold 50 g of salt. How many pots can be filled from a $\frac{1}{2}$ kg bag?

3 A shopper buys $\frac{1}{2}$ kg of carrots, 2.5 kg of potatoes, $\frac{1}{4}$ kg of onions and a bag of mushrooms weighing 360 g. How heavy is the shopping?

4 A tin of beans weighs 440 g. How much do 6 tins of beans weigh in kilograms?

5 A box weighs 420 g. A second box is 4 times as heavy. What is the total weight of the 2 boxes? Give the answer in kilograms.

6 A bag of 12 oranges weighs 4.2 kg. Find the average weight per orange.

7 How many 300 g bags of sweets can be made from 4.5 kg of sweets?

8 360 g of cake mix is divided equally between 24 small cake papers.
How much cake mix is in each cake paper?

9 2.5 kg of tobacco is put into 50 g packets. How many packets are there?

10 A crate holds 17 kg biscuits. The biscuits are in packets each holding 340 g.
How many packets are there in the crate?

c Kilograms and tonnes

> Remember 1000 kg = 1 tonne

Practice

1 A hospital uses 125 kg potatoes each day. How long will a 2 tonne delivery of potatoes last?

2 A tonne of grain is divided into 2.5 kg bags. How many bags are there?

3 A diesel locomotive weighs 100 metric tonnes.
A car transporter train carries 60 cars. Each car weighs 800 kg.
The cars travel on 15 car transporter trucks, each of which weighs 20 tonnes.

a) Find the total weight of the cars on the train in tonnes.
b) i) Which is the greater weight: the cars or the diesel locomotive?
   ii) Work out the difference in weight in tonnes.
c) What is the total weight of the car transporter trucks?
d) What is the total weight of the locomotive, transporter trucks and cars?

Weigh with metric and imperial units   MSS1/L1.4, MSS1/L2.3

Grams (g) and **kilograms** (kg) are **metric** units

▶ **1000 g = 1 kg**

Ounces (oz), **pounds** (lb) and **stones** (st) are **imperial** units.

▶ **16 oz = 1 lb   14 lbs = 1 st**

Many scales show both metric and imperial units.

Activities

1   Use grams, then ounces to weigh some items,
    e.g. a book, a dictionary, a pencil case or a bag.

2   Use kilograms, then pounds to weigh some items,
    e.g. a rubbish bin, 5 books, a chair, yourself.

Convert imperial weights   MSS1/L2.5

▶ **16 oz = 1 lb   14 lb = 1 st   2240 lb = 1 ton**

(Note: This is different from the metric tonne.)

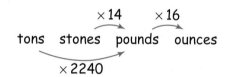

ounces   pounds   stones   tons   tons   stones   pounds   ounces

Examples   Convert:

3 st to lb	**2 lb 4 oz to oz**	**64 oz to lb**	$\frac{1}{2}$ **ton to lb.**
$3 \times 14 = $ **42 lb**	$2 \times 16 = 32$ oz	$64 \div 16 = $ **4 lb**	$2240 \div 2 = $ **1120 lb**
	$32 + 4 = $ **36 oz**		

Practice

Use a calculator where necessary.

1   Convert these to oz.

   a) 2 lb   b) 4 lb 6 oz   c) 6 lb 4 oz   d) 12 lb   e) 4 lb 8 oz
   f) 3 lb 7 oz   g) $\frac{1}{4}$ lb   h) $\frac{1}{2}$ lb   i) $\frac{3}{4}$ lb   j) $2\frac{1}{2}$ lb

2   How many lb are there in each of these?

   a) 16 oz   b) 96 oz   c) 48 oz   d) 128 oz   e) 80 oz
   f) 2 st   g) 4 st   h) $2\frac{1}{2}$ st   i) 2 tons   j) $1\frac{1}{2}$ tons

Take care with mixed imperial weights.

> **Example**   To convert 72 oz to lb and oz:
>
> Find how many whole pounds first $72 \div 16 = 4$ whole lbs
> Find how many ounces are used $4 \times 16 = 64$ oz
> Find how many ounces are left
>
> $$\begin{array}{r} 72\,oz \\ -\ 64\,oz \\ \hline 8\,oz \end{array}$$
>
> so 72 oz = **4 lb 8 oz**
>
> You can do this by long division, but take care if you use a calculator.
> This gives 4.5 lb. The 0.5 is half of a pound, **not** 5 ounces.

3   How may lb and oz are there in each of these?

    a) 56 oz   **b)** 82 oz   **c)** 38 oz   **d)** 73 oz

    e) 44 oz   **f)** 20 oz   **g)** 55 oz   **h)** 81 oz

Calculate using imperial weights

MSS1/L2.5

Practice

Use a calculator where necessary.

1   Brian weighs 183 lb. What is this in stones and pounds?

2   Bill weighs 115 lb. What is this in stones and pounds?

3   How much lighter is Bill than Brian:

    a) in pounds   **b)** in stones and pounds?

4   Paul's weight is half way between Brian and Bill. How much does Paul weigh:

    a) in pounds   **b)** in stones and pounds?

5   What is the average weight of Brian, Bill and Paul?

6   **a)** Brian decides to lose weight. He starts at 183 lb and loses 4 lbs in the first week of his diet. Copy and complete his weight chart.

	Week 1	Week 2	Week 3	Week 4	Week 5	Week 6
Weight lost	4 lb	4 lb		3 lb		2 lb
New weight in stones and pounds					11 st 11 lb	
New weight in pounds	179 lb		171 lb			

    b) How much did Brian lose altogether during these six weeks?

    c) How much more weight must he lose to reach his target of 10 st 7 lb?

Convert between metric and imperial weights

MSS1/L2.6

a Approximate metric/imperial conversions

Activity

Essential

Weigh some items in ounces and then in grams.
(Some suggestions are given in the table.)

Item	Weight in oz	Weight in g
Calculator		
5 pens		
A pad of paper		
Pencil case		

▶ **1 oz ≈ 25 g   1 lb ≈ 450 g**

An ounce is about the same as 25 grams.
If your pencil case weighs 4 oz, it should weigh about 100 g ($4 \times 25 = 100$).
Check your measurements in the table in the same way.

Approximate **conversion factors** between metric and imperial units of weight are given below.
Use them to answer the questions that follow.

Approximate conversion factors

▶ 1 lb is just under $\frac{1}{2}$ kg (about 450 g)   1 kg is just over 2 lb   1 oz ≈ 25 g

Remember this sign means 'approximately equal to'.

Examples $4\frac{1}{2}$ kg ≈ $4\frac{1}{2} \times 2 = 9$ lb   360 g ≈ 360 ÷ 25 = **14 oz** (nearest oz)

Practice

1 Which is the correct answer?

a) 4 oz is approximately _____.   5 g   2 kg   100 g   50 g
b) 1 kg is _____ than 2 lb?   **more**   **less**

2 A recipe has the following ingredients.
Convert each weight into ounces (to the nearest ounce).

a) 375 g fresh suet   b) 250 g currants   c) 560 g raisins
d) 280 g peel   e) 480 g breadcrumbs   f) 720 g flour

3 Pineapple cake has the following ingredients.
Give approximate conversions into grams (to the nearest 5 grams).

a) 14 oz pineapple rings in juice

b) 1 oz coconut

c) 5 oz chopped butter

d) 8 oz self-raising flour

e) 4 oz brown sugar

f) $3\frac{1}{2}$ oz melted butter

g) $2\frac{1}{2}$ oz plain flour

h) $7\frac{1}{2}$ oz caster sugar

b More accurate conversions

> **1 kg = 2.2 lb   1 oz = 28 g**

To convert, multiply or divide by the conversion factor.

Examples	Convert 5 kg to lb.	Convert 6 lb to kg.

$5 \text{ kg} = 5 \times 2.2 = \textbf{11 lb}$

$6 \text{ lb} = 6 \div 2.2 = \textbf{2.73 kg}$ (to 2 dp)
(using a calculator)

You can check using inverse calculations:
$11 \div 2.2 = 5$

$2.73 \times 2.2 = 6.006$
(near 6, the difference caused by rounding)

Practice

1 Use the conversion 1 kg = 2.2 lb to find the missing values.

a) 4 kg = ___ lb

b) 22 lb = ___ kg

c) 9 kg = ___ lb

d) 12 lb = ___ kg

2 Use the conversion 1 oz = 28 g to find these missing values.

a) 4 oz = ___ g

b) 140 g = ___ oz

c) 15 oz = ___ g

d) 70 g = ___ oz

Activity

Use conversions to check weights in recipe books.

Measures, Shape and Space

5 Capacity

Measure capacity and volume
MSS1/E3.7, MSS1/L1.4, MSS1/L2.3

a Know metric measures

The amount of liquid a container holds when full is its **capacity**.
Volume is the amount of space something fills.
The metric unit we use for capacity and volume of liquids is a **litre** (ℓ).

Many fruit juice drinks and squashes are sold in bottles that hold 1 litre.
Even large volumes, like the amount of water in a swimming pool are measured
in litres.

Activities

1 You need:

- a 5 litre plastic bottle (the sort used by campers)
- a $\frac{1}{2}$ litre measuring jug.

Measure $\frac{1}{2}$ litre water and pour it into the bottle.

Mark the $\frac{1}{2}$ litre level on the bottle.

Measure another $\frac{1}{2}$ litre water and pour it into the bottle.

Mark and label this level as 1 litre.
Repeat this until you have the bottle marked every $\frac{1}{2}$ litre up to 5 litres.

2 Collect together some large containers, e.g. bowls, saucepans, tins.
Estimate, then measure, how much each container holds.
Put your results in a table with the headings shown below.

Container	Estimate to nearest $\frac{1}{2}$ litre	Measure to nearest $\frac{1}{2}$ litre

---

We measure small amounts of liquid in **millilitres** (mℓ).

1 mℓ is just a few drops.
5 mℓ is the capacity of a medicine spoon/teaspoon.

Measuring jugs are often labelled in millilitres.

> **1000 mℓ = 1 litre**

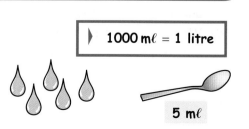

5 mℓ

Practice

1

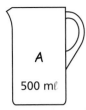

A
500 mℓ

B
200 mℓ

C
100 mℓ

1 litre
jug

1000 mℓ

a) How many jugs A are needed to fill this litre jug?
b) How many jugs B are needed to fill this litre jug?
c) How many jugs C are needed to fill this litre jug?

2 How many millilitres are there in:

a) 1 litre   **b)** $\frac{1}{2}$ litre   **c)** $\frac{1}{4}$ litre   **d)** $\frac{3}{4}$ litre   **e)** 1.5 litres?

3 How many litres are equal to:

a) 3000 mℓ   **b)** 5000 mℓ   **c)** 9000 mℓ   **d)** 12 000 mℓ   **e)** 4500 mℓ?

b Measure in litres and millilitres

Practice

1 This jug measures 2 litres.
Each division represents 200 mℓ.

a) Write down the amount shown by each arrow.
Give your answers in mℓ.

> 200 mℓ is the same as 0.2 litres, so each division on the
> scale measures 0.2 litres.
> The volume shown by arrow v can be written as 1.6 litres.

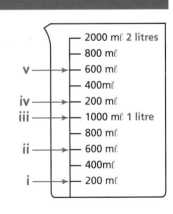

b) Write the amount shown by each of the other arrows in litres using decimals.

2 a) What is the maximum volume this jug can measure?
b) How many millilitres are represented by the large labelled divisions?
c) How many millilitres are represented by the small unlabelled divisions?
d) Write the measurements shown by each arrow.
e) How many millilitres are there in $\frac{1}{2}$ litre?

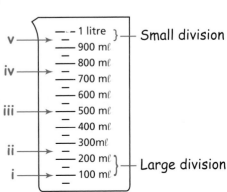

3 a) What is the greatest volume this cylinder can measure?
  b) How many large divisions are there between 100 mℓ and
     200 mℓ?
  c) How many mℓ does each large division represent?
  d) How many mℓ does each small division represent?
  e) Write the measures shown by the arrows.

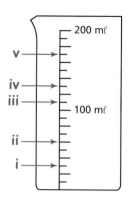

Estimate between marked divisions

MSS1/L2.3

Sometimes you need to estimate between the
divisions on a scale.

Arrow **(i)** is about half way between 350 mℓ and
400 mℓ.

An estimate of this reading is **375 mℓ**.

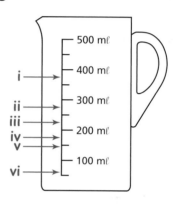

Practice

1 Estimate the readings shown by the other arrows as accurately as you can.

2 Here is another measuring jug.

  a) What is the greatest volume this jug measures?
  b) Write the measures shown by the arrows as
     accurately as you can.

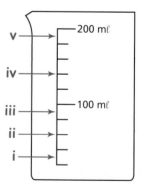

Activities

1 You will need a measuring jug marked in hundreds of millilitres to 1 litre.
   You also need a cup and some water.

  a) How many cupfuls of water do you need to fill the jug to the 1 litre mark?
  b) How many cupfuls do you need to fill to the $\frac{1}{2}$ litre level?

L1

2 Collect some bottles, jars and pots.
Estimate then measure how much each container holds.
Put your results in a table.

3 Pour 300 mℓ water into a measuring jug.
Put a stone in the jug.
Make sure it is covered with water.
Measure the level of the water again as accurately as
you can.
The difference between the two measurements is the
volume of the stone.

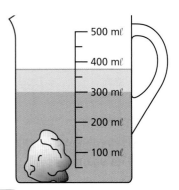

> **Example**
>
> The 1st water level was 300 mℓ.
> The 2nd water level is 375 mℓ. The difference is 75 mℓ.
> This is the same as **75 cm³**.
> (The volume of solid objects is usually measured in cm³.)

1 mℓ = 1 cm³

Find a selection of small objects or stones.
Find the volume of each.

4 a) Find the capacity of:
   i a teaspoon   ii) a dessertspoon   iii) a tablespoon.
  b) i)   How many teaspoons of water are needed to make $\frac{1}{2}$ litre?
   ii)   How many dessertspoons of water are needed to
      make $\frac{1}{2}$ litre?
   iii)   How many tablespoons of water are needed to make $\frac{1}{2}$ litre?

5 Weigh an empty measuring jug. Pour in 1 litre of water then weigh it again.
If you have measured and weighed accurately, you will have found that

   1 litre of water weighs 1 kg.

How much does 1 mℓ weigh?

Choose units and instruments

MSS1/E3.8, MSS1/L1.4, MSS1/L2.3

Practice

1 Which unit, litres or millilitres, would you use to measure these?

   a) The volume of water in a bath   b) The capacity of a baby bottle
   c) The capacity of a wine glass   d) The volume of water used to wash a car
   e) The capacity of a water butt   f) The water a rabbit drinks in a day

2 Which measuring container is most suitable to measure each amount?

A    **B** 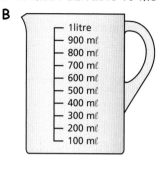   **C**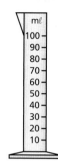

a) The volume of car shampoo to wash a car
b) The volume of gin in a gin and tonic
c) The capacity of a jam jar
d) The volume of paint to decorate a bedroom
e) The capacity of a saucepan

3 Which measuring container would you use to measure each volume?

A    **B**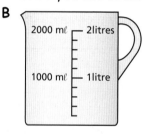

a) 1.8 litres   b) 300 mℓ   c) 600 mℓ   d) 1500 mℓ
e) $3\frac{1}{2}$ litres   f) 2.5 litres   g) 1400 mℓ   h) 500 mℓ

4 Which level of accuracy would you use to measure the volume of each item?
Choose from the nearest **1 mℓ**, **10 mℓ**, **100 mℓ** or **1 litre**.

a) Kettle   b) Indoor watering can   c) Medicine
d) Sauce jar   e) Fish tank   f) Saucepan
g) Concentrated plant food to mix with 1 litre of water

Convert between millilitres, litres and centilitres

MSS1/L1.7, MSS1/L2.5

Litres and millilitres are the most common metric units for capacity.

> **1 litre = 1000 mℓ**

To convert between litres and millilitres, multiply or divide by 1000.
You can use millilitres, litres or a mixture of these.
For example, 1400 mℓ is the same as 1.4 litres or 1 litre 400 mℓ

$$litres \xrightarrow{\times 1000} mℓ$$
$$mℓ \xrightarrow[\div 1000]{} litres$$

Examples

Convert: **4 litres to ml**
$4 \times 1000 = 4000\,\text{ml}$

2.5 litres to ml
$2.5 \times 1000 = 2500\,\text{ml}$

4875 ml to litres
$4875 \div 1000 = 4.875$ litres

or 4 litres and 875 millilitres (the first 3 dp are ml)

Practice

1 Write these in millilitres.

a) 5ℓ   b) 2.6ℓ   c) 12.5ℓ   d) 3.625ℓ   e) 0.682ℓ

f) 1.85ℓ   g) 5.063ℓ   h) 0.75ℓ   i) 0.083ℓ   j) 0.04ℓ

2 Write these in litres, using a decimal point where necessary.

a) 6000mℓ   b) 3500mℓ   c) 11000mℓ   d) 1100mℓ   e) 65400mℓ

f) 330mℓ   g) 250mℓ   h) 425mℓ   i) 7300mℓ   j) 5320mℓ

3 Write these in millilitres.

a) 3 litres 400mℓ   b) 6 litres 250mℓ   c) 3 litres 630mℓ   d) 8 litres 760mℓ

4 Write these as: i) litres and millilitres   ii) litres using a decimal point.

a) 1200mℓ   b) 3830mℓ   c) 4935mℓ   d) 5769mℓ   e) 15800mℓ   f) 5098mℓ

5 Find the matching volumes.

a) 100mℓ   b) 10mℓ   c) 250mℓ   d) 50mℓ

e) 2550mℓ   f) 500mℓ   g) 0.1 litre   h) 0.05 litre

i) 0.5 litre   j) 2.55 litres   k) 0.01 litre   l) 0.25 litre

6 Put the bottles in order of size, starting with the **greatest** capacity.

A    B   C   D

1.1 litres   94 mℓ   1.650 ℓ   875 mℓ

L2 Another unit that is used for measuring wine or spirits is the **centilitre** (cℓ).

> **1cℓ = 10mℓ**   (like cm and mm)   **1 litre = 100cℓ**   (like m and cm)

To convert between units you need to multiply or divide by 10, 100 or 1000.

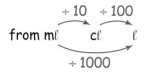

Examples	Convert:

320 mℓ to cℓ	**45 cℓ to litres**	**3.45 litres to cℓ**
$320 \div 10 = 32\,c\ell$	$45 \div 100 = 0.45$ litres	$3.45 \times 100 = 345\,c\ell$

---

7 How many litres are the same as:

   a) 300 cℓ   **b)** 250 cℓ   **c)** 3050 cℓ   **d)** 740 cℓ

   e) 50 cℓ   **f)** 33 cℓ   **g)** 75 cℓ   **h)** 3210 cℓ?

8 How many centilitres are the same as:

   a) 30 mℓ   **b)** 500 mℓ   **c)** 250 mℓ   **d)** 85 mℓ

   e) 3.5 litres   **f)** 1.75 litres   **g)** $\frac{1}{2}$ litre   **h)** 0.46 litres?

Calculate using metric units of capacity

MSS1/L1.6, MSS1/L2.5

a Litres and millilitres

Parts of a litre

Units	.	Tenths	Hundredths	Thousandths
litres	.	$\frac{1}{10}$ litre	$\frac{1}{100}$ litre	$\frac{1}{1000}$ litre
1000 mℓ	.	100 mℓ	10 mℓ	1 mℓ

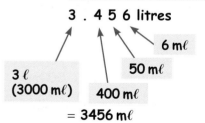

The first 3 dp are mℓ.

3 . 4 5 6 litres

3 ℓ (3000 mℓ)
400 mℓ
50 mℓ
6 mℓ
= 3456 mℓ

Example

Either **work in millilitres:**

$2.4\,\ell + 200\,m\ell + 15\,m\ell = 2400 + 200 + 15 = \mathbf{2615\,m\ell}$ or $\mathbf{2.615\,\ell}$

        × 1000 = 2400 mℓ   **Remember the units.**

```
2400 +
 200
  15
2615
```

or **work in litres using decimals:**

$2.4\,\ell + 200\,m\ell + 15\,m\ell = 2.4\,\ell + 0.2\,\ell + 0.015\,\ell = \mathbf{2.615\,\ell}$

      2 × 100 mℓ   **10 mℓ**   **5 mℓ**

               Use a calculator to check.

```
 2.4   +
 0.2
 0.015
 2.615
```

Practice

1 Work these out. Give the answer in litres if it is more than 1000 mℓ.
Check your answers with a calculator.

    a) 3 litres + 650 mℓ + 50 mℓ   **b)** 1 litre + 2 litres + 400 mℓ + 25 mℓ
    c) 7 litres + 2 litres + 75 mℓ   **d)** 3.6 litres + 850 mℓ + 80 mℓ
    e) 400 mℓ + 1.64 litres + 40 mℓ   **f)** 4.7 litres + 2 litres + 50 mℓ
    g) 150 mℓ + 5 mℓ + 50 mℓ   **h)** 10 litres − 4 litres − 40 mℓ
    i) 5 litres − 2 litres − 340 mℓ   **j)** $7\frac{1}{2}$ litres − 3.2 litres + 75 mℓ

2 A caterer mixes together 2.5 litres lemonade, 2 litres orange juice, 1.25 litres apple juice and 750 mℓ pineapple juice. How much liquid is this altogether?

3 A student buys 2 litres milk. He drinks 350 mℓ then gives 30 mℓ to the cat. How much milk is left?

4 Give the answers to these in litres. Check with a calculator.
    a) 6.2 litres + 600 mℓ + 3 cℓ   **b)** 50 mℓ + 43 cℓ + 0.6 litre
    c) 5 litres − 43 cℓ   **d)** 10 litres − 1 litre − 1 cℓ

5 For each part assume you start with 1 litre of fruit juice.
How much is left after pouring out the drinks described?

    a) 2 cups holding 250 mℓ each   **b)** 3 glasses holding 200 mℓ each
    c) 4 glasses holding 150 mℓ each   **d)** 2 tumblers holding 420 mℓ each

6

       A   **B**   **C**   **D**

    a) What is the total capacity of the 4 jars?
    b) What is the difference in capacity between the largest and the smallest jars?
    c) Which 2 jars have a total capacity of 2.12 ℓ?
    d) A new jar has half the capacity of A? How much does it hold?
    e) Another jar holds 4 times as much as D. What is its capacity?

b Mixed calculations

Practice

1 a) A 5mℓ dose of medicine is taken 4 times a day. What is the total dose per day?
  b) The medicine bottle holds 140mℓ. How many days will the medicine last?

2 A water butt holds 100 litres of water. A watering can holds 5 litres.
  How many times can you fill the can from the water butt?

3 Water drips from a tap at a rate of 400mℓ every 5 minutes.
  How much water drips from the tap in an hour?

Questions 4 to 9 are about a child's birthday party.

4 a) 24 children go to the party. Each child drinks two 200mℓ glasses of cola.
  How much cola do they drink altogether?
  b) A bottle holds 1.5 litres. How many bottles are used at the party?

5 The cake recipe needs 3 tablespoons of oil. The cook doesn't have a tablespoon, but she
  does have a teaspoon. She knows a tablespoon holds 15mℓ and a teaspoon holds 5mℓ.
  How many teaspoons of oil does she need to use?

6 a) Each packet of chocolate whip needs 300mℓ milk.
  How much milk is needed for 6 packets?
  b) The whip is divided between the 24 children.
  How much milk does each child get in their whip?

7 Each jelly dish holds 850mℓ. How much will 5 jelly dishes hold?

8 One party game has a large bucket of water. A 750mℓ jug is used to fill
  the bucket. The jug is filled 8 times. How much water is in the bucket?

 9 A cocktail is made for the adults. The recipe for 4 needs
  60mℓ orange juice,   80mℓ lemonade,   30mℓ rum   and   20mℓ tequila.

  a) There are 10 adults at the party. How much of each ingredient is needed?
  b) How much fruit cocktail is made altogether?

10 a) How many people are in your maths group?
  b) If everyone has a 200mℓ drink, how much liquid is needed altogether?
  c) A drink recipe is: $\frac{1}{3}$ lemonade, $\frac{1}{4}$ orange juice, $\frac{1}{6}$ apple juice
  $\frac{1}{6}$ grapefruit juice and $\frac{1}{12}$ lemon juice

  How much of each do you need to make the drink for your group?

Measure with metric and imperial units

Many measuring jugs and other containers show both metric and imperial measures.

The common metric units are millilitres and litres.

> **1000 mℓ = 1 litre**

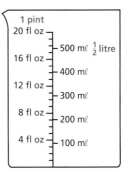

The common imperial units are fluid ounces (fl oz), pints (pt) and gallons (gal).

> **1 pt = 20 fl oz   1 gal = 8 pt**

Activities

1 Fill a pint milk bottle with water.
Pour the water into a measuring jug similar to the one shown above.
Does the milk bottle hold: **a)** more or less than 1 litre
b) more or less than $\frac{1}{2}$ litre?

2 Collect some containers that hold less than 1 pint, e.g. cup, mug, small jug, egg cup.
Estimate, then measure, the capacity of each container in appropriate metric and imperial units. Record your results in a table like this:

Container	Imperial		Metric	
	Estimate	Measurement	Estimate	Measurement

3 Collect some containers that hold more than 1 pint, e.g. tin, bowl, saucepan.
Add these to your table and compare the imperial and metric measurements.
Can you see that a litre is a bit less than 2 pints?

Convert between imperial units of capacity

> **1 pint = 20 fluid ounces   1 gallon = 8 pints**

To convert between units multiply or divide by the conversion factor.

$$\text{fl oz} \xrightarrow{\div 20} \text{pt} \xrightarrow{\div 8} \text{gal} \qquad\qquad \text{gal} \xrightarrow{\times 8} \text{pt} \xrightarrow{\times 20} \text{fl oz}$$

Examples   Convert:

80 fl oz to pints   **2.5 gallons to pints**

$80 \div 20 = 4$ pints $2.5 \times 8 = 20$ pints

Practice

1 Convert these to pints.

   a) 60 fl oz   **b)** 40 fl oz   **c)** 70 fl oz   **d)** 150 fl oz

   e) 4 gallons   **f)** 1.5 gallons   **g)** $\frac{1}{2}$ gallon   **h)** $\frac{1}{4}$ gallon

2 Convert these to gallons.

   a) 24 pints   **b)** 2 pints   **c)** 60 pints   **d)** 140 pints

3 Convert these to fluid ounces.

   a) 2 pints   **b)** 0.1 pint   **c)** $\frac{1}{2}$ pt   **d)** $\frac{1}{4}$ pt

Calculate using imperial units of volume
MSS1/L2.5

Practice

1 A recipe needs the following ingredients:

   1 pt lemonade, $\frac{1}{2}$ pt orange juice, $\frac{1}{4}$ pt apple juice,

   $\frac{1}{8}$ pt lemon juice, $\frac{1}{10}$ pt tequila.

   a) Write each volume in fluid ounces.
   b) What is the total volume in fluid ounces?

2 Sixty guests have $\frac{1}{10}$ pint punch each. How many gallons is this?

3 **a)** Five gallons of milk are poured into pint bottles.
       How many bottles are filled?
   b) Pints of milk are sold for 44p each. At the end of the day only 6 pints remain unsold.
       How much money is taken?

4 A bottle of concentrated disinfectant holds $1\frac{1}{5}$ pints. The disinfectant is diluted by mixing
   4 fluid ounces with 1 gallon of water.
   How many gallons of water is mixed with the concentrated disinfectant?

5 A day centre orders milk. There are 5 groups in the centre with 16 people in each group.
   Each person drinks 1 pint of milk per day.

   a) How many pints of milk are needed:
       i) each day   **ii)** each week   **iii)** each year?   (Assume 365 days in 1 year)
   b) Convert each of your answers to part **a)** into gallons.

Convert between metric and imperial measures

a Approximate metric/imperial conversions

There's a rhyme about the approximate conversion between litres and pints.

'A litre of water's a pint and three quarters.'

This means

> **1 litre $\approx 1\frac{3}{4}$ pints**

Other useful approximations are:

> **1 fl oz $\approx 30\,\text{m}\ell$**   **1 pint $\approx \frac{2}{3}$ litre**   **1 gallon $\approx 4\frac{1}{2}$ litres**

Practice

1 Which is larger: **a)**   a pint or a litre   **b)** $\frac{1}{2}$ litre or 1 pint?

2 Copy and complete these.

   a) 2 litres $\approx$ ____ pt   **b)** 3 pt $\approx$ ____ litres   **c)** 120 mℓ $\approx$ ____ fl oz
   d)   45 ℓ $\approx$ ____ gal   **e)** 4 gal $\approx$ ____ litres   **f)** 8 fl oz $\approx$ ____ mℓ

3 $\frac{1}{2}$ pint is poured out of a litre bottle. How much is left?

4 **a)** Which petrol can holds more: 2 gallon   or   10 litre?
   b) Which yoghurt pot holds more: 6 fl oz   or   120 mℓ?

b More accurate conversions

> **1 fl oz = 28.4 mℓ**   **1 pt = 568 mℓ**   **1 gallon = 4.55 litres**   **1 litre = 0.22 gallons**

To convert units, multiply and/or divide by the conversion factor(s).
Use a calculator when necessary

Example	Convert 5 fl oz to mℓ	Convert 330 mℓ to fl oz.

5 fl oz $= 5 \times 28.4 = \textbf{142 m}\ell$ 330 m$\ell = 330 \div 28.4 = \textbf{11.6 fl oz}$ (to 1 dp)

Check: 142 $\div$ 28.4 = 5   Check: 11.6 $\times$ 28.4 = 329.44
(nearly 330 – the difference is due to rounding)

Example   A water tank holds 500 litres of water. Use the conversion
10 litres = 2.2 gallons to find its capacity in gallons.

Capacity $= 500 \div 10 \times 2.2 = 50 \times 2.2 = \textbf{110 litres}$

Practice

Use a calculator when the calculations are very difficult.

1 Use the conversion 1 pint = 0.57 litres to find the missing values.

a) 2 pt = ____ litres   **b)** 2.85 litres = ____ pt
c) 8 pt = ____ litres   **d)** 4 litres = ____ pt

2 Use the conversion 1 gallon = 4.55 litres to find the missing values.

a) 5 gal = ____ litres   **b)** 500 litres = ____ gal
c) 9 gal = ____ litres   **d)** 36 litres = ____ gal

3 Use 2.2 gallons = 10 litres for each part.

a) A water container holds 20 litres. How many gallons is this?
b) How many litres does a 12 gallon petrol tank hold?

4 A medicine bottle holds 3 fl oz. How many mℓ is this? (1 fl oz = 28.4 mℓ)

5 a) A pond holds 80 gallons water. How many litres is this? (1 gallon = 4.55 litres)
b) The pond loses 2.5 litres each day from a leak at the bottom of the pond, how many gallons of water are lost each week?
c) How long will it take for the pond to be empty? Give your answer to the nearest week.

Use conversion scales and tables

MSS1/L2.6

There are other ways to convert between metric and imperial units of length, weight and capacity.
Side by side scales like this are often given on maps. They can be used to convert distances between miles and kilometres.

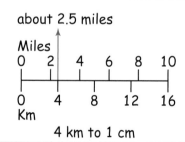

| **Example** | This scale shows 4 km ≈ 2.5 miles |

Practice

1 a) Use the scale above to complete these.
     i) 10 miles ≈ ____ km   **ii)** 8 km ≈ ____ miles
     iii) 8 miles ≈ ____ km   **iv)** 12 km ≈ ____ miles

A map of the Shetland Islands uses the scale shown above (4 km to 1 cm).

b) On the map the islands are 17 cm north to south.
Write the actual distance in: **i)** km   **ii)** miles.
c) On the map the islands are 14.8 cm east to west.
Write the actual distance in: **i)** km   **ii)** miles.

2 Copy the following table. Use the ruler to complete the table using the nearest markers.

Length in inches	1	2		$3\frac{1}{2}$			$4\frac{3}{4}$	$5\frac{1}{4}$		
Length in cm	2.5		8		10	11.5			13.7	14.3

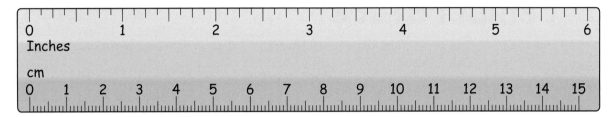

3 Write the value marked by each **orange** line or arrow in both metric and imperial units.

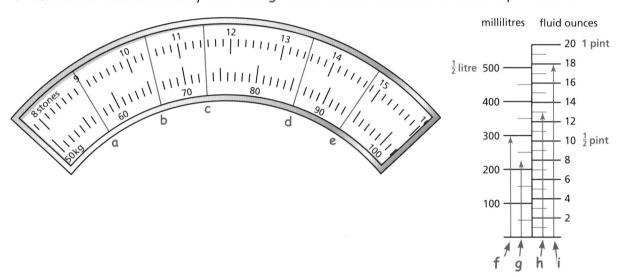

4 You can also use this table to convert lengths.
For example, 3 miles = 4.828 km and
3 km = 1.864 miles.

a) A taxi journey is 5 km.
How many miles is this?

b) A student cycles 6 miles to college and 6 miles back each day.
How many km is this in a 5 day week?

c) A postman walks 8 miles each day.
How many km is this in a 6 day week?

d) A delivery person makes three calls.
She travels 6 km to the first call,
3 km to the second and 9 km to the third.
How many miles is this altogether?

km	miles/km	miles
1.609	1	0.621
3.219	2	1.243
4.828	3	1.864
6.437	4	2.485
8.047	5	3.107
9.656	6	3.728
11.265	7	4.350
12.875	8	4.971
14.484	9	5.592
16.093	10	6.214

Measures, Shape and Space

6 Temperature

Measure temperature

MSS1/E3.9, MSS1/L1.4, MSS1/L2.4

Temperature is measured using a thermometer.
There are many different types.
Temperature is usually measured in degrees Celsius (°C).

This is a **medical thermometer**.
It is placed under a patient's arm or tongue.
The level of the liquid in the tube shows the patient's temperature.

The large divisions on the thermometer go up in full degrees Celsius.

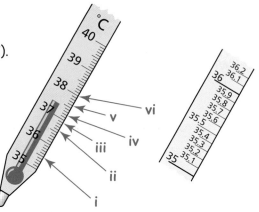

The enlarged section shows each degree is divided into 10 small divisions.
After 35°C, the markers show 35.1°, 35.2°, 35.3°, ..., up to 35.9° (just before 36°C).

Practice

1 a) What is the highest temperature the medical thermometer above can measure?
   b) What is the lowest temperature it can measure?
   c) Write down the temperatures shown by each arrow.

2 **Room thermometers** often look like this.
   a) What is the highest temperature that it can show?
   b) What is the lowest temperature it can show?
   c) Write down the temperatures shown by each arrow.

3 a) What is the highest temperature that this thermometer measures?
   b) What is the lowest temperature that it measures?
   c) What does each small division on the scale show?
   d) Write down the temperature shown by each arrow.

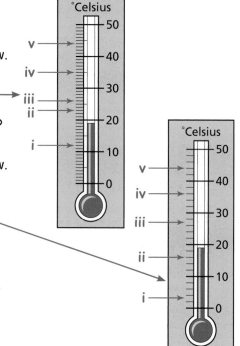

4 For each part, find the number of degrees the temperature **rises**.

a) 15°C → 24°C   **b)** 25°C → 32°C   **c)** 3°C → 11°C
d) 5°C → 21°C   **e)** 9°C → 33°C   **f)** 12°C → 39°C

5 For each part, find the number of degrees the temperature **falls**.

a) 21°C → 12°C   **b)** 27°C → 9°C   **c)** 34°C → 27°C

6 For each part, choose a likely temperature from the list below.

a) sauna   **b)** cool workroom   **c)** the surface of the sun
d) hotel room   **e)** ice cube   **f)** sunny summer day

0°C,   11°C,   20°C,   28°C,   54°C,   6000°C

Temperatures below freezing point   N1/L1.2, MSS1/L1.4

The Celsius scale has **negative** numbers **below the freezing point of water**.

0 = freezing point   −1 = getting colder   −2 = colder still   −3 = even colder

Practice

1 a) What is the highest temperature this thermometer measures?
b) What is the lowest temperature it measures?
c) Write down the temperature shown by each arrow.
d) From your answers to part **c)** write down:
   i) the warmest temperature
   ii) the coldest temperature.

2 The cool fridge in the supermarket should be at −4°C.
Which phrase describes the cool fridge with the
temperature shown?

a) too cold
b) the right temperature
c) too warm

3 For each part put the temperatures in order, starting with
the **warmest**.

a) −21°C   21°C   2°C   −12°C   −2°C   12°C
b) −4°C   23°C   0°C   −2°C   14°C   4°C
c) 7°C   1°C   −2°C   5°C   −7°C   3°C

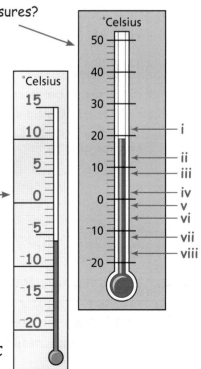

4 For each part, find the number of degrees the temperature rises.

a) –5°C to 4°C   b) –2°C to 12°C   c) –3°C to 11°C   d) –6°C to 10°C

e) –9°C to 13°C   f) –12°C to 29°C   g) –7°C to –2°C   h) –20°C to –3°C

5 For each part, find the number of degrees the temperature falls.

a) 21°C to –2°C   b) 17°C to –9°C   c) 14°C to –7°C   d) 15°C to –3°C

e) 22°C to –4°C   f) 18°C to –1°C   g) –1°C to –10°C   h) –4°C to –18°C

6 This chart shows the temperature over a period of 10 days.

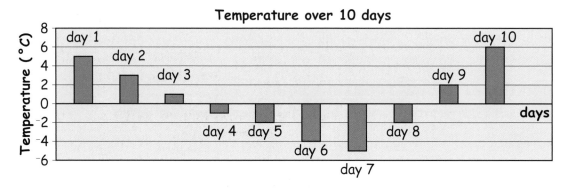

Temperature over 10 days

a) What was: **i)** the highest temperature   **ii)** the lowest temperature?

b) On which days was the temperature below freezing point?

c) What was the range of the temperatures over the 10 days?

For range see page 309.

7 The table below shows the central heating settings in a house.
The differences between the inside temperatures and those outside the house are missing.
Fill in these missing values. The sitting room has been done for you.

Central heating settings		Difference between the inside and outside temperature when the outside temperature is:		
Room	Temperature	11°C	6°C	–3°C
Sitting room	21°C	10°C	15°C	24°C
Dining room	20°C			
Kitchen	16°C			
Bedroom 1	19°C			
Bedroom 2	16°C			
Bedroom 3	13°C			

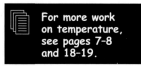

For more work on temperature, see pages 7–8 and 18–19.

Activities

1 a) Use a medical thermometer to take your temperature.
   b) Take the temperature of some other people. Make sure you wash the
       thermometer after each person has used it. Record your results in a table.

2 Make a copy of a thermometer.
   Mark and label the scale to show the following descriptions that weather forecasters
   use: Very hot 30°C, Warm 20°C, Mild 15°C, Cold 5°C.

3 Estimate the temperature for different places, e.g. room, corridor, outside, window
   sill or the top of a heater.
   Place a room thermometer in each place for two minutes.
   Record your estimates and the temperatures in a table.

4 **Digital thermometers** like this one are often used in hospital to take a patient's
   temperature. The units used on this thermometer are **degrees Fahrenheit (°F)**.

   Generally only the numbers are shown, not the °C or °F sign.
   Use a digital thermometer to find temperatures, e.g. a warm drink, cup of tea, your
   temperature, the room, a sock or outside. Record your results in a table.

Celsius and Fahrenheit

MSS1/E3.9, MSS1/L1.4, MSS1/L2.4

Ovens measure temperatures in Celsius, Fahrenheit or Gas marks.
This table gives equivalent temperatures and a description for cooking.

°C	°F	Gas mark	Description
110	225	$\frac{1}{4}$	Cool
120/130	250	$\frac{1}{2}$	
140	275	1	Very low
150	300	2	
160/170	325	3	Low/moderate
180	350	4	Moderate/warm
190	375	5	Moderately hot
200	400	6	Hot
210/220	425	7	
230	450	8	
240	475	9	Very hot

Practice

1 Look at the table at the bottom of page 229.
   a) Which °C settings should you use for:
      i) a hot oven   ii) a warm oven   iii) a cool oven?
   b) Repeat part a) for °F.

2 a) A roasting temperature for meat is 180°C. What is this in °F?
   b) The roasting temperature for poultry is 50°F higher than for other meats. What is the temperature in °F for roasting poultry?
   c) Draw dials like those shown below.

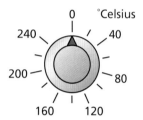 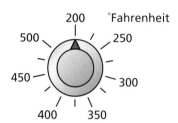

On each dial mark the temperature for cooking: i) meat   ii) poultry.

Use scales to convert temperatures

MSS1/L2.4

Many thermometers like the one on the right show temperatures in both Celsius and Fahrenheit. You can use the scales to compare and convert temperatures.

Office temperatures should be at least 16°C.
The scales show that this is also 61°F.

The table gives these and other useful temperatures:

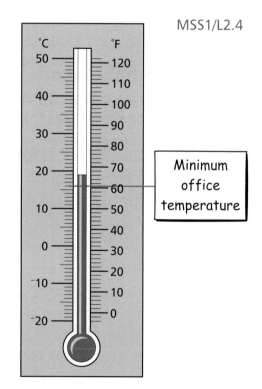

Description	Celsius	Fahrenheit
Office temperatures should not fall below	16°C	61°F
Chilled food in shops should be below	8°C	46°F
Frozen food in shops should be kept below	−18°C	0°F
Most house plants like a temperature over	18°C	64°F
Body temperature	37°C	98.6°F

Use the scales to check the other temperatures in the table.

Practice

1 The freezing point for water is 0°C. What is this in °F?
2 Use the scales to convert these temperatures to Fahrenheit.
Give your answers to the nearest degree.

    a) 30°C   **b)** 45°C   **c)** 23°C   **d)** –10°C   **e)** –15°C   **f)** –6°C

3 Use the scales to convert these temperatures to Celsius.
Give your answers to the nearest degree.

    a) 80°F   **b)** 50°F   **c)** 42°F   **d)** 18°F   **e)** 105°F   **f)** –4°C

Use formulae to convert temperatures
N1/L2.4

This rhyme might help you remember how to convert between Celsius and Fahrenheit.

If C to F confuses you, $\dfrac{9C}{5}$ + 32 is all you have to do.
▶ $F = \dfrac{9C}{5} + 32$

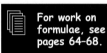
For work on formulae, see pages 64–68.

Example   Convert: 40°C to °F

$$F = \frac{9C}{5} + 5 = \frac{9 \times 40}{5} + 32 = \frac{360}{5} + 32 = 72 + 32 = \mathbf{104°F}$$

Practice

1 Below are some of the temperatures taken in capital cities around the world.
Convert each temperature to °F.

    a) Alexandria   20°C   **b)** Barbados   30°C   **c)** Toronto   5°C
    d) Belgrade   15°C   **e)** Berlin   10°C   **f)** Algiers   25°C
    g) London   12°C   **h)** New York   14°C

To convert from °F to °C use: ▶ $C = \dfrac{5(F - 32)}{9}$

Example   Convert: 86°F to °C

$$C = \frac{5(86 - 32)}{9} = \frac{5 \times 54}{9} = \frac{270}{9} = \mathbf{30°C}$$

2 Here are some more temperatures from capital cities. Convert these temperatures into
Celsius. (Round to the nearest degree.)

    a) Amsterdam   52°F   **b)** Athens   66°F   **c)** Bahrain   93°F
    d) Belfast   48°F   **e)** Hong Kong   81°F   **f)** Moscow   55°F
    g) Paris   59°F

Measures, Shape and Space

7 Perimeter, Area and Volume

Perimeter

MSS2/E3.2, MSS1/L1.8, MSS1/L2.7

The **perimeter** of a shape is the distance all the way round it.

Measuring the sides of this shape gives the lengths shown.

Perimeter of the shape
= 2.5 + 5 + 6.5 + 3
= **17 cm**

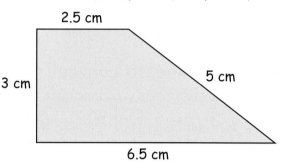

2.5 cm

3 cm

5 cm

6.5 cm

Practice

1 For each shape: **i)** use a ruler to measure each side **ii)** work out the perimeter.

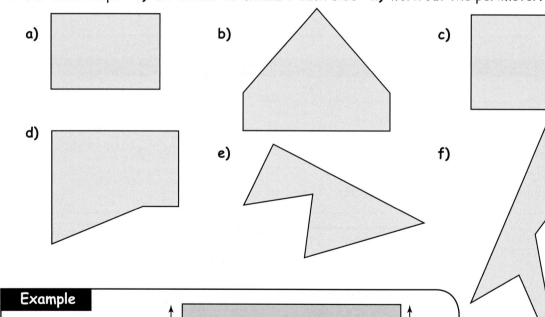

a)

b)

c)

d)

e)

f)

Example	
A gardener wants to put lawn edging around the perimeter of this lawn. Work out the total lengths he needs.	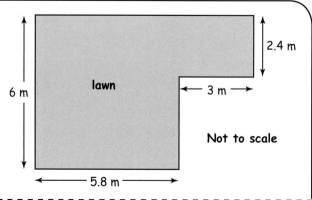

6 m

2.4 m

lawn

3 m

Not to scale

5.8 m

The lengths of two sides are not given on the diagram.
These must be worked out first.

Length needed (perimeter)
= 8.8 + 2.4 + 3 + 3.6 + 5.8 + 6
= 29.6 m

Check by rounding:
9 + 2 + 3 + 4 + 6 + 6 = 30

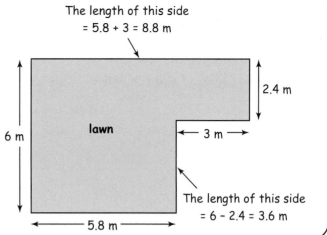

The length of this side
= 5.8 + 3 = 8.8 m

2.4 m

lawn

6 m

3 m

The length of this side
= 6 – 2.4 = 3.6 m

5.8 m

2 A builder needs coving to go round these ceilings.
   Work out the total length he needs for each ceiling.

Diagrams not to scale

a)

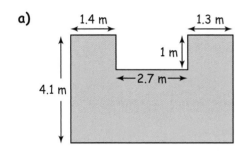

1.4 m   1.3 m

1 m

4.1 m   2.7 m

b)

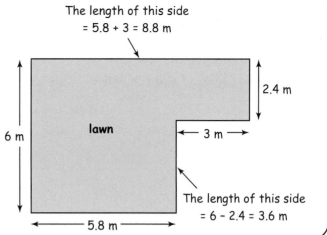

6.2 m

1.2 m

4.3 m

4.1 m

c)

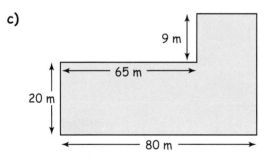

9 m

65 m

20 m

80 m

d)

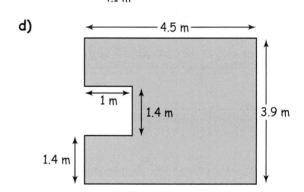

4.5 m

1 m

1.4 m   3.9 m

1.4 m

3 Find the length of sealant needed to go round the
   base of the shower shown in the diagram.

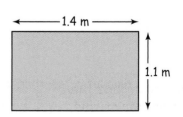

1.4 m

1.1 m

4 The sketch shows a loft hatch.
   Find the length of draft excluder
   needed to go round it.

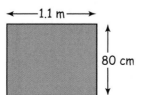

1.1 m

80 cm

**Note: You will need to
use the same units for
each side.**

5   The table gives the lengths and widths of some rectangles. Copy and complete the table.

Length	Width	Perimeter
6 cm	2 cm	
10 mm	6 mm	
20 cm	6.2 cm	
4.5 m	3 m	
20 mm	15 mm	
2.7 m	2.25 m	

6   Copy and complete the table below. Give the length and width of two possible rectangles for each perimeter.

Perimeter	Rectangle 1		Rectangle 2	
	length	width	length	width
a) 14 cm				
b) 20 cm				
c) 18 cm				
d) 28 cm				
e) 32 cm				

7   A rectangular field is 400 m long and 250 m wide. What is its perimeter?

8   Lace is needed to go round the perimeter of a 80 cm square table cloth. What length of lace is needed?

9   A rectangular rabbit run is 2 m 40 cm long and $1\frac{1}{2}$ m wide. What is its perimeter?

To find the perimeter of the rectangular shapes in questions 3 and 4 you have added together 2 lengths and 2 widths. This can be written as a formula:

> **Perimeter of rectangle = 2 × (length + width)**

Note that length and width are sometimes called **dimensions**.

 In letters the perimeter formula is:

> **P = 2(l + w)**

For more on formulae, see pages 64–68.

Use this formula to check your answers to the previous questions.

Area
MSS1/L1.9, MSS1/L2.7

a Measuring in square units

Area is the amount of surface inside a 2-D shape (flat shape).
We measure area in **square units**, e.g. **square centimetres (cm²)** or **square metres (m²)**.

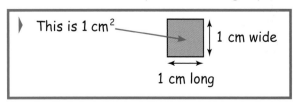

> This is 1 cm². 1 cm wide, 1 cm long

The area of this shape is 4 cm².

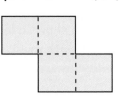

Practice

1 Write down the area of these shapes.

a) 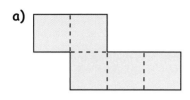   b) 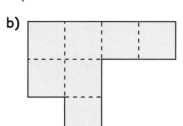   c)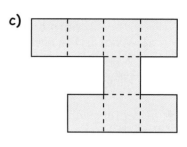

2 On cm squared paper, draw 4 different shapes, each with an area of 12 cm².

b Area of a rectangle

Example   Find the area of a rectangular room that is 6m long and 3m wide.

You can sketch a rectangle and count the squares.
The area of the room is 18 m².

You can also find the area without drawing the squares.
The area of the room is 6 × 3 = 18 m².

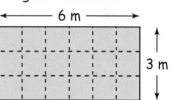

Practice

1 Find the area of these rectangles.

> **Use the same units for both sides.**

a) 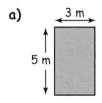   b) 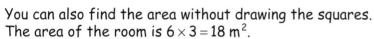   c)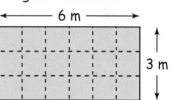

d) 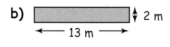   e) 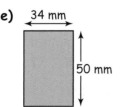   f)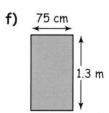

2 Give possible dimensions for rectangles with these areas.

a) 24 cm²   b) 18 m²   c) 35 cm²   d) 80 mm²   e) 75 m²   f) 320 cm²

The formula for the area of a rectangle is:
In letters this is:

> **Area = length × width**
> **A = l × w or A = lw**

3 A rectangular lawn is 9 metres long and 6 metres wide.

a) What is the area of the lawn?

b) A gardener spreads fertiliser on the lawn. He uses 20 grams per square metre. How much fertiliser does he use?

4 A rectangular bathroom is being decorated. It is 2.2 m long and 2 m wide.

a) A piece of lino 2 m wide and 2.2 m long is bought for the floor at a price of £9.99 per square metre. What is the cost of the lino?

b) The wall is re-tiled along the length and width of the bath.
The length of the bath is 1.7 m.
The width of the bath is 0.7 m.
The height of the wall above the bath is 1.8 m.
What is the area that is re-tiled?

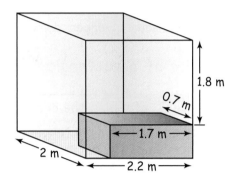

Circles

MSS1/L2.7

a Circumference

A **radius** is a straight line from the centre of a circle to the edge.

A **diameter** is a straight line from one side of the circle to the other that passes through its centre.
The diameter is twice the length of the radius.

The perimeter of a circle is called its **circumference**.

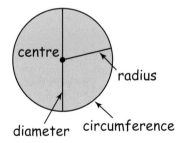

Activity

You can find the circumference of a coin by rolling it along a ruler.

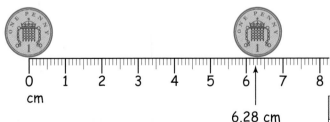

6.28 cm

Do this with the coins listed in the table.
Check all the measurements given.

Coin	Diameter	Circumference
1p	2 cm	6.3 cm
2p	2.5 cm	7.9 cm
5p	1.7 cm	5.3 cm

Divide the circumference of each coin by its diameter.
What do you find?

You should find that circumference ÷ diameter is always a bit more than 3.
A more accurate result is 3.14 (to 2 decimal places) or 3.142 (to 3 decimal places).
The exact answer is a never-ending decimal called **pi**, which is usually written as π.
Your calculator might have a key for π. If so, press it – you should get the number 3.14159265 ...

The relationship between the diameter and the circumference gives these formulae:

▶ **Circumference = $\pi \times$ diameter**
$$C = \pi d$$

or

▶ **Circumference = $2 \times \pi \times$ radius**
$$C = 2\pi r$$

Example   Find the circumference of a circular pond with a radius of 3 m.

$C = 2\pi r = 2 \times \pi \times 3 = 18.84955 ... = 18.8$ m (using the π key on a calculator)
or $2 \times 3.14 \times 3 = 3.14 \times 6 = 18.84 = 18.8$ m (without a calculator)
(rounded to 1 decimal place)

Practice

Give your answers to questions 1 to 3 to 1 decimal place.

1 a) Measure the diameter of these coins.   **i)** 10p   **ii)** £1   (in centimetres)
  b) Multiply the diameter by π to find the circumference of each coin.
  c) Check your answers to **b)** by measuring.

2 Find the circumference of each circle. Use 3.14 for π.

  a) diameter 10 cm   **b)** radius 7 cm   **c)** diameter 1.5 m
  d) radius 0.8 m   **e)** diameter 45 m   **f)** diameter 30 m.

  Check using your calculator key for π.

3 Copy and complete the table.

  Use 3.14 or your calculator key for π.

Radius	Diameter	Circumference
2 cm		
	30 mm	
	6 m	
1.2 m		
		15.7 cm
		75.4 mm

4 A wheel has a radius of 24 cm.
  How far does the wheel travel in one full turn (to the nearest cm)?

5 A circular window needs insulating. The diameter of the window is 80 cm.
  What length of insulation tape is needed to go around the window (to the nearest cm)?

b Area of a circle

The formula for the area of a circle in letters is: $A = \pi r^2$ r^2 **means** $r \times r$

In words this means: **Area of a circle** $= \pi \times$ **radius** $\times$ **radius**

Example   Find the area of a circular pond with a radius of 3 m.

$A = \pi \times 3^2 = \pi \times 3 \times 3 = 28.2743 \ldots = $ **28.3 m²** (using the π key on a calculator)
or $3.14 \times 9 = 28.26 = $ **28.3 m²** (without a calculator)
(rounded to 1 decimal place)

Practice

In each question use 3.14 or your calculator key for π.
Round answers for questions 1 to 3 to 1 decimal place.

1 Find the area of each circle.

a) 1 m

b) 3 cm

c) 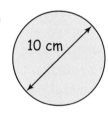 10 cm

d) 4.4 m

2 Find the area of a circle with a radius of:

a) 7 cm   b) 2 m   c) 8 mm   d) 12 mm   e) 15 cm   f) 1.4 m.

3 Find the area of a circle with a diameter of:

a) 8 cm   b) 20 mm   c) 11 cm   d) 2.5 m   e) 14 mm   f) 80 cm.

4 A circular flower bed has a diameter 0.75 m.
Find the area in cm² to the nearest 10 cm².

5 a) What is the diameter of the mirror including the frame?
b) Find the area of the mirror including the frame.
c) Find the area of the glass.
d) What is the area of the frame to the nearest cm²?

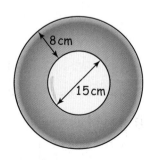
8 cm
15 cm

Areas of composite shapes

MSS1/L2.8

Sometimes the area of a shape can be found by splitting it into parts.

Example

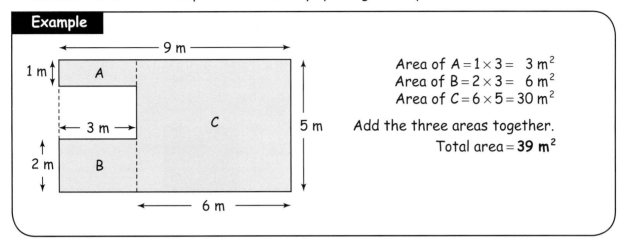

Area of $A = 1 \times 3 = 3$ m^2
Area of $B = 2 \times 3 = 6$ m^2
Area of $C = 6 \times 5 = 30$ m^2

Add the three areas together.
Total area $= $ **39 m^2**

Sometimes you can make a rectangle around the shape and subtract part of it.

Example Here is the same shape again.

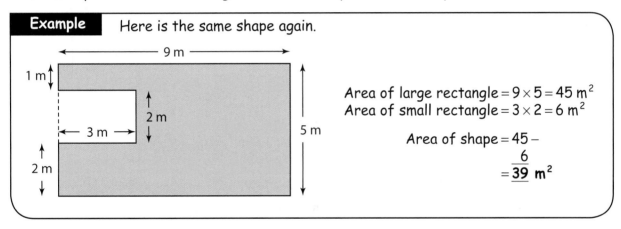

Area of large rectangle $= 9 \times 5 = 45$ m^2
Area of small rectangle $= 3 \times 2 = 6$ m^2

Area of shape $= 45 -$
$ \underline{6}$
$= $ **39 m^2**

Practice

1 Find the area of each room.

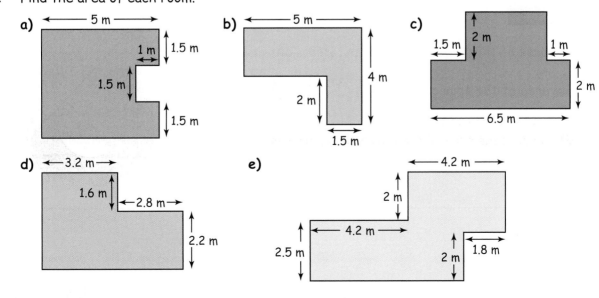

2 The sketch shows the layout of a garden.

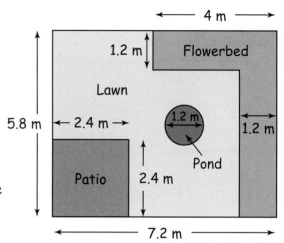

a) Find the area of:
i) the patio
ii) the flowerbed
iii) the pond (to 2 dp)
iv) the lawn.

b) Fertiliser is spread onto the flowerbed at a rate of 20 g per square metre. How much fertiliser is used to the nearest 10 grams?

c) Weedkiller is spread onto the lawn at a rate of 50 g per square metre. How much weedkiller is used to the nearest 10 grams?

3 A room is 3.6 m long and 3 m wide. The walls are 2.2 m high. The door is 0.8 m by 2 m and the window is 1.5 m by 1.2 m.

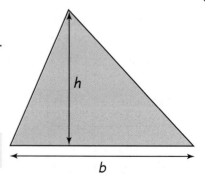

a) What is the area of the ceiling?

b) Find the total area of the walls by adding together the areas of the walls and then subtracting the area of the door and window.

c) The walls and ceiling need painting. Each litre of paint covers 12 m². Is a 2.5 litre tin of paint enough to paint this room?

Use formulae to find areas

MSS1/L2.8

The formula for the area of a triangle is:

where b is the base of the triangle and h is its height.

$$A = \frac{bh}{2}$$

Example

The base of this triangle is 5 cm and its height is 3.6 cm.

The area of the triangle is $A = \dfrac{5 \times 3.6}{2} = \dfrac{18}{2} = \textbf{9 cm}^2$

There must be a right angle between the base and height.

Practice

1 Use the formula given on page 240 to find the area of these triangles.

a)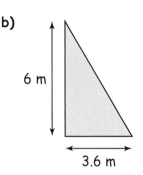

3 cm

4 cm

b)

6 m

3.6 m

c)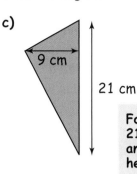

9 cm

21 cm

For part c), use 21 as the base and 9 as the height.

2 A triangular sign has a base 1.5 m long and its height is 1.4 m.
What is the area of the sign?

3 The formula for the area of a kite is $A = 0.5wl$
where w is the width and l is the length.
A kite is 80 cm wide and 1.2 m long.
What is its area? (Remember to use the same units for each length.)

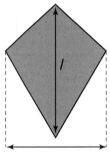

l

w

4 This shape is called a trapezium.

The formula for its area is $A = \dfrac{h(a+b)}{2}$.

Find the area when $h = 12$, $a = 11$ and $b = 19$ (all in cm).

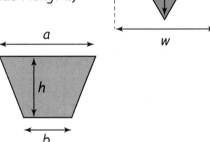

a

h

b

5 The formula for the area of this shape is $A = \dfrac{\pi ab}{4}$.

A pond of this shape is 2.2 m long and 1.3 m wide.
Use your calculator to find the area of the pond.
(Use 3.14 or the special key on your calculator for π.)
Give your answer to 1 decimal place.

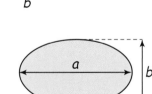

a

b

6 A formula that estimates the area of this window is

$$A = hd + \frac{3d^2}{8}$$

Find the area when h is 1.6 (metres) and d is 2 (metres).

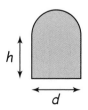

h

d

Volume of a cuboid

MSS1/L1.10, MSS1/L2.9

Volume is the space taken up by a 3-D shape (a solid) or liquid.
We measure volume in **cubic units**. This is 1 cubic centimetre.

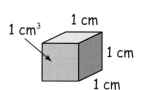

To find the volume of a shape, you need to
find the number of unit cubes that fill it.
The shape of a box is called a **cuboid**.
The volume of this cuboid is 3 cm³.

Practice

1   Write down the volume of each cuboid. Assume each of the cubes is 1 cm³.

a)   b)   c)   d)

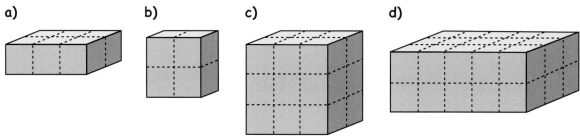

You can find the volume of a cuboid without drawing and counting the cubes.

Look at the cuboid in question 1 d) again. It is 5 cm long, 3 cm wide and 2 cm high.
Its volume is $5 \times 3 \times 2 = 30$ cm³.

> **Volume of a cuboid = length × width × height**

---

2   Find the volume of each cuboid.

> **Use the same units for all the sides.**

a)
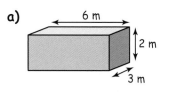
6 m
2 m
3 m

b)
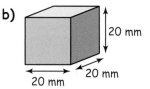
20 mm
20 mm
20 mm

c)
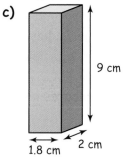
9 cm
1.8 cm
2 cm

d)
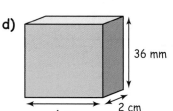
36 mm
4 cm
2 cm

3 Copy and complete this table:

Length	Width	Height	Volume
2 cm	4 cm	5 cm	
3 m	3.4 m	3.5 m	
4 mm	2.5 mm	6 mm	
9 cm	6.2 cm	8 cm	

4 A box has a 1 m square base and it is $\frac{1}{2}$ m high. What is the volume of the box?

5 What is the volume of a box that is 50 cm long, 10 cm wide and 8 cm high?

6 A carton of fruit juice has a volume of 1 litre (i.e. 1000 cm^3).
   Find possible dimensions for the length, width and height of the carton.

7 A sand pit is 100 cm long, 75 cm wide and 15 cm high.
   What is the volume of sand it holds when full?

8 Each side of a dice is 25 mm long. What is its volume?

9 The diagram shows a cube with edges of length 1 metre.

   a) What is the length of each edge in cm?
   b) What is the area of each face in cm^2?
   c) What is the volume of the cube in cm^3?

 The formula for the volume of a cuboid is:

The formula for the volume of a cube is:

> $V = l \times w \times h = lwh$
>
> $V = l \times l \times l = l^3$ (say this as 'l cubed')

---

10 The diagram shows a podium made from three
    cuboids. Each cuboid is 40 cm wide.

    a) Find the volume of each cuboid in m^3.
    b) What is the total volume of the podium?

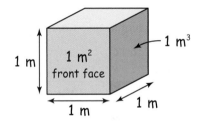

To find one dimension, divide the volume by the product of the others.

> **Example**
>
> If a cuboid has length 5 cm, width 4 cm and volume 120 cm³, its height is:
> 120 ÷ (5 × 4) = 120 ÷ 20 = **6 cm**

11 Copy and complete this table.

Length	Width	Height	Volume
4 m	2 m		24 m³
3 cm		8 cm	48 cm³
	3 m	9 m	54 m³
4 cm	4 cm		64 cm³

Volume of a cylinder

MSS1/L2.9

The formula for the **volume of a cylinder** is:

> $V = \pi r^2 h$

This means $V = \pi \times r \times r \times h$

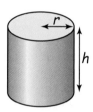

When finding volumes remember:

- The same units must be used for each measurement.

- The volume units are cubed, e.g. 4 **cm³** or 35.6 **m³**.

- You can convert volume to litres using **1 litre = 1000 cm³**.

> **Example**   This pipe is 0.8 m long and has a diameter of 40 mm.
>
> To find the volume we need r and h in the same units.
> Here it is sensible to use cm:
>
> $r = 20$ mm $= 2$ cm
> $h = 0.8$ m $= 80$ cm
>
> The volume is $V = \pi \times r \times r \times h$
> $\qquad\qquad\quad = \pi \times 2 \times 2 \times 80$
> $\qquad\qquad\quad = 1005$ cm³ (using a calculator).
>
> If the pipe is filled with water it will hold just over 1 litre.

Practice

In each question use the π key on your calculator or π = 3.14.
Give the answers to questions 1 and 2 to 1 dp.

▶ **Use the same units for *r* and *h*.**

1 Find the volume of the following cylinders.

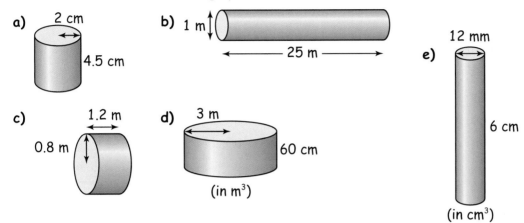

a) 2 cm 4.5 cm

b) 1 m 25 m

e) 12 mm 6 cm (in cm³)

c) 1.2 m 0.8 m

d) 3 m 60 cm (in m³)

2 Calculate the volume of a cylinder with:

a) radius 1.2 m and height 2 m
b) radius 6 cm and height 9 cm
c) radius 3 m and height 2.2 m
d) radius 3 m and height 1.6 m
e) diameter 12 cm and height 2.5 cm
f) diameter 1 m and height 0.5 m.

3 Find the volume of each cylinder.
   (Give your answers to **b)** and **c)** in cm³ and the rest in m³ to 2 decimal places.)

a) diameter 4 m and height 30 cm
b) diameter 2.5 cm and height 16 mm
c) radius 1.2 cm and height 6 mm
d) radius 70 cm and height 0.6 m
e) radius 0.3 m and height 40 cm
f) diameter 1.2 m and height 50 cm

4 A paddling pool has a radius of 60 cm and a depth of 30 cm.
   How much water does it hold when full? Give your answer in m³.

Use formulae to find volumes

MSS1/L2.9

The volume formulae for a cuboid and cylinder are similar. In both formulae:

▶ **Volume = area of base × height**

Shapes like these are called **prisms**.
Some of the shapes in the following
questions are also prisms.

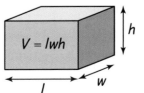

$V = lwh$

**Area of rectangular
base = *lw***

$V = \pi r^2 h$

**Area of circular
base = *πr*²***

Practice

1 The sketch shows a triangular prism.

The formula $V = \dfrac{bhl}{2}$ gives its volume.

Find the volume when $b = 10$, $h = 8$ and $l = 30$ (all measured in cm).

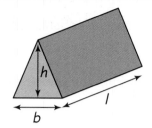

2 The diagram shows the shape of a plant box.

The formula $V = \dfrac{hl(a + b)}{2}$ gives its volume.

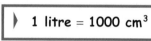

a) Find V if $h = 25$ cm, $l = 120$ cm, $a = 20$ cm and $b = 30$ cm.
b) A gardener buys compost in 40 litre bags.
   How many bags does she need to fill the plant box?

▶ 1 litre = 1000 cm³

3 This shape is called a square-based pyramid.

Its volume formula is $V = \dfrac{x^2 h}{3}$.

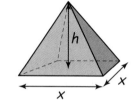

a) What is the volume of a pyramid with $x = 6$ cm and $h = 5$ cm?
b) How many paperweights of this shape can be made from 25 litres
   of molten metal?

4 The formula that estimates the volume of this section of gutter
is $V = \dfrac{3d^2 l}{8}$.

How many litres of water will it hold if $d = 12$ cm and $l = 5$ m?

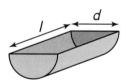

Measures, Shape and Space

8 Shape

Angles

MSS2/E3.1, MSS2/L1.1

Polygons have straight **sides**.
Where two sides meet, an **angle** is formed.

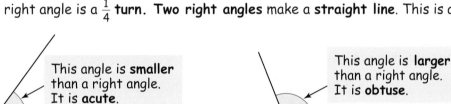

side → polygon ← angle

Get a sheet of paper.

Fit a corner into this drawing of a **right angle**.

Does it fit exactly?

If so, the corner is a right angle.

Try the other corners.

A right angle is shown like this.

A right angle is a $\frac{1}{4}$ **turn**. **Two right angles** make a **straight line**. This is also a $\frac{1}{2}$ **turn**.

This angle is **smaller** than a right angle. It is **acute**.

This angle is **larger** than a right angle. It is **obtuse**.

Practice

1 These letters all have right angles. How many right angles are there in each?

a)   b)   c)   d)   e)   f)

2 Write down ten objects in the room that have right angles.

 L1

Angles are measured in **degrees** using a **protractor**.

▸ A right angle is **90°**.

▸ **2 right angles** = **180°**.

You will use a protractor to measure angles in pie charts (see pages 276 and 291).

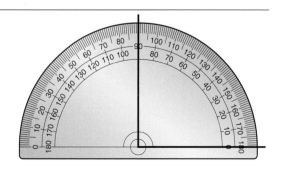

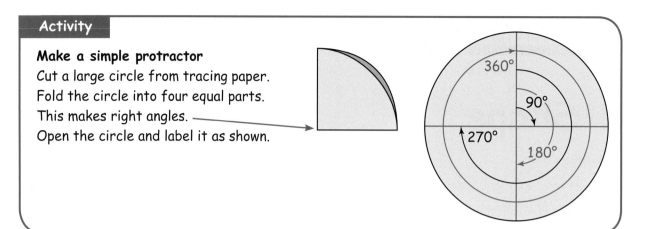

Activity

Make a simple protractor
Cut a large circle from tracing paper.
Fold the circle into four equal parts.
This makes right angles.
Open the circle and label it as shown.

---

3 Use your protractor.

a) How many degrees are there in these angles?

i)   **ii)**    **iii)**

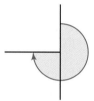

b) How many degrees are there in a full turn?

4 The points of a compass are: N (north), NE (north-east), E (east), SE (south-east),
S (south), SW (south-west), W (west) and NW (north west).

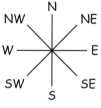

How many degrees are there between:
a) N and S   **b)** N and E   **c)** N and NE
d) S and W   **e)** S and SW   **f)** N and SE
g) E and W   **h)** W and NW   **i)** E and SW?

Parallel lines

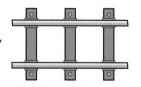

MSS2/E3.1, MSS2/L2.2

Parallel lines are always the same distance apart,
like railway tracks or the sides of a door.

Practice

L2 **1** At different points, measure the distance between each of these pairs of lines.
Which pairs of lines are parallel?

A    B   C    D   E

2 List examples of parallel lines from everyday life.

Describe positions

Look at the items on these shelves.
You can describe their positions using words like:
above, below, top, bottom, middle, right, left, between

Examples

The jam is at the **right**-hand end of the **top** shelf.
The coffee is on the **top** shelf, **between** the soup and the jam.
The fruit juice is on the **bottom** shelf, **below** the soup and to the **left** of the sugar.

Practice

1 Copy and complete the following.

    a) The cornflakes are on the .... shelf, to the .... of the sugar.

    b) The soup is at the .... -hand end of the .... shelf.

    c) The sugar is on the .... shelf, ... the fruit juice and the cornflakes.

2 Describe the position of each fruit and vegetable.

3 Draw two shelves. Use these descriptions to label the positions of objects on your shelves.

    a) A tube of **toothpaste** is at the **right**-hand end of the **bottom** shelf.

    b) A **toothbrush** is on the **bottom** shelf to the **left** of the toothpaste.

    c) The **shampoo** is on the **top** shelf **above** the **toothpaste**.

    d) The **conditioner** is at the **left**-hand end of the **top** shelf.

    e) The **soap** is in the **middle** of the **top** shelf **between** the conditioner and the shampoo.

    f) A **razor** is **below** the conditioner.

Give directions

MSS2/E3.3, MSS2/L1.3, MSS2/L2.4

Here is a sketch map of a town centre.

You can use the sketch to describe where things are. To give directions, you need to imagine walking along the route.

Sometimes you can use the compass points: **N**orth, **S**outh, **E**ast, **W**est.

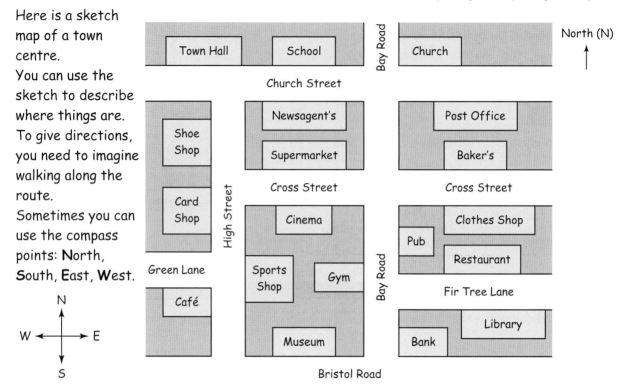

Example

- The cinema is on Cross Street **opposite** the supermarket.
- To go from the post office to the café, walk west along Church Street to High Street. Turn left (south) and walk down High Street past Cross Street. The café is on the right on the corner where High Street meets Green Lane.

Practice

1 Use the map to answer the questions.
   a) What is on the corner where Bay Road meets Bristol Road?
   b) I start from the café and walk north up High Street.
      (i) What is the first shop I pass on my left?
      (ii) Which way must I turn to go to the cinema?
   c) From the baker's, I walk west along Cross Street to Bay Road.
      Which way must I turn to go to the bank?
   d) Copy and fill in the gaps to give directions from the school to the library.

   Walk .... along Church Street to Bay Road and turn ....
   Walk .... down Bay Road, cross over .... and turn ....
   into .... The library is on the ....

2 Work in pairs. Use the sketch map to make up more questions.
   Ask your partner to answer them, then check the answers they give.

L1 Here is a sketch map of another town centre.

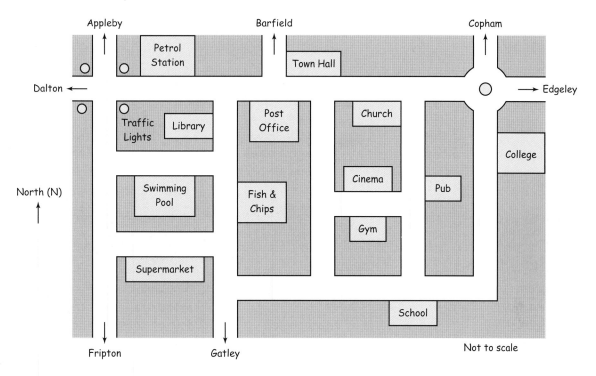

Example

A student from Fripton asks for directions to the college.
One possible route is given below:
Coming into town from Fripton, turn right (east) after the supermarket, turn right again (south) at the T-junction (taking the road to Gatley), then first left.
Follow this road past the school and round a left-hand bend.
The college is further along on the right.

3 Write down directions to the college for a student from:
   a) Copham   b) Edgeley   c) Dalton
   d) Gatley   e) Barfield   f) Appleby.

4 Write down directions to go from:
   a) the college to the post office   b) the school to the petrol station
   c) the gym to the fish and chip shop   d) the cinema to the supermarket
   e) the swimming pool to the pub   f) the cinema to the swimming pool.

L2 5 There is more than one way to go from the college to the swimming pool.
   a) Give the directions for two different routes.
   b) Which route do you think is best? Why?

6 Work in pairs. Use the sketch to make up more questions.
   Ask your partner to answer them, then check the answers they give.

Two-dimensional (2-D) shapes

Apart from the circle, all the following shapes are **polygons**.

Circle

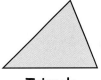

Triangle
3 sides
3 angles

Square
4 equal sides
4 right angles
2 pairs of parallel sides

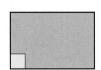

Rectangle
2 equal long sides
2 equal short sides
4 right angles
2 pairs of parallel sides

Parallelogram
2 equal long sides
2 equal short sides
2 pairs of equal angles
2 pairs of parallel sides

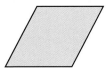

Rhombus
4 equal sides
2 pairs of equal angles
2 pairs of parallel sides

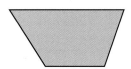

Trapezium
4 sides
1 pair of parallel sides

> **Note**
> **Any** polygon with **5 sides** is a **pentagon**.
> **Any** polygon with **6 sides** is a **hexagon**, etc.
> **Regular** means the sides are all the same length and the angles are all the same size.

Regular pentagon
5 equal sides
5 equal angles

Regular hexagon
6 equal sides
6 equal angles

Regular octagon
8 equal sides
8 equal angles

Practice

1 Look at the polygons. Compare each angle with a right angle.
   Copy and complete the following table.

| | Number of angles that are: | | |
	Right angles	Smaller than a right angle	Larger than a right angle
Square			
Rectangle			
Regular hexagon			
Regular octagon			
Regular pentagon			
Parallelogram			
Rhombus			
Trapezium			

2 Look at the polygons. Many have four sides.
Some of the shapes have parallel sides.
Copy the following table. Tick where appropriate.

	4 sides equal	4 right angles	Opposite sides equal	Opposite angles equal	No parallel sides	1 pair of parallel sides	2 pairs of parallel sides
Square							
Rectangle							
Parallelogram							
Rhombus							
Trapezium							

3 Use a protractor to measure the angles in these shapes on page 252.
a) trapezium
b) regular pentagon
c) regular hexagon
d) regular octagon

Symmetry

MSS2/E3.1, MSS2/L1.1, MSS2/L1.2, MSS2/L2.3

Line symmetry

When a shape can be folded in half so that
the halves match, it has **line symmetry**.

The fold line is a **line of symmetry**.
Sometimes it is called a **mirror line**
because each side is like the **reflection** of the other side in a mirror.

Some shapes have more than one line of symmetry.

Butterflies are symmetrical.

Line of symmetry

**This triangle has
3 lines of symmetry.**

Rotational symmetry

When a shape **looks exactly the same** when it is **rotated** by less than a full turn to a **new**
position, then the shape has **rotational symmetry**.

The shape's **order** of rotational symmetry is the number of different positions in which it looks
the same during a complete turn.

The triangle above has **rotational symmetry of order 3**.
The butterfly does not have rotational symmetry.

**You can use tracing paper
to check line symmetry
and rotational symmetry.**

L1

Example

Complete this shape so that the
dotted line is a line of symmetry.

Drawing the reflection in the dotted line completes the shape.
(Use tracing paper to check this matches when you fold it.)

Example

Shade 2 more squares so that the result
has rotational symmetry of order 2.

Two more squares have been shaded so that the pattern
looks the same in two different positions during a turn.
(Use tracing paper to check this.)

Practice

1 How many lines of symmetry does each of these letters have?
   a) T   b) H   c) E   d) A   e) D   f) O
   g) B   h) Z   i) M   j) S   k) X   l) V

2 Which letters in question 1 have rotational symmetry?
   Write down the order of rotational symmetry for each of these letters.

3 Draw each of the following shapes on squared or isometric paper:
   square, rectangle, rhombus, regular hexagon, parallelogram.
   Cut out each shape.
   By folding, find which of the shapes have:
   a) no lines of symmetry   b) one line of symmetry
   c) two lines of symmetry   d) more than two lines of symmetry.

4 Use tracing paper to copy each of the following shapes from page 252:
   square, rectangle, parallelogram, rhombus, regular pentagon, regular octagon.
   Rotate the tracing paper to find the order of rotational symmetry of each shape.

L1 5 Copy and complete each shape. The dotted line is a line of symmetry.

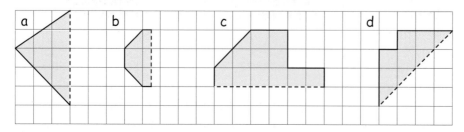

6 Copy each grid onto squared paper. Shade more squares so that the results have rotational symmetry of order 2.

a)   b)   c)

7 Copy and complete each shape. Each dotted line is a line of symmetry.

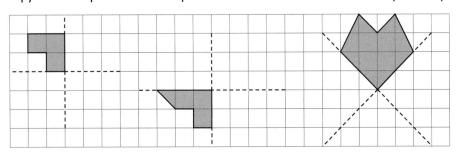

8 On squared paper draw some shapes that have:

  a) 1 line of symmetry **b)** 2 lines of symmetry **c)** 4 lines of symmetry.

9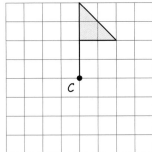

When the flag is rotated clockwise through 90°, 180° and 270° about the point C, the result is this pattern.

For this pattern, write down:
  i) the order of rotational symmetry
  ii) the number of lines of symmetry.

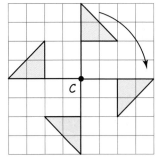

10 For each shape below:
  i) copy the shape and rotate it clockwise by the given angle(s) about point C
  ii) write down the order of rotational symmetry of the resulting pattern
  iii) write down the number of lines of symmetry of the resulting pattern.

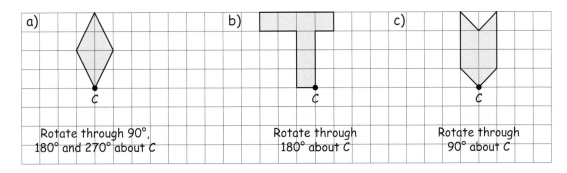

11 On squared paper draw some patterns that have:
  a) rotational symmetry of order 2 **b)** rotational symmetry of order 4.

Tessellations

Shapes **tessellate** if they fit together leaving no gaps.

Circles do not
tessellate. They
leave gaps.

Hexagonal cells
tessellate in a
honeycomb.

Practice

1 How many triangles fit together to make each rectangle?

a)   b)   c)   d)

2 Name the shapes in these tessellations.   a)   b)

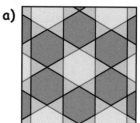 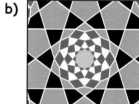

Activities

1 Cut out 10 crosses like the one labelled a) (below) from squared paper.
   Fit them together to form a tessellation. Repeat this with shapes b), c) and d).

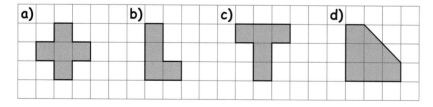

a)   b)   c)   d)

2 Use tracing paper to copy a hexagon, an octagon, a pentagon and a parallelogram.
   Cut out several of each shape.

   a) Which shapes tessellate by themselves?
   b) Which shapes leave triangles or squares when you try to tessellate them?

3 Many houses and other buildings have walls and floors covered with tiles.
   Find and draw some interesting tiling patterns. Name the shapes used.

Three-dimensional (3-D) shapes
MSS2/E3.1, MSS2/L2.1

3-D (3-dimensional) shapes have: **edges**, **vertices** (i.e. corners) and **faces**.

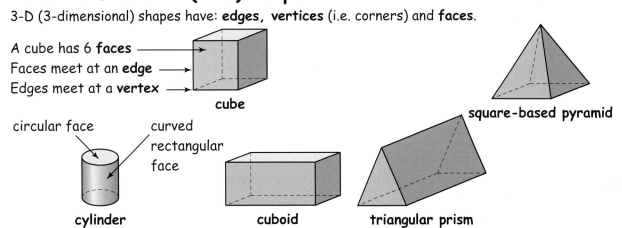

A cube has 6 **faces**
Faces meet at an **edge**
Edges meet at a **vertex**

cube

square-based pyramid

circular face
curved rectangular face

cylinder

cuboid

triangular prism

Practice

1 Copy and complete the following table.

	Shape of faces	Number of faces	Number of edges	Number of vertices
Cube				
Cuboid				
Cylinder				
Square-based pyramid				
Triangular prism				

Activity

Collect together a number of each 3-D shapes.

a) Which shapes stack together easily on a shelf?
b) Which shapes are difficult to stack? Why?

Nets
MSS2/L2.1

Folding the 2-D patterns gives the 3-D shapes. The patterns are called **nets**.

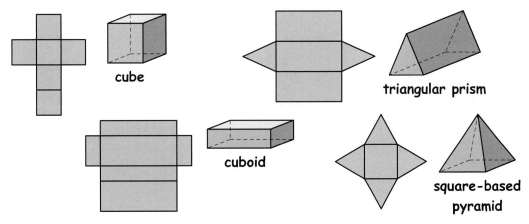

cube

triangular prism

cuboid

square-based pyramid

Activities

Use graph paper.

1 Draw four different nets for a cube.

2 Draw three different nets for a triangular prism.

3 Draw three different nets for a cuboid.

4 Make a net for:
   a) a pyramid with a 5 cm square base and triangular faces of height 7 cm
   b) a prism of length 12 cm, with triangular ends of width 6 cm and height 5 cm.

5 Make a net for a cube with a pyramid on top.

Solve problems involving 2-D and 3-D shapes   MSS2/L2.1, MSS2/L2.2

When some dimensions of an item are given on a diagram, you can often work
out other dimensions by adding, subtracting, multiplying or dividing the given lengths.

Example

From this diagram of storage
units we can work out that the
total width of the cupboards and
drawers = 50 + 50 + 60
          = 160 cm = **1.6 m**

This means the top is 10 cm wider
than the units. Assuming the top
overhangs the units equally
at each side, the overhang
at each side is 10 ÷ 2 = **5 cm**.

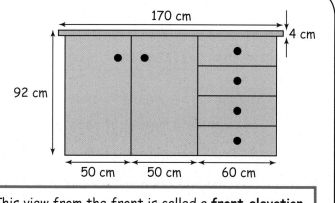

▶ This view from the front is called a **front elevation**.

Practice

1 For the storage units shown above, calculate:

   a) the height of the cupboards
   b) the height of each drawer
      (assuming they are identical).

2 Here is a bedroom unit.
   Find in metres:

   a) the total length of the unit
   b) the total height of the unit.

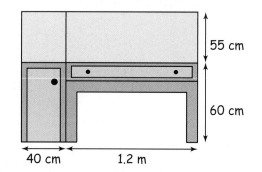

Example

Plan means view from above.

This is the plan of a kitchen.
The work surface is 2.75 m long and 60 cm wide.
Square tiles with sides 20 cm long are used to cover
the work surface.

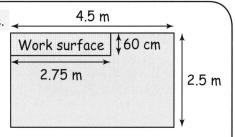

To find out how many tiles are needed look at the length then the width.
Length = 2.75 m = 275 cm.
Number of tiles = 275 ÷ 20 = 14 tiles (rounded up)
Width = 60 cm
Number of tiles = 60 ÷ 20 = 3 tiles
Total number of tiles = number along length × number along width
= 14 × 3 = **42 tiles**

---

3 a) Square tiles with sides 50 cm long are used to cover the ceiling of the kitchen.
How many tiles are needed?

b) Square tiles with sides 25 cm long are used to cover the whole floor of the kitchen
(including under the worktop). How many floor tiles are needed?

4 The walls are tiled above where the worksurface meets
them with square tiles whose sides are 20 cm long. The
height of these parts of the walls is 1.6 m.

a) Find the number of tiles needed
  i) above the width of the work surface
  ii) above the length of the work surface.
b) How many tiles are needed altogether for
this part of the walls?

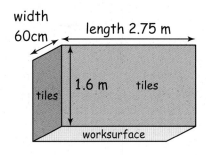

5 Cans of soup have diameter 7.5 cm and height 10 cm.
What is the **maximum** number of cans that you could pack
upright in a box that is 90 cm long,
75 cm wide and 30 cm high?

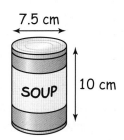

6 a) What is the greatest number of packs of butter that will fit inside the box?.

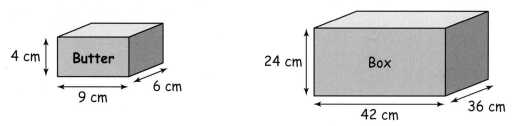

b) Draw a net of the box and give its dimensions.

7 This tent comes in a 2-man, 3-man and 4-man design.

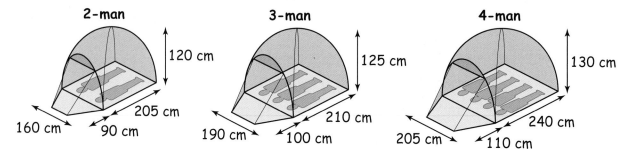

a) Give the total length of each tent.
b) Which tent is the highest?
c) What is the difference in height between the 2-man tent and the 4-man tent?
d) Which is the widest tent?
e) The tents are sold in cylindrical bags 45 cm long with a diameter of 20 cm.
How many bags fit in a box 60 cm wide, 60 cm high and 50 cm deep?

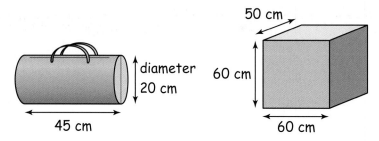

Handling Data

1 Extracting and Interpreting Information

Information from lists, tables, pictograms and bar charts

HD1/E3.1, HD1/E3.2, HD1/L1.1, HD1/L2.1

a Find information from lists

Lists are often used to give information. Examples can be found in books, catalogues, newspapers, magazines, maps and websites.

Lists can be arranged:

- in alphabetical order – as in a phone book
- numerically – a list of instructions is often numbered
- using dates – as in bank statements
- in categories, i.e. grouped by type – as in sales catalogues
- randomly – some shopping lists are written this way.

Practice

1 Look at this list.

   a) How is the list ordered?
   b) Why do you think it is ordered like this?
   c) Write down the school's phone number.
   d) Write down the dentist's phone number.
   e) Whose phone number is 778887?

Name	Phone number
After school club	453521
Child minder	797654
Dentist	787652
Doctor	778887
Fire station	773774
Police station	776431
Railway station	768998
School	791455
Work	798976

2 Here is a list of important events.

School play	17/03
Party	11/04
Joe's birthday	13/04
Dental appointment	15/04
Dinner date	17/04
Wedding anniversary	20/04

a) How is this list ordered?
b) What is happening on April 17th?
c) On what date is the wedding anniversary?
d) When is the school play?
e) The parent's evening is on April 19th.
 Where would this go in the list?

3 Here is a shopping list.

> $\frac{1}{2}$ dozen eggs
> Shampoo
> Washing powder
> Beans x 4
> Family pack crisps
> Ham
> Semi-skimmed milk

a) How is this list arranged?
b) How many cans of beans are on the list?
c) How many loaves of bread are on the list?
d) What kind of milk do they want?

4 a) How are words in a dictionary arranged?
 b) How are events ordered in a diary?
 c) How are items ordered on a credit card or bank statement?
 d) How are items ordered in an index?
 e) How are items ordered in the Yellow Pages?

Lists on the Internet

A **search engine** looks for information on the Internet. When you type in a subject the search engine gives you a list of websites in order of relevance. **Relevant** means closely linked to your search word or words. The list you get depends on the words you use.

Suppose you wanted to find some information about Kirk Michael in the Isle of Man.

Typing Kirk Michael into a search engine gave these websites:

Which of these websites do you think are relevant?
Which are not relevant? Why?

Typing in 'Kirk Michael Isle of Man' gives more relevant results – try this if you can.

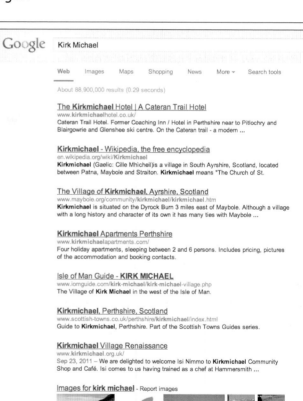

Activities

(If necessary ask your tutor or teacher for help.)

1   Use a dictionary to find the following words.
    evaluate   proportion   integer   frequency

2   Use the Yellow Pages to find the address and phone number of a plumber, a gardener, a taxi and a vet.

3   Use a search engine to find a list of websites that have information on a subject you find interesting. Type in just one word and look at the list of the first ten websites the search engine finds.
    Do you think the first website is the most relevant on the list?
    Do you think the tenth website is the least relevant on the list?

    Carry out another search using more than one word.
    Do you get more relevant results?
    Investigate further, using different combinations of words.
    Find out how to use quotation marks ("...").

b Find information from tables

This table shows the number of men and women in some college evening classes.

This table has a **title** at the top to tell us what the table is about.

Number of students in evening classes ← title

heading →

Evening class	No. of women	No. of men
Local History	10	7
Cookery	11	5
Italian	8	9
Do-It-Yourself	4	15

row →

column

Each column has a **heading**. The first column's heading tells us that this column gives the evening classes. The other column headings show that these columns give the number of women and men in each class. Without the title and headings we would not understand what the table is about.

Practice

1 Use the table on page 263 to answer these questions.

   a) How many women are there in the Cookery class?
   b) How many men are there in the Local History class?
   c) How many more men than women are there in the Do-It-Yourself class?
   d) Altogether, how many students are in the Italian class?
   e) Which class has the most students?

2 Use the Film Prices table to answer these questions.

   a) How much is a Thriller from the Internet?
   b) How much is a Western from the shop?
   c) Is Comedy cheaper from the Internet or shop?
   d) Which is more expensive from the shop – a Horror film or a Comedy film?
   e) How much more is a Sci-fi film from the shop than from the Internet?

Film Prices

Genre	Internet price	Shop price
Action	£12.00	£15.99
Sci-fi	£10.99	£14.00
Thriller	£11.00	£13.99
Comedy	£9.00	£12.00
Horror	£11.99	£15.00
Western	£9.00	£11.99

L1

3 Use the information in the Car Hire table to answer the questions.

Car Hire

Vehicle type	Hire period		
	1 day	3 days	7 days
Ford Focus	£41	£114	£227
Mercedes C200	£89	£259	£565
Nissan Qashqai	£49	£138	£284
Toyota Aygo	£29	£79	£147
VW Golf	£55	£156	£324
VW Tiguan	£72	£207	£445

   a) How much does it cost to hire:
      i) a VW Tiguan for 3 days   ii) a Toyota Aygo for 7 days?
   b) You pay £138 to hire a car.
      i) Which car is it?   ii) How long have you hired it for?
   c) What is the cheapest car you can hire for 1 day?
   d) How much more does it cost to hire a Toyota Aygo for 7 days then for 3 days?
   e) Which car is the most expensive to hire?
   f) How much more does it cost to hire a VW Golf than a Ford Focus for 3 days?

4   **Flight Information**

Route	Flight number	Departure–Return Days	Outward times		Return times	
			Depart	Arrive	Depart	Arrive
Heathrow – Split	453/65S	Sat–Sat	16:50	21:00	14:30	17:40
Gatwick – Split	789/45S	Fri–Fri	10:00	13:50	21:00	23:50
Manchester – Split	791/75S	Sat–Sat	19:00	22:50	18:10	21:00
Heathrow – Pula	465/71P	Thurs–Thurs	11:30	15:20	12:20	15:30
Gatwick – Pula	781/31P	Fri–Fri	07:00	11:10	06:30	09:20
Manchester – Pula	867/91P	Sat–Sat	18:00	21:50	17:10	20:00

a) What is the flight number of the flight from Manchester to Split?

b) What time is the flight from Gatwick to Pula due to arrive?

c) What time does the return flight for flight number 781/31P leave?

d) Which route does flight number 781/31P fly?

e) What day of the week does the flight from Manchester to Pula fly?

5 a) What is missing from the table below?

Altea	H/Esplendia	7	HB	750	FB	1010	14	HB	1076	FB	1568
Benidorm	H/Fantastica	7	HB	600	FB	950	14	HB	989	FB	1300
Malaga	H/Luxuriosa	7	HB	570	FB	900	14	HB	976	FB	1286
Salou	H/Fabulosa	7	HB	500	FB	845	14	HB	855	FB	1015

b) Redraw the table. Put in the missing items.

c) How much does it cost to stay in Hotel Luxuriosa for 14 nights full board?

d) How much does it cost to stay in Hotel Fabulosa for 7 nights half board?

e) Your hotel costs £1010.

   i)   Where are you staying?

   ii) Is it half board or full board?

   iii) How many nights are you staying there?

f) How much more does it cost to stay full board for 7 nights in Hotel Fantastica in Benidorm than in Hotel Fabulosa in Salou?

6 The table below gives insurance premiums for winter sport travel insurance. Children under 2 years old are covered for free.

Age	2–15 years		16–64		65 or over	
No. of nights	Europe	Outside Europe	Europe	Outside Europe	Europe	Outside Europe
6–9	£18.20	£33.90	£24.30	£44.50	£30.60	£55.25
10–17	£21.80	£38.25	£28.40	£51.20	£35.25	£63.75
18–23	£24.75	£42.50	£33.20	£56.25	£41.75	£70.50

Use the table at the bottom of page 265. Find the total premium for each of these families who are going on a skiing holiday.

a) 14 nights in Europe

Mr Bell	Age 28
Mrs Bell	Age 26
Ben	Age 4
Julie	Age 1

b) 21 nights outside Europe

Mr Khan	Age 38
Mrs Khan	Age 39
Imran	Age 14
Mr Khan (senior)	Age 67

Activities

1 Choose a holiday from a brochure.
   Use information from the brochure to find out the total cost and the flight times.

2 Look up prices in tables in catalogues.

3 Find the distance between cities from a road atlas table. (See page 174.)

---

c Find information from pictograms

A **pictogram** uses pictures or symbols to represent data.

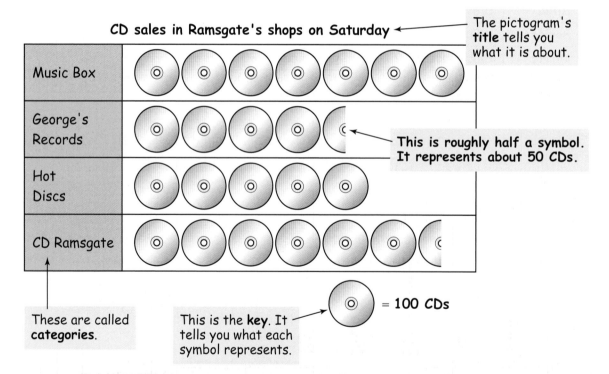

CD sales in Ramsgate's shops on Saturday — The pictogram's **title** tells you what it is about.

This is roughly half a symbol. It represents about 50 CDs.

These are called **categories**.

This is the **key**. It tells you what each symbol represents.

= 100 CDs

When each symbol represents a number of items, you can only estimate how many are in each category when parts of symbols are used.

Practice

1 Use the pictogram of CD sales to answer these questions.

   a) How many CDs were sold in George's Records?
   b) Which shop sold: i) the most CDs   ii) the least CDs?
   c) How many more CDs did CD Ramsgate sell than Hot Discs?
   d) How many CDs were sold altogether?

2 The pictogram below shows how much a householder spends on his bedroom.

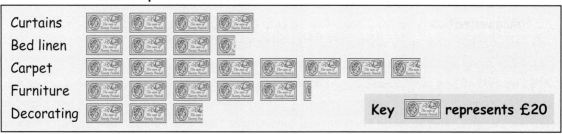

Amount spent on bedroom

Curtains	Key ☺£20 represents £20

   a) Which category of goods cost: i) the most   ii) the least?
   b) Estimate how much he spends: i) on each category   ii) altogether.

d Find information from bar charts

Bar charts use vertical or horizontal bars to give data.
The bar chart below shows the type of morning drinks chosen by the workers in a factory.
The main features are labelled. Note 'decaff' means decaffeinated.

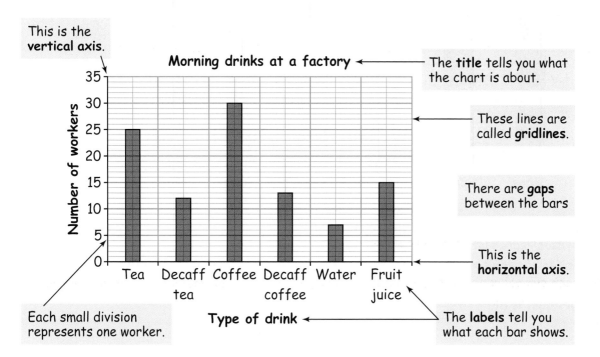

This is the **vertical axis**.

The **title** tells you what the chart is about.

These lines are called **gridlines**.

There are **gaps** between the bars

This is the **horizontal axis**.

Each small division represents one worker.

The **labels** tell you what each bar shows.

The **scale** on the vertical axis tells you how many workers had each type of drink. The first bar shows 25 workers had tea. The second bar shows 12 workers had decaffeinated tea.

Practice

1 a) Copy this table. Use the bar chart on page 267 to fill in the missing numbers.

Use the table to answer these questions then use the bar chart to check.

Type of drink	Number of workers
Tea	25
Decaff tea	12
Coffee	
Decaff coffee	
Water	
Fruit Juice	

b) How many more workers had tea than decaffeinated tea?

c) How many more workers had fruit juice than water?

d) Altogether how many workers had decaffeinated drinks?

e) How many workers were there altogether?

In a survey people are asked which type of TV programmes they like best. The bar chart below shows the results. The label on the vertical axis shows that the number of people is also called the **frequency**. Frequency means how often something happens – in this case how often people chose each type of TV programme.

Can you see what each division on the vertical axis represents this time?

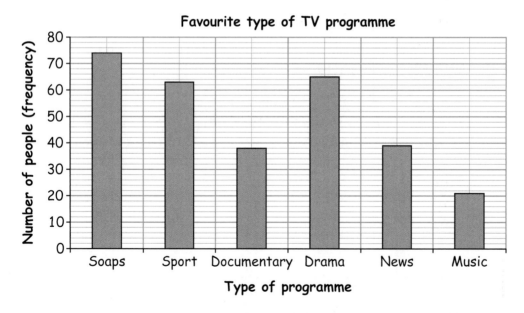

Each division represents 2 people.

The number of people who like soaps best is 74.

The top of the bar for sport lies between 62 and 64 – this means the number of people who like sport best is 63.

2 Use the bar chart on page 268 to answer these questions.

   a) How many people chose:

      i) documentaries   **ii)** drama   **iii)** news   **iv)** music?

   b) Which is:

      i)   the most popular type   **ii)** the least popular type?

   c) How many more people chose documentaries than music?

   d) Altogether, how many people took part in the survey?

This bar chart shows the population of some countries in Europe.
Can you see what each division on the horizontal axis represents?

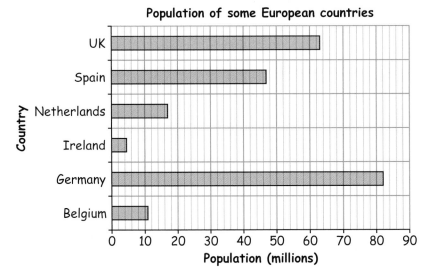

Population of some European countries

Each division represents 2 **million** people. It is not possible to read the populations accurately, but you can estimate them to the nearest million.

3 Use the bar chart above to answer these questions.

   a) Which of the countries has:

      i) the largest population   **ii)** the smallest population?

   b) Write down the population of each country. Give your answers to the nearest million.

   c) **i)**   How many more people live in Germany than in the UK?

      ii)   How many more people live in the UK than in the Netherlands?

   d) What is the total population of the UK and Ireland?

The bar chart below shows the height of each student in a class. Their names are given in a **key** because it is difficult to fit them along the horizontal axis.
What does each division on the vertical axis represent?

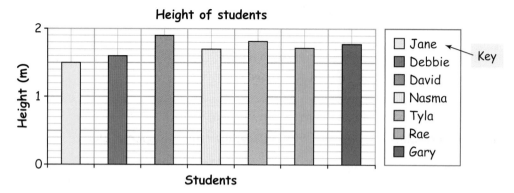

Height of students

Each division represents 0.1 metres (or 10 cm).
Some of the tops of the bars lie between two gridlines.
Using the **nearest gridline** gives the height to the **nearest 0.1 m**.
For example, Gary's height is 1.8 m to the nearest 0.1 m.

4 Look at the bar charts above.
   a) Copy this table.
       Use the bar chart to fill in the missing heights, correct to the nearest 0.1 m.
   b) Who is the shortest student?
   c) Who is the tallest student?
   d) Who is taller: Nasma or Rae?
   e) Who is shorter: Tyla or Gary?

Student	Height (m)
Jane	
Debbie	
David	
Nasma	
Tyla	
Rae	
Gary	1.8

5 A café sells baked potatoes with a choice of fillings.
   The bar chart shows the number of each type of filling bought in one day.

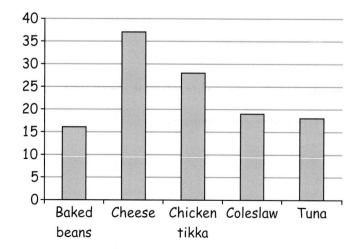

a) What is missing from the bar chart?
b) Which was the **most** popular filling?
c) Which was the **least** popular filling?
d) Draw a table giving the frequency for each type of filling.
   Give your answers as accurately as you can.
e) How many baked potatoes were sold altogether?
f) How many more potatoes were filled with cheese than were filled with tuna?

---

5 These two charts both show the same information.

Chart A

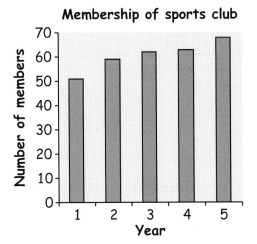

Chart B

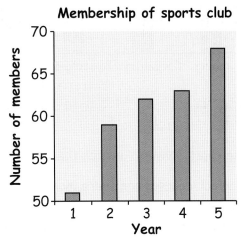

Changing the scale on an axis changes the impression a chart gives.
Starting at a non-zero value also makes it look different.

---

6 a) Look at the charts above. Which chart makes the
   increase in the number of members look more dramatic?
b) The charts do not have any gridlines, but you can
   estimate the number of members.
   Look at Chart A, the number of members in Year 1 is
   just above 50. A reasonable estimate is 51. This has
   been entered into the table.

Year	Chart A	Chart B
1	51	
2		
3		
4		
5		

i) Copy the table and use Chart A to complete the
   second column.
ii) Use Chart B to complete the last column of your table.
iii) Which chart was easier to use when estimating the number of members?

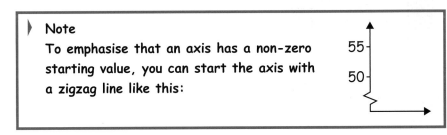

▶ **Note**
To emphasise that an axis has a non-zero
starting value, you can start the axis with
a zigzag line like this:

The chart below is a **comparative** bar chart. It gives a comparison between the number of full-time and part-time students on vocational courses at a college.

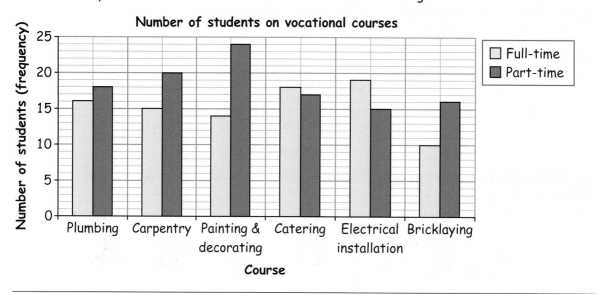

7 Use the chart above to answer these questions.

a) On which courses are there more full-time than part-time students?

b) How many more part-time students than full-time students are doing:
   i) carpentry   ii) plumbing   iii) painting and decorating?

c) Find the total number of students who are doing:
   i) bricklaying   ii) catering   iii) electrical installation.

d) Find the total number of:
   i) full-time students   ii) part-time students.

The chart below is a **component** bar chart. It shows the amount each salesman in a car showroom earns in a week. The amount earned is the total of basic pay and commission.

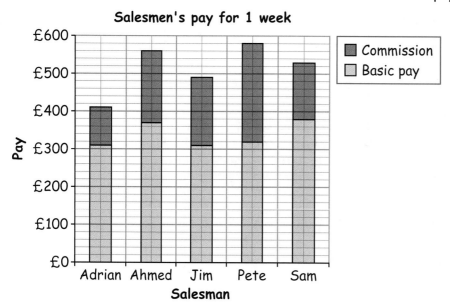

8 Use the bar chart at the bottom of page 272 to answer these questions.

    a) i)   Who earned the **most**?   **ii)** How much did he earn?

    b) i)   Who earned the **least**?   **ii)** How much did he earn?

    c) Write down the amount earned by each of the other salesmen.

    d) i)   Who has the highest basic pay?   **ii)** How much is this?

    e) i)   Which salesmen have the same basic pay?   **ii)** How much is it?

    f) i)   Find the commission earned by each salesman.

       ii)   List the salesmen in order of their commission, starting with the salesman who earned the **most** commission.

You have met many different types of data in this chapter.

Qualitative data is not numerical – examples include colours, types of drink.

Quantitative (numerical) data can be discrete or continuous. **Discrete** data takes only particular values, for example shoe sizes or number of children in families.

Continuous data can take any value in a given range, for example heights, weights, time to complete a task and other things that you measure.

---

9 Give three examples of each of:

    a) qualitative data   **b)** discrete data   **c)** continuous data.

10 Each year a National Readership Survey is carried out to find out what people have read. The chart shows estimates of the number of adults who read a selection of TV magazines in one year.

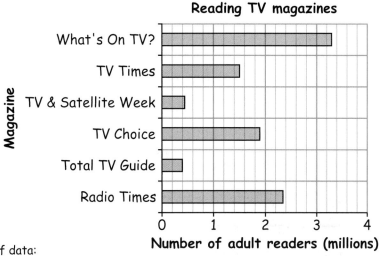

Reading TV magazines

Source of data:
www.nrs.co.uk

Estimate the number of adult readers of each TV magazine.

Give each answer:

a) to the nearest million   **b)** to the nearest 100 000.

11 The chart shows the grades achieved by the candidates in some A level exams one year.

a) Which subject had the biggest % of grade As?

b) Every grade except U is a pass grade.
Which subject had the best pass rate?

c) Draw up a table to give the % in each grade for each subject. Estimate values to the nearest %.

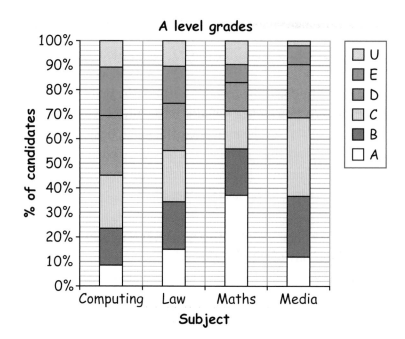

Source of data: www.qca.org.uk

Information from pie charts and line graphs   HD1/L1.1, HD1/L2.1

a Find information from pie charts

Pie charts show how something is divided into parts.
The pie chart below shows how a student spends a typical day.

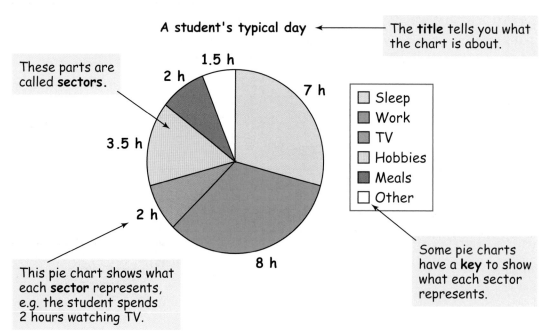

A student's typical day ← The **title** tells you what the chart is about.

These parts are called **sectors**.

Some pie charts have a **key** to show what each sector represents.

This pie chart shows what each **sector** represents, e.g. the student spends 2 hours watching TV.

Practice

1 Use the pie chart on page 274 to answer these questions.

   a) Which activity took the most time?

   b) How much longer did the student spend on hobbies than on watching TV?

   c) What is the total time shown on the pie chart? Why?

> **Some pie charts give information as percentages.**

See page 126 for a reminder.

2 This pie chart shows the results of a survey about how people travel to work.

   a) Which is:
      i)   the most popular way of getting to work
      ii) the least popular way of getting to work?

   b) What fraction of the people:
      i) travel to work by bus   ii) walk to work
      iii) travel to work by train   iv) drive to work
      v)   cycle to work?

   c) Altogether, 400 people took part in the survey. Find how many travelled by each method.

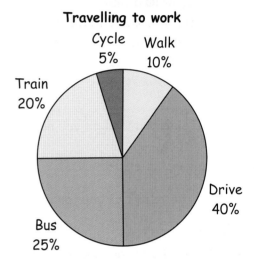

Travelling to work

Cycle 5% Walk 10% Train 20% Drive 40% Bus 25%

Sometimes pie charts do not give the number or the percentage of items in each sector.
You can **estimate** the fraction or percentage from the size of the sector.
If you are told the total number of items you can estimate how many are in each category.
For example, if a pie chart represents altogether 100 people, then half of the circle represents 50 people.

---

3 This pie chart shows the grades given to students for an exam.

   a) Estimate the fraction of the students who got each grade.

   b) Altogether, 48 students took the exam.
       Estimate how many got each grade.

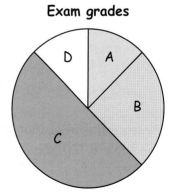

Exam grades

D A B C

4   The pie chart shows the proportion
of each type of property a developer
builds on a new estate.

Types of property on development

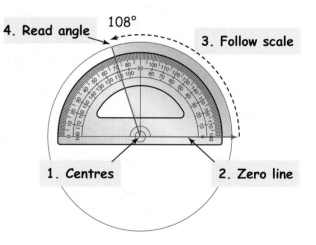

☐	Semi-detached
▨	Detached
▨	Flats
☐	Terraced
■	Bungalows

a)   Say which of the following statements are true and which are
false.
  i)   More than a quarter of the
  properties are flats.
  ii)   There are more bungalows
  than flats.
  iii)   There are approximately the same number of bungalows as detached houses.
  iv)   There are more terraced properties than any other type.

b)   Put the types of property in order of the numbers built by the developer, starting with
the largest.

For more accurate values you need to measure the angles of the pie chart with a **protractor**.

To measure an angle:

1   Put the centre of your protractor on the
centre of the pie chart.

2   Lie the zero line of the protractor over one
side of the angle.

3   Follow the scale round the edge of the protractor to the other side of the angle.

4   Read the size of the angle.
Take care to use the correct scale!

4. Read angle   108°

3. Follow scale

1. Centres   2. Zero line

There are 360° in a circle. To find accurate information from a pie chart you must find the
connection between the angles and the data.

For example, the following pie chart shows how a telephone bill of £180 is shared between
five flatmates.

$360°$ represents £180   so $1°$ represents $\dfrac{£180}{360} = £0.50$ (i.e. 50p)

Measure the angle that shows Bev's share.
The angle is 120°. So Bev's share of the bill $= 120 \times £0.50 = £60$.

5 a) Copy and complete this table.

Sharing the telephone bill

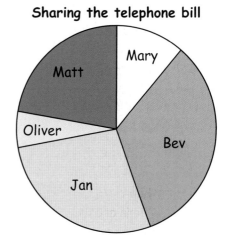

Flatmate	Angle	Amount to pay
Bev	120°	120 × £0.50 = £60
Mary		
Jan		
Oliver		
Matt		

b) Check that the total of the amounts in the last column is £180.

6 The pie chart below shows how a day's takings were divided between different types of goods at a do-it-yourself store.

a) On which type of goods did customers spend the most?

b) Did customers spend more or less on gardening tools than decorating equipment?

c) The manager says that more than half of his takings were from goods for decorating.
Is this true or false?

Takings at a DIY store

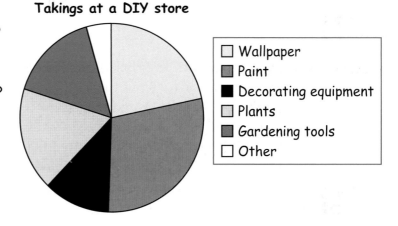

- ☐ Wallpaper
- ◩ Paint
- ■ Decorating equipment
- ☐ Plants
- ◩ Gardening tools
- ☐ Other

d) The total amount taken in the day was £8100.
Measure the angles and calculate how much was spent on each type of goods.

7 This pie chart shows the employees in each department of a company.
There are 90 **office workers**.

a) Work out what 1° represents.

b) Find the number of employees in each of the other departments.

c) How many employees are there altogether?

Employees of a company

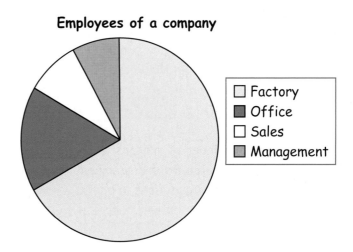

- ☐ Factory
- ◼ Office
- ☐ Sales
- ◩ Management

8 The pie charts show the people who visit a leisure centre on two days.

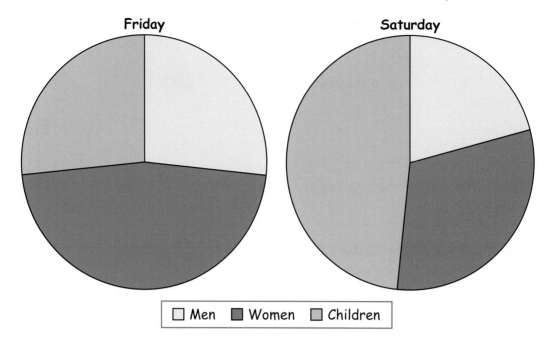

Friday   Saturday

☐ Men   ■ Women   ▨ Children

a) Which of the statements given below **must** be true?
   i)   More women than men visited the leisure centre on Friday.
   ii) The number of men who visited the leisure centre on Friday was the same as the number of children.
   iii) More than half of the people who visited the leisure centre on Friday were women.
   iv) More than a quarter of the people who visited the leisure centre on Saturday were women.
   v)   More men visited the leisure centre on Friday than on Saturday.
   vi) More than twice as many children as men visited the leisure centre on Saturday.

b) In fact, 240 children visited the leisure centre on Friday.
   i)   Measure the angle of the children sector on the Friday pie chart.
   ii) Work out how many people are represented by each degree on the Friday pie chart.
   iii) Copy the table and complete the column for Friday.

	Fri	Sat
Men		
Women		
Children		
Total		

c) The total number of visitors to the leisure centre on Saturday was 1200.
   i)   Complete the Saturday column in your table.
   ii) Answer the questions in part **a)** from your table.
        Do all of your answers agree with your answers for part **a)**?

b Find information from line graphs

Line graphs are often used to show how something changes as time goes by.

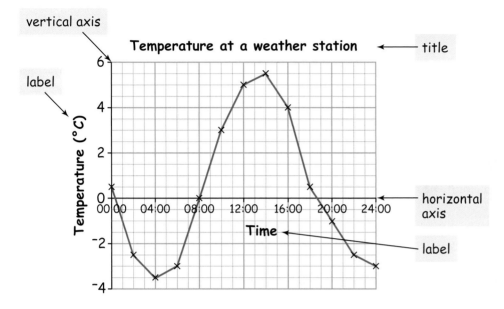

The **title** and **labels** on this line graph show that it is about the temperature at a weather station during a day.

The **crosses** show the temperatures that were recorded.
The **lines** that join them show how the temperature went up and down during the day.

Practice

1 Use the line graph above to answer these questions.

   a) What does each small square represent on:
      i)   the horizontal axis
      ii) the vertical axis?
   b) How often was the temperature measured?
   c) i)   What was the **highest** temperature measured?
      ii) When was this reading taken?
   d) i)   What was the **lowest** temperature measured?
      ii) When was this reading taken?
   e) i)   When was the temperature **measured** as 0°C?
      ii) **Estimate** the other times when the temperature was 0°C.
   f) Between what times was the temperature:
      i)   rising
      ii) falling?
   g) When was the temperature rising most quickly?

2 Use this line graph to answer the questions that follow.

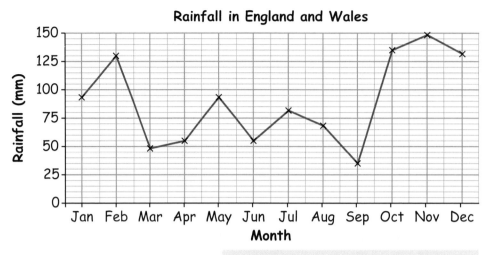

Source of data www.met-office.gov.uk

a) What does each small division represent on the vertical axis?
b) i)   In which month did the **most** rain fall?
   ii) Estimate how much rain fell in this month.
c) i)   In which month did the **least** rain fall?
   ii) Estimate how much rain fell in this month.
d) How much more rain fell in December than in June?
e) Estimate the total rainfall for the year.

3 These graphs both show the profit made by a company over a 5-year period.

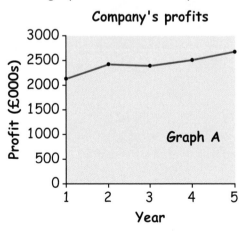

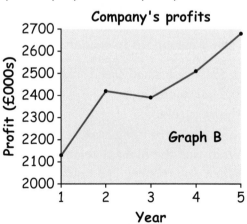

a) i)   Which graph gives the best estimates of the yearly profits?
   ii) Use this graph to estimate the profit each year.
b) Explain why graph B may be misleading.

This line graph is a **conversion graph**. You can use it to convert gallons to litres and vice versa.

Look at the scales.
What does a small square
represent on each axis?

The arrows show that:
14 gallons ≈ 64 litres
16 litres ≈ 3.5 gallons.

> **Note:** ≈ means
> 'approximately equal to'.

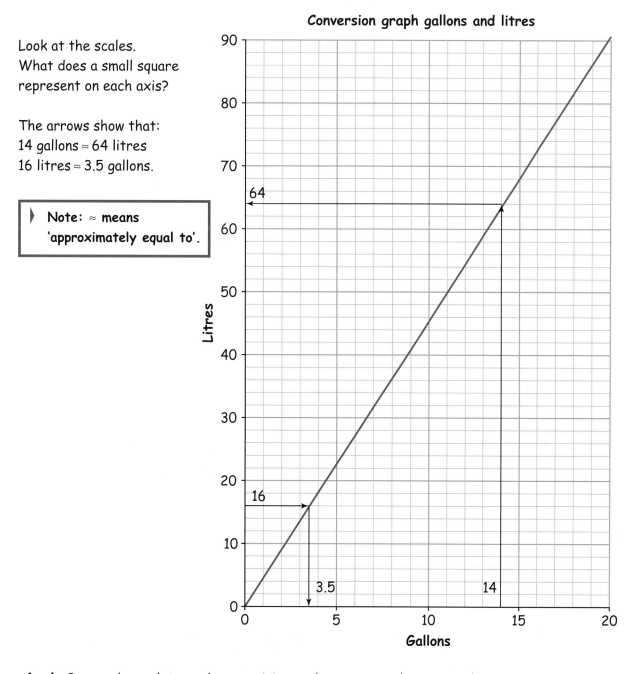

Conversion graph gallons and litres

4 a) Copy and complete each part, giving each answer to the nearest litre.
   i)   10 gallons ≈ ____ litres   **ii)** 5 gallons ≈ ____ litres
   iii) 18 gallons ≈ ____ litres   **iv)** 7.5 gallons ≈ ____ litres
   b) Copy and complete each part, giving each answer to the nearest 0.5 gallon.
   i)   40 litres ≈ ____ gallons   **ii)** 72 litres ≈ ____ gallons
   iii) 24 litres ≈ ____ gallons   **iv)** 67 litres ≈ ____ gallons
   c) The petrol tank of a car holds 15 gallons. How many litres is this?
   d) There are 82 litres of water left in a tank. How many gallons is this?

5 The graph below converts temperatures from Celsius to Fahrenheit and vice versa.

a) What does each small square represent on:
  i) the horizontal axis   ii) the vertical axis?

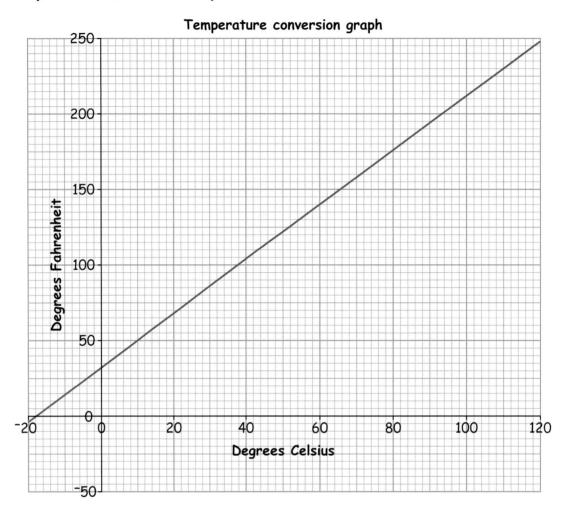

Temperature conversion graph

Copy and complete these. Estimate each answer as accurately as you can.

b) i)   20°C ≈ ____°F   ii) 48°C ≈ ____°F   iii) 75°C ≈ ____°F
   iv) 27°C ≈ ____°F   v) −10°C ≈ ____°F   vi) −7°C ≈ ____°F

c) i)   50°F ≈ ____°C   ii) 140°F ≈ ____°C   iii) 75°F ≈ ____°C
   iv) 205°F ≈ ____°C   v) 184°F ≈ ____°C   vi) 28°F ≈ ____°C

d) Water freezes at 0°C and boils at 100°C.
   What are these temperatures in Fahrenheit?

> Line graphs can be used to show **more than one set of data**.

6 The graph below shows the average number of hours of sunshine per day over the course of a year in Scotland and in England.

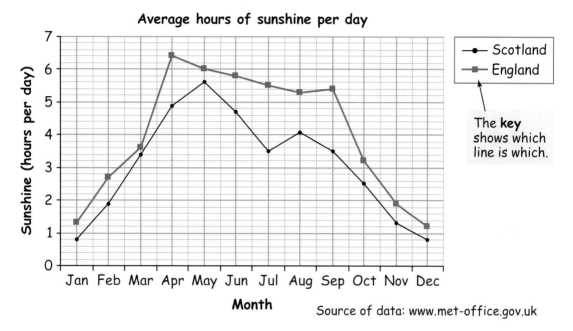

Average hours of sunshine per day

Source of data: www.met-office.gov.uk

The **key** shows which line is which.

a) Which month had the **most** sunshine in:
   i) England
   ii) Scotland?

b) Which month(s) had the **least** sunshine in:
   i) England
   ii) Scotland?

c) How much more sunshine was there in England than in Scotland in:
   i) January
   ii) July?

d) How much more sunshine was there in August than November in:
   i) England
   ii) Scotland?

e) Write a paragraph to explain what the graph shows.

7 The graph shows the amount of solid fuel, petroleum and gas consumed by households in the UK during the years 1970–2010.

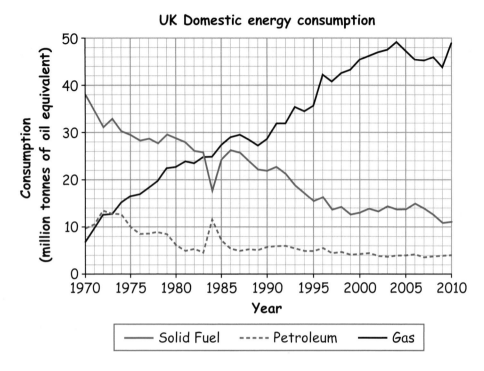

a) In which year was the consumption of solid fuel the lowest?

b) In which year was the consumption of petroleum the highest?

c) Estimate the increase in gas consumption between 1970 and 2010.

d) Estimate the decrease in the consumption of solid fuel between 1980 and 2000.

e) Estimate the total consumption of solid fuel, petroleum and gas in 1990.

f) Describe what happened to solid fuel and petroleum consumption in 1984.

g) Describe the way in which the consumption of each type of fuel changed between 1970 and 2010.

Activities

1 Look for examples of charts, graphs and diagrams in newspapers, books and magazines. Explain what they show.

2 Find the latest edition of Social Trends on the government's statistics website at www.statistics.gov.uk.
Print some of the information given in charts and graphs and discuss what they show.
Find other information including charts and graphs on the internet.

Handling Data

2 Collecting and Illustrating Data

Record information using a tally chart   HD1/E3.3, HD1/L1.2, HD1/L2.2

How could you find out which type of pet is most popular?

One way is to carry out a survey asking people which pets they have.

The more people you ask, the more representative the results.

Before starting the survey you need to think of the possible answers and how to organise them.

The best way of organising the information you collect is in a **tally chart**.

Here is a tally chart from a survey about pets.

The question asked was: **Which types of pet do you own?**

Type of Pet	Tally	Frequency
Dog	JHT JHT JHT I	16
Cat	JHT JHT JHT IIII	
Rabbit	JHT	
Bird	JHT II	
Fish	JHT I	
Reptile	II	
Other	JHT JHT I	
None	JHT JHT III	

A **tally mark** | is put into the table whenever a person says they have a particular type of pet.

The fifth tally mark is drawn across the first four to make a group of five. This makes it easier to count the total.

The total number of people for each pet is the **frequency**.

Practice

1 a) The frequency for dogs is 16.
       Write down the frequency for each other type of pet.
   b) Which is the most popular type of pet?
   c) List some types of pet that could be in the 'Other' category.

2 The pet survey also asked people how many pets they had. The answers were:

0, 1, 3, 1, 2, 1, 0, 6, 4, 3, 1, 1, 0, 1, 3, 1, 1, 0, 1, 6, 1, 2, 1, 3, 1, 0, 0, 1, 2, 1,
1, 2, 4, 1, 0, 3, 2, 1, 1, 0, 3, 1, 0, 2, 5, 1, 1, 2, 1, 3, 1, 0, 0, 1, 2, 1, 0, 1, 2, 0

a) This tally chart has been started to show the results. It shows the first four answers.
Copy and complete the tally chart.

Number of pets	Tally	Frequency
0	I	
1	II	
2		
3	I	
4		
5		
6		

b) How many people had:
   i) no pets   **ii)** 3 pets   **iii)** more than 3 pets?
c) How many people took part in the survey?

3 The tally chart below shows the grades awarded to students on a course.

a) Copy and complete the table.
b) How many students were on the course?
c) What fraction of the students got:
   i) a distinction   **ii)** a merit?
d) What percentage of the students failed the course?

Grade	Tally	Frequency
Distinction	IHI II	
Merit	IHI IHI IIII	
Pass	IHI IHI IHI IIII	
Fail	IHI IHI	

4 A college library's records show the number of books taken by each student who borrowed books during one lunch hour. This data is given below:

1	3	4	1	2	5	4	1	3	1	2	5	1	3	1	2	2	4	1	1	4
2	1	1	6	3	6	1	5	2	1	3	1	2	1	1	4	1	3	1	5	2
1	4	3	3	1	1	2	1	6	4	1	3	4	2	5	6	3	2	3	1	1
5	2	1	3	4	1	5	1	1	2	3	2	1	4	1	5	3	2	1	4	2

a) Draw a tally chart to show these results.
b) What was the most common number of books borrowed?
c) How many students borrowed books during this lunch hour?
d) What % of students borrowed more than 4 books?

5   A garage records the time taken (to the nearest minute) for car mechanics to find the cause of a particular engine problem.
The results are:

11	23	30	22	9	25	14	19	23	18	12
25	10	29	21	16	18	8	13	21	24	28
13	18	26	20	13	19	15	21	27	17	22

a) Copy and complete the tally chart.
b) What was the most common time group?
c) How many times were recorded?
d) What fraction of the recorded times were less than 11 minutes?
e) What % of the recorded times were more than 20 minutes?

Time (minutes)	Tally	Frequency
6–10		
11–15		
16–20		
21–25		
26–30		

Illustrate data using a pictogram or bar chart   HD1/E3.4, HD1/L1.2, HD1/L2.2

a Using a pictogram

A **pictogram** uses simple pictures or symbols to illustrate data (see page 266).
When drawing a pictogram you should:

- Choose a **simple symbol** that is easy to draw.
- Decide how many items each symbol stands for.
- Work out how many symbols you need for each category.
- Draw the pictogram. Use squared paper – this makes it easier to line up the symbols. Try to keep the symbols the same size (except when only part of a symbol is needed).
- Give the pictogram a **title** saying what it is about.
- Include a **key** showing what each symbol stands for.

Practice

1   The table shows the number of hours of sunshine at a seaside resort each day during a week in August. Draw a pictogram to show this data.

Use ⚪ to represent 2 hours of sunshine.

Use ◖ to represent 1 hour of sunshine.

Day	Hours of sunshine
Monday	6
Tuesday	4
Wednesday	3
Thursday	5
Friday	8
Saturday	10
Sunday	9

2   The table shows the number of houses sold by an estate agent in the first six months of a year. Draw a pictogram to show this information.

Use to represent 10 houses.

You will have to decide how much of the symbol to use when you need to show less than 10 houses.

Month	No. of houses
January	25
February	30
March	45
April	52
May	60
June	57

3   A car manufacturer keeps records of the colours of the cars sold. The table gives the results for one model.
Choose a symbol to represent 100 cars and draw a pictogram.

Colour	No. of cars sold
Red	650
Black	425
White	386
Blue	541
Green	150

b Using a bar chart

A **bar chart** uses vertical or horizontal bars to illustrate data (see page 267).

When drawing a bar chart you should:

- Use **squared** or **graph** paper.
- Use an **easy scale**, i.e. make each small square an easy value to use. (e.g. 1, 2, 5, 10, 20, 50, 100, 200, 500 are easier than 3, 4, 7, 8, 9, 30, 40, ...)
- Aim for a large chart that is easy to read.
- Use a ruler and sharp pencil to draw the chart. Leave gaps between the bars.
- Give the bar chart a **title** saying what it is about.
- **Label** the axes (give **units** if necessary). Use a pen to do this if you wish.

Practice

1   The pizza toppings ordered in a snack bar during one day are shown in the table.

a) Draw a bar chart to show this information. Use vertical bars.
b) Use your chart to answer these questions.
    i)   Which topping is **most** popular?
    ii) Which topping is **least** popular?

Topping	No. of pizzas
Cheese & tomato	15
Seafood	8
Ham & mushroom	12
Pepperoni	18
Roasted vegetable	11

2 The table gives the type and number of fruit trees in an orchard.

Type of tree	No. of trees
Apple	36
Cherry	12
Pear	18
Plum	25

a) Draw a bar chart to show this information. Use horizontal bars.

b) Use your chart to answer these questions.
  i) Which is the **most common** type of fruit tree?
  ii) How many more plum trees than pear trees are there?

3 A company offers holiday homes for rent in some European countries.
The table shows the number of properties it has in each country.

Country	No. of properties
Belgium	15
France	63
Germany	42
Greece	27
Italy	36
Spain	54

a) Draw a bar chart to show this data.

b) In which country does the company have the **most** properties?

c) How many more properties does the company have
  i) in France than in Belgium
  ii) in France than in Spain?

Also draw bar charts of the data given in the tally chart and pictogram Practice questions on pages 285–288.

4 A shop records the number of films rented from different categories.
The table shows the results for one week.

Film type	No. of films
Adventure	340
Children's	276
Comedy	125
Horror	205
Musical	83
Sci-fi	107
Thriller	194

a) Draw a bar chart to show this data.
b) Which category was **most** popular?
c) Which category was **least** popular?
d) List the categories in order of popularity, starting with the most popular.

5 A newsagent records the number of newspapers delivered to the houses he supplies. The results for one day are given in the table.

No. of newspapers	No. of houses
1	324
2	191
3	54
4	27
5	8

a) Draw a bar chart to show this data.
b) Describe briefly what the chart shows.

6 The table shows estimates of the number of people who read some daily newspapers.

a) Draw a bar chart to show this data.

b) Which of these newspapers is:
   i) **most** popular   ii) **least** popular?

c) List the newspapers in order of popularity, starting with the most popular.

Source: National Readership Survey 2011/2012 (www.nrs.co.uk)

Newspaper	No. of readers (thousands)
Daily Express	1304
Daily Mail	4385
Daily Star	1439
Daily Telegraph	1387
Guardian	1081
Mirror	3178
Sun	7244
Times	1302

7 The table gives the life expectancy of men and women in eight countries.

a) Draw a comparative bar chart to illustrate this data.

b) In which of these countries is
   i)   male life expectancy **greatest**
   ii) female life expectancy **lowest**?

c) i)   Briefly compare male and female life expectancy.
   ii) In which country is the difference between male and female life expectancy greatest?

d) Compare life expectancy in the eight countries.

Country	Life expectancy (years)	
	Male	Female
Angola	51	53
Brazil	70	77
China	72	76
India	63	66
Malawi	44	51
Russia	62	74
UK	78	82
USA	76	81

Source: World Health Report 2012

8 The table shows the number of men and women employed in UK industries.

a) Copy and complete the table by filling in the totals.

b) Draw a component bar chart to show the data.

Industry	Number of employees (thousands)		
	Male	Female	Total
Construction	1905	236	
Education	865	2195	
Health and social work	818	3009	
Manufacturing	2139	689	
Wholesale and retail	2090	1901	

Source: Labour Force Survey 2012

c) Which industry has:
   i)   the greatest total number of employees
   ii) the greatest number of male employees
   iii) the greatest number of female employees?

d) Briefly compare the number of men and women employed in each industry.

Activities

Collect some data. Then use a pictogram or bar chart to illustrate it and describe what your illustration shows. Here are some suggestions:

1 Use a tally chart to record the colours (or ages) of cars in a car park.

2 Use a tally chart to record the type of vehicles travelling along a road.

3 Keep a record of how long you watch each TV channel during a week.

4 Record how much time you spend on different activities
   (e.g. hobbies, shopping, housework, work) during a week.

5 Carry out a survey using tally charts to record the information you collect.
   Ask students questions such as 'What is your favourite sport?',
   'What type of programmes do you like most?', 'What is your favourite hobby?'

6 Take some measurements from students in your group
   (e.g. height, length of right hand, length of stride, time taken to recite the alphabet).

Use a comparative or component bar chart to show the results for male and female students from activity 5.

Illustrate data using a pie chart
HD1/L1.2, HD1/L2.2

Pie charts show how something is divided into parts. (See page 274.)
A pie chart is the best way to show the **proportion** (or fraction) of the data that is in each category.

To draw a pie chart:

- Find the **total** of the data (unless it has been given).

- **Divide 360° by the total** to find how many degrees represent each item.

- **Multiply this by the number of items** in each category to find the **angle** for each category.
  Check the total angle is 360°.

- Write a **title** to say what information the pie chart gives.

- Use a **compass** to draw a circle and a **protractor** to measure the angles needed.

- **Label** each sector of the pie chart or give a **key** to show what each sector represents.

Example

The table shows the percentage of students on a course who got each grade.

Grade	% of students
Distinction	13%
Merit	31%
Pass	45%
Fail	11%

As the data are percentages, the total must be 100% (but check to make sure).

Each % will be represented by:

$$\frac{360°}{100} = 3.6°$$

The table shows how the angle is worked out for each grade.

Use your calculator to check the working in the table. Here the total angle is 361°. The rounding has given an

Grade	% of students	Angle (nearest °)
Distinction	13%	$13 \times 3.6 = 47°$
Merit	31%	$31 \times 3.6 = 112°$
Pass	45%	$45 \times 3.6 = 162°$
Fail	11%	$11 \times 3.6 = 40°$
Total	100%	361°

extra degree. When this happens, take 1° away from the largest angle, here the angle for the Pass grade should be changed to 161°.

Practice

1 a) Draw a pie chart to show the proportion of students who got each grade.
   b) According to your chart, which grade had:
      i) the smallest proportion of students   ii) the largest proportion?

2 A group of students have lunch at a snack bar. The table shows what the students eat.

   a) How many students are there altogether?
   b) In a pie chart how many degrees represent each student?
   c) Copy the table and add an extra column to show how you work out the angles.
   d) Draw a pie chart to illustrate the data.

Lunch	No. of students
Sandwich	6
Pasta	5
Baked potato	8
Salad	4
Pizza	1
Total	

3 The table shows the number of cars owned by households in the UK.

   a) Draw a pie chart to illustrate this.
   b) What is the most common number of cars owned by households?
   c) What fraction of households have no car?

Number of cars	% of households
0	25
1	43
2 or more	32

Source: National Travel Survey

4   The table gives estimates of the number of
    cars produced in each region during a year.

    a) Use the data to draw a pie chart.
    b) Which region produces the greatest
        proportion of cars?

Region	No. of cars (millions)
Europe	17.0
North America	5.1
South America	3.4
Asia	32.4
Other	0.3

Source: International Organisation of Motor
Vehicle Manufacturers

5   The total amount spent on different types of advertising is given in the table.

    a) Draw a pie chart to illustrate this data.
    b) Use your chart to answer these questions.
        i)   What type of advertising takes the
             greatest share of the money?
        ii) Does advertising on television earn
             more or less than advertising in the
             press (newspapers and magazines)?
        iii) Is it true that more than a third of
             the total amount spent on advertising
             is spent on the internet?
    c) Suggest one type of advertising that could
        be in the other category.

Type of advertising	Amount (£billions)
Internet	5.3
Cinema	0.2
Radio	0.4
Television	4.2
Press	3.7
Other	3.0

Source: www.adassoc.org.uk

6   The table gives the population and area of
    each country in the UK.
    a) Draw a pie chart for each set of data
    b) Write a paragraph explaining what your
        charts show.

Country	Population (millions)	Area (000 km^2)
England	53.1	130.4
N. Ireland	1.8	13.6
Scotland	5.3	78.1
Wales	3.1	20.8

Source: Office for National Statistics

Draw graphs   HD1/L1.2, HD1/L2.2

a Draw a line graph

A **line graph** is often used to show how something changes as time goes by. (See page 279.)

To draw a line graph:

- Use **squared** or **graph** paper.
- Use an **easy scale**, i.e. make each small square an easy value to use by making
  (e.g. 1, 2, 5, 10, 20, 50, 100, 200, 500 are easier than 3, 4, 7, 8, 9, 30, 40, ...).

- Aim for a **large graph** that fills most of the page.
- Use a ruler and sharp pencil to draw the graph.
- Give the graph a **title** saying what it is about.
- **Label** the axes (give **units** if necessary). Use a pen to do this if you wish.
- (L2) If your graph has more than one line, label each line or use a key to show what each line represents.

Practice

1 Students on a college course can attend an extra drop-in class for help.
The number of students who go to this class each week is shown in the table.

Week	1	2	3	4	5	6	7	8	9	10
No. of students	1	0	6	9	5	8	12	15	10	7

a) Draw a line graph to illustrate this data. Put the week number on the horizontal axis and the number of students on the vertical axis.

b) Use your graph to answer these questions.
   i)   In which week did the **greatest** number of students attend the class?
   ii) In which week did the **least** number of students attend the class?
   iii) Describe how the attendance varied over the weeks shown.

2 The table shows how the number of computers owned by a school varied between 1980 and 2010.

Year	1980	1985	1990	1995	2000	2005	2010
No. of computers	5	12	30	60	120	170	200

a) Draw a line graph to illustrate this data.

b) Estimate how many computers the school had in: **i)** 1982 **ii)** 1994 **iii)** 2007.

c) Estimate the year when the school had:
   i) 20 computers   **ii)** 100 computers   **iii)** 150 computers.

3 The table below shows how the number of motoring offences detected by camera increased between 2000 and 2009.

Year	2000	2001	2002	2003	2004	2005	2006	2007	2008	2009
No. of offences (thousand)	651	924	1206	1785	1900	1948	1808	1406	1147	1027

Source: Social Trends 41

a) Draw a line graph to illustrate this data.

b) Describe what your line graph shows.

c) If the trend continued, how many motoring offences would be detected by camera in 2010?

> It is often difficult to plot large but accurate numbers like these. You may find it easier if you round the number of thousands first (perhaps to the nearest 20 or 50 depending on your scale).

L2

4 A health and safety officer records the noise level in a factory in decibels (dB) every hour. The results are given in the table below.

Time	8 am	9 am	10 am	11 am	noon	1 pm	2 pm	3 pm	4 pm	5 pm
Noise (dB)	40	65	85	89	82	74	84	83	72	54

a) Use these results to draw a line graph.
b) Use your graph to estimate the times when the noise level is exactly 80 decibels.
c) Estimate the total time when the noise level is over 70 decibels.
d) Describe briefly how the noise level varies during the day.
   Suggest reasons for high and low values.

5 The table below gives the number of licensed motorcycles in Great Britain between 2001 and 2011, and the number of deaths and serious injuries to motorcyclists in this period.

Year	Number of motorcycles (thousands)	Number of deaths and serious injuries
2001	1010	7305
2002	1070	7500
2003	1135	7652
2004	1191	6648
2005	1206	6508
2006	1210	6484
2007	1248	6737
2008	1275	6049
2009	1276	5822
2010	1234	5183
2011	1238	5609

Source: www.dft.gov.uk

a) Draw line graphs to show:
   i)   the number of licensed motorcycles between 2001 and 2011
   ii) the number of deaths and serious injuries to motorcyclists between 2001 and 2011.
b) Describe what your graphs show.

6 A clothing company has shops in London and Manchester.
   The table below gives the value of sales at these shops during the months of a year.

Month		Jan	Feb	Mar	Apr	May	Jun	Jul	Aug	Sep	Oct	Nov	Dec
Sales	Lon	385	143	162	189	221	254	207	198	240	262	376	437
(£000s)	Man	258	124	138	142	196	205	212	186	193	187	256	328

   a) Draw a graph with two lines showing the sales in London and Manchester.
   b) For each shop write down:
       i)   the month when sales were **highest**
       ii)   the month when sales were **lowest**
       iii) the months in which sales were above £300 000.
   c) In which month was the difference between the London and Manchester sales:
       i) the **greatest**   ii) the **least**?
   d) Briefly describe how sales varied in each shop over the course of the year.
       Describe any similarities and differences in the sales patterns.

7 The table below gives the average temperature in England and Wales, Scotland and
   Northern Ireland during each month of a year.

Month		Jan	Feb	Mar	Apr	May	Jun	Jul	Aug	Sep	Oct	Nov	Dec
Average temp (°C)	E&W	5.3	6.6	7.2	8.8	11.4	13.9	15.4	16.5	13.9	9.8	8.2	5.4
	Scot	4.2	3.6	5.0	7.0	9.5	11.8	12.5	14.0	11.8	6.8	6.2	3.5
	NI	6.1	5.6	6.8	8.3	10.6	12.9	13.6	15.0	13.0	8.6	7.8	4.9

   Source: www.met-office.gov.uk

   a) Draw a graph with a line for each set of data.
   b) For each set of data write down the months when the average temperature was
       below 6°C.
   c) Describe how the temperature varied in these countries over the course of the year.
       Describe any similarities and differences.

8 A nurse measures the temperature of a patient every three hours.
   The results are given in the table below.

Time	7 am	10 am	1 pm	4 pm	7 pm	10 pm	1 am	4 am
Temp (°F)	99.8	101.2	102.9	103.7	103.6	102.9	101.1	99.3

   a) Draw a line graph to illustrate these results. Show times from 7 am to 4 am on the
       horizontal axis and temperatures from 98°F to 104°F on the vertical axis.
   b) A patient has a fever when their temperature is 100°F or more.
       Use your graph to estimate how long this patient's fever lasted.
   c) This patient has a normal temperature of 98.5°F. Use your graph to estimate the time
       when this patient's temperature returns to normal.

Activities

Collect some data and use a pie chart or line graph to illustrate it.
Then describe what your chart or graph shows. Here are some suggestions:

Pie charts

- Keep a record of how you spent the time during one day.
- Collect data from your class, e.g. hair colour, eye colour, method of travel to college.

Line graphs

- Record the outside temperature over the course of a day.
- Use the Internet to find out how currency exchange rates or share prices change over a period of time.

Visit government (and other) websites

Carry out searches to find some information you are interested in (e.g. sport).
The addresses of some useful government websites are given below.

www.statistics.gov.uk   All sorts of data on population, employment, leisure and tourism, etc.

www.homeoffice.gov.uk   Home Office website giving crime data.

www.gov.uk   For information on communities and local government.

---

b Draw and use conversion graphs   MSS1/L2.6

A **conversion graph** is used to convert units from one system to another (see page 281). When drawing a conversion graph you should use graph paper, choose scales and label in the same way as for other line graphs. But before you can do this you need to work out values for some points on the graph.

Example

Using 1 inch = 2.54 cm gives this table:

inches	0	5	10
cm	0	12.7	25.4

Note: only 3 points have been found – 2 points are enough to give the line; the 3rd point acts as a check.

5×2.54 10×2.54

Now a sensible scale is needed on both axes.
One possible graph is shown below.

Practice

1 a) Write down what a small square represents on each of the axes on this graph.

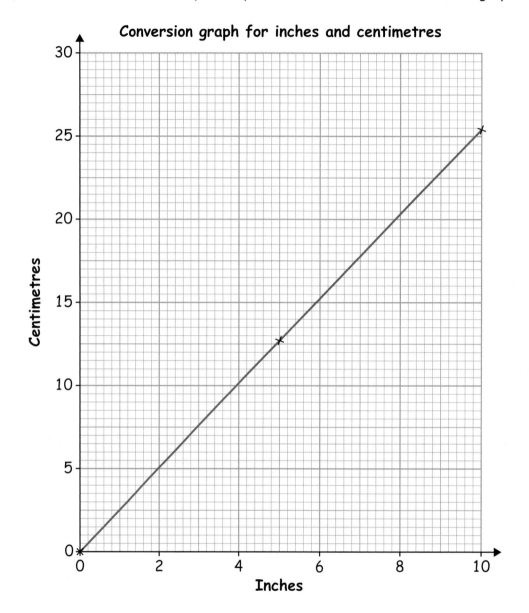

Conversion graph for inches and centimetres

b) Check that the points shown on the graph agree with the values given in the table above.

c) Use the graph to complete these.

i) 6 in ≈ ____ cm   **ii)** 9 in ≈ ____ cm

iii) 2.6 in ≈ ____ cm   **iv)** 4.3 in ≈ ____ cm

v) 10 cm ≈ ____ in   **vi)** 25 cm ≈ ____ in

vii) 18 cm ≈ ____ in   **viii)** 12.5 cm ≈ ____ in

Check your answers by calculation.

2 a) Use the approximation 1 kg ≈ 2.2 lb to draw a conversion graph for kilograms and pounds up to 20 kg.

b) Use your graph to complete the following, then check by calculation.

i) 15 kg ≈ ___ lb   **ii)** 9 kg ≈ ___ lb   **iii)** 12.5 kg ≈ ___ lb

iv) 30 lb ≈ ___ kg   **v)** 14 lb ≈ ___ kg   **vi)** 38.5 lb ≈ ___ kg

---

Activities

1 Draw a conversion graph for miles and kilometres. Find the distance between your home town and major cities such as Birmingham and Manchester from a road atlas. Use your conversion graph to convert the distances to kilometres.

2 Draw a conversion graph for grams and ounces and/or pints and litres. Use your graph(s) to convert the amounts given in recipes.

3 Use a spreadsheet to draw charts and graphs – use the data given in the questions or that you find on the Internet. Much of the data from government websites is available in spreadsheet form.

Choosing a statistical diagram
HD1/L1.2, HD1/L2.2

The choice of statistical diagram depends on what you want to show.

For example:
- If you want to show how something is divided into parts, a pie chart is probably the best.
- If you want to show how something varies over time, then a line graph is likely to be the most appropriate.
  You can draw two or more lines on the graph to compare how different things vary over time.
- A bar chart can be used in many different situations, particularly when you want to compare two sets of data.
  A comparative bar chart emphasises the difference between the datasets, whereas a component bar chart emphasises the total for each category or group while also showing how this total is made up.

Often you can illustrate data in a variety of effective ways.

Practice

For each question below:
- Choose the diagram that you think will illustrate the data best (choose from bar chart, pie chart or line graph).
- Draw the diagram.
- Describe briefly what your diagram shows.

1 Cinema attendances in the months of a year:

Month	Jan	Feb	Mar	Apr	May	Jun	Jul	Aug	Sep	Oct	Nov	Dec
Attendance (millions)	15.2	17.2	11.1	11.2	13.3	12.7	17.8	21.4	11.5	13.6	12.9	13.6

Source: British Film Institute, www.bfi.org.uk

2 The percentage of last month's earnings that Tracy spent on different things:

	Food	Rent	Transport	Clothes	Entertainment	Other
Percentage of earnings	29%	32%	9%	14%	10%	6%

3 The numbers of men and women who read different magazines:

Magazine	Gardeners' World	Reader's Digest	What Car?	Xbox 360
No. of men (thousands)	445	588	601	889
No. of women (thousands)	738	673	102	150

4 The number of drink drive accidents reported in one year at different times of day:

Time of day	12 am	1 am	2 am	3 am	4 am	5 am	6 am	7 am	8 am	9 am	10 am	11 am
No. of accidents	854	804	700	582	412	296	238	230	178	176	158	150

Time of day	12 pm	1 pm	2 pm	3 pm	4 pm	5 pm	6 pm	7 pm	8 pm	9 pm	10 pm	11 pm
No. of accidents	180	186	200	274	374	520	538	592	596	672	756	822

Source: Department for Transport, www.dft.gov.uk

5 A survey asks people who are moving house to give the main reason.
The table shows the results.

Reason	No. of people
Need different size of house	90
Personal (e.g. marriage, divorce)	63
To move to a better area	30
Job-related reason	96
Other	21

6 This table shows the number of students in a class who achieved each grade in the two papers of an exam.

Grade	Paper 1	Paper 2
A	2	1
B	5	5
C	8	5
D	6	6
E	7	9
F	2	4

7 Population of continents in 2011:

Region	Population (millions)
Asia	4140
Africa	995
Europe	739
North America	529
South America	386
Oceania	36

8 Medals won by different countries in the London 2012 Olympics:

Country	Gold	Silver	Bronze
United States of America (USA)	46	29	29
China	38	27	23
Great Britain (GB)	29	17	19
Russia	24	26	32

Handling Data

3 Averages and Range

Find the mean

An **average** is a representative or 'typical' value. There is more than one type of average.

The **mean** is the **arithmetical average**.

To find the mean, add all the values together, then divide by how many there are.

> **Mean** = $\dfrac{\textbf{Sum of data values}}{\textbf{Number of values}}$

Sum means total
(add up the values).

Example

The number of goals scored by a footballer in the games he played this season were:

2 0 1 1 0 3 1 2 3 1

Sum of data values $= 2+0+1+1+0+3+1+2+3+1=14$

Number of values $=10$ Mean score $=\dfrac{14}{10}=\textbf{1.4}$ goals per match

Check by reverse calculation: mean × no. of items should equal the total
$1.4 \times 10 = 14$

Practice

1 Find the mean of each set of data. Check each answer.

a) The number of students absent from a class each week of a term:

0 1 2 2 1 4 3 0 2 3

b) The number of calls a student makes on her mobile phone each day in a week:

5 2 4 0 8 12 11

c) The ages of the children in a family: 12 10 6 4 1

d) The number of letters delivered to each house in a street:

4 5 2 0 4 5 8 6

e) The weights of puppies in a litter:

0.41 kg, 0.36 kg, 0.32 kg, 0.36 kg, 0.41 kg, 0.39 kg

f) Prizes in a lottery:

£10 £10 £10 £250 £10 £10 £10 £10

Note: in part **f)** the mean is **distorted** by the large win of £250. *It is not a good representative value.*

2 The table shows the scores of five
   batsmen in four cricket matches.

   a) Find the mean score for each
      batsman. Check each answer.
   b) Which batsman has:
      i)   the **best** mean score
      ii) the **worst** mean score?

Player	Match			
	1	2	3	4
Andy	112	16	3	49
Imran	0	–	68	43
Mark	–	11	–	37
Stuart	40	33	17	25
Tim	35	27	54	43

Note: – means 'did not play'.

3 Four clerks in an office each earn £9 per hour.
   Their manager earns £30 per hour.

   a) Find the mean wage.   b) Is this mean a 'typical' value?

4 The table below shows the number of tickets sold for a concert.

Day	Weekdays					Weekend	
	Mon	Tues	Wed	Thurs	Fri	Sat	Sun
Ticket sales	126	205	167	152	298	324	282

Find the mean number of tickets sold:

a) per day at the weekend   b) per weekday (Mon to Fri).

5 The bar chart shows the rates of
   pay earned by six students in
   part-time work.

   a) Write down the rate of pay for
      each student.
   b) Calculate the mean rate of pay.
   c) Which students get more than
      the mean rate of pay?

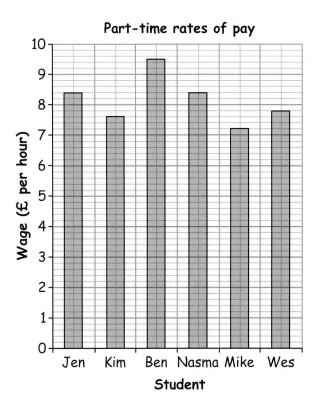
Part-time rates of pay

For extra practice, find the mean of
each set of data in these questions:

*Using a pictogram, pages 287–288,
questions 1, 2*
*Using a bar chart, pages 288–290,
questions 1, 2, 3, 4, 6*
*Draw a line graph, pages 294–296,
questions 1, 3, 4, 5, 6*

L2 When data is given in a frequency table it is more difficult to find the mean.

Example

A survey investigates the number of occupants in a block of flats.
The table gives the results.

Number of occupants	Number of flats
0	1
1	6
2	12
3	7
4	3
5	0
6	1

The mean number of occupants per flat

$$= \frac{\text{total number of occupants}}{\text{total number of flats}}$$

The total number of flats is the sum of the 2nd column:
$1+6+12+7+3+0+1=30$

Take care with the number of occupants, the total is **not** just $0+1+2+...$
Think about what each row in the table means:

No occupants in 1 flat	$=0$ occupants	
1 occupant in each of 6 flats	$=6$ occupants	
2 occupants in each of 12 flats	$=24$ occupants	
3 occupants in each of 7 flats	$=21$ occupants	and so on.

Total number of occupants $=0 \times 1 + 1 \times 6 + 2 \times 12 + 3 \times 7 + 4 \times 3 + 5 \times 0 + 6 \times 1$
$ = 0 + 6 + 24 + 21 + 12 + 0 + 6 = 69$

A table can be used to show this working neatly.

x	f	Multiply	xf
0	1	0×1	0
1	6	1×6	6
2	12	2×12	24
3	7	3×7	21
4	3	4×3	12
5	0	5×0	0
6	1	6×1	6
Total	30		69

Here, x represents the number of occupants and f, the frequency, is the number of flats.

Mean number of occupants per flat $= \dfrac{69}{30} = \mathbf{2.3}$

Total number of flats

Total number of occupants

The method can be written as a formula:

> **Mean from a frequency table** $= \dfrac{\text{Sum of (value} \times \text{frequency)}}{\text{Sum of frequencies}}$
>
> or in letters **mean** $= \dfrac{\sum xf}{\sum f}$ where $\sum$ means 'add up'.

6 A trainee secretary has her work checked to see how many errors she makes.
The table shows the results for one document she prepares.

   a) How many pages are in the document?
   b) Find the mean number of errors per page.

Number of errors	Number of pages
0	9
1	8
2	0
3	2
4	1

7 A health visitor notes the number of children in each of the families he visits.
The table shows her results.

   a) Find the mean number of children.
   b) How many families had more than the mean number of children?

Number of children	Number of families
1	2
2	8
3	4
4	4
5	1
6	1

8 A rail company records the number of trains on a particular route that arrive late each day.
The table gives the results for June.

   a) Find the mean number of late trains per day.
   b) Five trains travel on this route per day. What % of the trains were late in June?

Number of late trains	Number of days
0	8
1	11
2	9
3	1
4	1

9 The table gives the ages of students on a carpentry course.

   a) How many students are over 20 years old?
   b) Find the mean age.

Age (years)	Number of students
16	5
17	7
18	3
19	4
20	3
21	2
22	1

10 The table below gives the number of goals scored by a football team in home and away matches.

No. of goals		0	1	2	3	4	5	6
No. of games	Home	2	4	3	0	2	0	1
	Away	4	6	0	1	1	0	0

a) Find the mean number of goals scored per match   i) at home   ii) away.

b) Is the team better at scoring goals at home or away?

Find the mode and median

HD1/L2.3

The mean is not the only type of average – there are two others.

> The **mode** is the **most common value** in a data set.

Sometimes there is more than one mode (when 2 or more values are equally common).
Sometimes there is no mode (when all values occur just once).

> The **median** is the **middle value** in an ordered list.

If there are two middle values, add them together and divide by 2.
This gives the value halfway between them.

Example

The number of tracks on a band's CDs are:

   15   12   14   12   12   14   10   13

Putting these in order gives: ⟵ **middle values**

   10   12   12 | 12   13 | 14   14   15

The modal number of tracks is **12** – this is the value that occurs most often.

There are 2 middle values, 12 and 13. The median number of CDs is **12.5**.

Also, the mean $= \dfrac{15 + 12 + 14 + 12 + 12 + 14 + 10 + 13}{8} = \dfrac{102}{8} = \textbf{12.75}$

Often one of the averages is a better representative of the data than the others.
In this case, the mode is the only average that gives a number of tracks that is possible, so you could argue that the mode is the best average to use.

Some of the advantages and disadvantages of each average are given below.

- The mean has the advantage of **using all the data values**.
  A disadvantage is that **abnormally high or low data values can make it 'distorted'**.
  Also, the mean is **not usually one of the values in the dataset**.
- The mode is **always one of the values in the dataset**. It is **useful to manufacturers** who want to know the most common size.
  However, sometimes there is **no mode** or a **lot of modes**.
  The mode **may be at one end of the distribution**.
- Being in the middle means the median is **not affected by abnormal values**.
  However, if there are 2 middle values, the median **may not be one of the values in the dataset**.

Which average you choose to use will depend on which of these characteristics you think are most important. Opinions may differ about this.

Practice

1 Find the mode, median and mean of each set of data.
   Decide which is the best representative value: the mean, the mode or the median.
   In each case explain your choice.

   a) The number of pints of milk delivered to a house each day in a week:
       2   1   1   0   1   6   0
   b) The number of students attending a class each week during a term:
       18   17   17   17   18   16   16   15   12   16
   c) The ages of the players on a football team:
       19   18   18   20   29   34   18   18   18   21   19
   d) The time taken in minutes by students to complete a test:
       32   28   28   30   27   25   36   27   31   33   30   28
   e) Heights of a group of children:
       1.24 m   1.26 m   1.18 m   1.21 m   1.17 m   1.19 m

Find the mode and median for each set of data given in this question:
Find the mean, page 302, question 1

Look back at the scores given for the batsmen in:
Find the mean, page 303, question 2
What happens if you try to find the mode for each batsman?
Find the median for each batsman.
Why do you think the mean is usually used to compare batsmen rather than the median or mode?

> To find the **mode** from a **frequency table**, look for the **value with the highest frequency.**

Example

The results of the survey done in a block of flats are shown again in the first two columns of the table below. The **highest frequency** (number of flats) is 12.
This means that the **modal number of occupants per flat** is **2**.
(More flats had 2 occupants than any other number of occupants.)

It is more difficult to find the **median.** It helps if you add another column to the table to show **cumulative frequency,** i.e. a 'running total' of the frequencies.

Number of occupants	Number of flats	Cumulative frequency	
0	1	1	
1	6	7	← $1 + 6 = 7$
2	12	19	← $7 + 12 = 19$
3	7	26	← $19 + 7 = 26$
4	3	29	
5	0	29	
6	1	30	← Total number of flats

There are 30 flats altogether. If the number of occupants for all of these flats were written in a line, there would be two middle values – the 15th and 16th:

$$15^{th} \quad 16^{th}$$
0 1 1 1 1 1 1 2 2 2 2 2 2 2 ②, ② 2 2 2 3 3 3 3 3 3 4 4 4 6
Middle

(Whenever there is an even number of data values, there are two middle values.)

Look at the cumulative frequency column in the table. The 15th and 16th values are both after the 7th value, but before the 19th. This means the 15th and 16th values are both 2. The median is **2 occupants**.

---

2 The table shows the ages of the children in a creche one morning.

a) Find:
  i) the mode
  ii) the median
  iii) the mean.

b) Which value from part a) do you think is the best representative value?

Age	Number of children
1	1
2	5
3	6
4	8

3 A clothes shop keeps records of the sizes of dresses it sells. The table shows the results for one style of dress.

Size	Number sold
8	2
10	8
12	10
14	13
16	5
20	2

a) Find:
   i) the mode   **ii)** the median
   iii) the mean.
b) Which of these averages do you think is most useful to the shop manager?

> Find the mode and median for each set of data given in these questions:
> *Find the mean, pages 303–306, questions 3, 6, 7, 8, 9, 10*
> Compare your answers with the means you found for these questions.
> In each question say which average you think is most appropriate and why.

4 In a survey, new drivers are asked how many attempts they needed to pass their driving test. The table gives the results.

Number of attempts	Number of drivers	
	Men	Women
1	9	7
2	7	8
3	3	2
4	0	2
5	1	1

a) What is the mode for:
   i) the men   ii) the women?
b) What is the median for:
   i) the men   ii) the women?
c) Calculate the mean for:
   i) the men   ii) the women.
d) Which group do you think were the better learners and why?

> Find the mean, mode and median for the data given in this question:
> *Using a bar chart, page 289, question 5*
> Which of the averages do you think is the best representative? Why?

Find the range

HD1/L1.4, HD1/L2.4

> ▶ **Range = highest value − lowest value**

The range gives a measure of how spread out the values are in a set of data.

Example

The table gives the marks achieved by two students in assignments.

Assignment	1	2	3	4
Bernie	76%	65%	54%	69%
Matt	71%	58%	45%	75%

The range of Bernie's marks = 76% − 54% = 22%
The range of Matt's marks = 75% − 45% = 30%
Matt's marks are more spread out than Bernie's.
Bernie is the more **consistent** student and Matt is the more **variable**.

Practice

1   The table gives the prices of tins of fruit in a number of supermarkets.

	Penny-Saver	Sam's Store	Low-Price	Super-Food
Apricots	78p	82p	75p	93p
Grapefruit	71p	77p	69p	81p
Peaches	72p	80p	72p	£1.05
Pineapple	85p	94p	90p	89p

a) Find the price range for each type of fruit.
b) Which type of fruit has:
  i)   the most consistent price
  ii) the most variable price?

2   The table below gives the scores of six snooker players in the last ten games they have played.

Player	Scores in the last 10 games									
Alex	38	0	56	83	12	64	21	34	54	71
Cilla	34	56	27	39	42	35	50	76	57	29
Kath	63	19	27	32	60	55	48	85	94	87
Paul	76	58	32	103	67	82	59	147	96	49
Rhona	25	92	147	120	45	62	0	76	15	83
Steve	93	21	47	56	109	74	23	37	88	43

a) Find the mean and range for each player.
b) Who do you think is the best player and why?
c) Who do you think is the most consistent player and why?

> Find the range for each set of data given in these questions:
> **Find the mean, pages 302–303, questions 1, 2, 3, 4, 5**
> **Find the mode and median, page 307, question 1**

3   A student has drawn the bar chart below to show how long he spent on his computer last week. Each time is shown to the nearest five minutes.

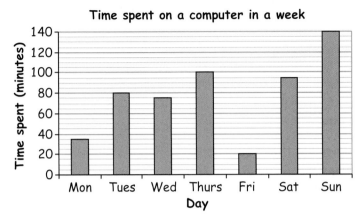

a) Find the mean time per day.
b) Find the range of the times.

 You can also find the range from a frequency table, but **take care to use the correct values**.

Example

The results of the survey carried out in a block of flats are shown again in the table.
The highest number of occupants is 6 and the lowest number is 0.

The range $= 6 - 0 = \mathbf{6}$

Note: When the data is given in a frequency table, **watch out for zero values** in the frequency column.

Number of occupants	Number of flats
0	1
1	6
2	12
3	7
4	3
5	0
6	1

If the last two values in the second column had been the other way round, the highest number of occupants would have been 5 rather than 6.

---

4   A doctor records the number of home visits she makes to patients each day. The results are given in the table.

a) What is the range of the number of visits?

b) What is the modal number of visits?

Number of visits	Number of days
0	0
1	4
2	6
3	10
4	7
5	2
6	1

5 The table shows the ages of the children on a summer camp.

a) What is the range of ages for:
i) boys   ii) girls?
b) What is the modal age for:
i) boys   ii) girls?
c) Use your answers to a) and b) to compare the age distribution of the girls and boys.

Find the range for each set of data given in these questions:
Find the mean, pages 305–306, questions 6, 7, 8, 9, 10
Find the mode and median, pages 308–309, questions 2, 3, 4

Age (years)	Number of:	
	Boys	Girls
8	3	0
9	5	2
10	12	5
11	6	8
12	5	6
13	2	2
14	0	2
15	1	0

6 The chart shows the number of licence penalty points awarded by a court to drivers convicted of driving offences.

a) For the data for men find:
i)   the mode
ii) the median
iii) the mean
iv) the range.
b) Repeat part a) for the women.
c) Compare your results for parts a) and b).

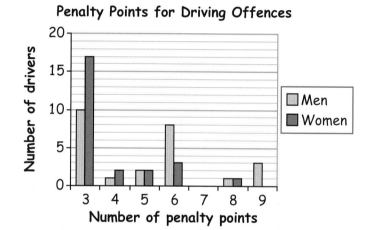

Penalty Points for Driving Offences

Activities

Collect data from your class or from the Internet, newspapers, etc.
Find and compare averages and ranges. Some suggestions are given below:

From your group: ages, number of pets, heights, time spent on homework, etc.
From other sources: temperature or rainfall at two or more places
                      prices of cars of different makes and/or ages
                      prices of houses, electrical equipment, CDs, DVDs, etc.

Use a spreadsheet to work out the MODE, MEDIAN, AVERAGE (mean), MAX – MIN (range)

Handling Data

4 Probability

Show that some events are more likely than others HD2/L1.1

Some things are **certain** to happen, e.g. 10 am is certain to come after 9 am.
Some things are **impossible**, e.g. if you empty a bucket the water won't flow uphill.
Other things may or may not happen – these '**events**' may be **likely** or **unlikely**.

Practice

1 Say whether each of these **events** is **impossible**, **unlikely**, **likely** or **certain**.

a) You will watch television tonight.
b) You will win a prize with the next raffle ticket you buy.
c) You will meet a dinosaur on your way home.
d) It will get dark tonight.
e) A relative will visit you next week.

Activities

1 Toss a penny ten times. Record your results in a table like the one shown below.
After each toss, work out the % of tosses that have been heads.

Toss number	1	2	3	4	5	6	7	8	9	10
Result										
Total number of heads										
% of heads										

Compare your results with those of other students in your group.
Add together the number of heads for all the students in your group together.
Find the % of all the tosses that were heads.

2 Toss a drawing pin ten times. Use a table like the one in Activity 1 to record whether the drawing pin lands point up or point down.
After each toss, work out the % of times the drawing pin has landed point up.
Again, compare your results with those of other students and find the % of tosses that were point up for the whole group.

3 Throw a dice 30 times and record the results in a tally table (see page 285)

Work out the % of throws that gave:

a) five   **b)** an even number

c) eight   **d)** a number less than 7.

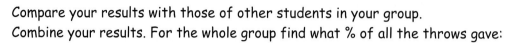

Compare your results with those of other students in your group.

Combine your results. For the whole group find what % of all the throws gave:

a) 5   **b)** an even number   **c)** 8   **d)** a number less than 7.

Put the events a, b, c, d in order of how often each occurred, starting with the event that occurred **least** often.

There are six different possibilities when a dice is thrown: 1, 2, 3, 4, 5, 6.

Assuming the dice is **not biased**, the possibilities are all **equally likely**.

One of the six numbers is a 5. We say there is a **1 in 6 chance** of getting a 5.

Three of the numbers are even numbers, so there is a **3 in 6 chance** of getting an even number.

There is no 8, so there is a **0 in 6 chance** of getting an 8. It is **impossible**.

All of the numbers are less than 7, so there is a **6 in 6 chance** of getting a number less than 7. This event is **certain**.

---

2 a) How many possibilities are there when a coin is tossed?

b) What is the chance of getting a head when you toss a coin?

3 A spinner is numbered 1 to 5 as shown.

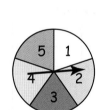

a) When it is spun, what is the chance of getting:

   i) 4   **ii)** an even number   **iii)** an odd number?

b) Which is more likely: an even number or an odd number?

4 A pack of cards is shuffled, then one is picked out at random.

a) How many possibilities are there?

b) What is the chance of getting:

   i) an ace   **ii)** a diamond   **iii)** the ace of diamonds?

5 The letters of the alphabet are written on pieces of paper and put into a hat.
A letter is taken out at random.

a) How many possibilities are there?

b) What is the chance of getting:

   i) Z   **ii)** a vowel (a, e, i, o or u)

   iii) a consonant (a letter that is not a vowel)?

Use fractions, decimals and % to measure probability   HD2/L1.2

Probability is a measure of how likely something is to happen.
If an event is **impossible**, we say the probability is **0**.
If an event is **certain**, we say the probability is **1**.

Other events have probabilities between 0 and 1. The probabilities of such events can be written as fractions, decimals or percentages.
To find the fraction use:

> **Probability of an event** = $\dfrac{\text{Number of ways the event can happen}}{\text{Total number of possibilities}}$

To find the decimal, divide the numerator by the denominator.
To find the %, multiply the decimal by 100 (see page 144).

Example

When a coin is tossed, the probability of a head is $\frac{1}{2}$, **0.5** or **50%**.   **Head is 1 out of 2 possibilities.**
You should get a head for about half, i.e. 50%, of the tosses.
Look back at your results from Activity 1 on page 313 How near were they to 50%?

Example

A set of cards is numbered from 1 to 20. If you take a card at random:

a) the probability of getting 5 is:
   1 in 20   or $\frac{1}{20}$ as a fraction   or   0.05 as a decimal   or   5% as a %

b) the probability of getting an odd number (1, 3, 5, 7, 9, 11, 13, 15, 17, 19) is:
   10 in 20   or $\frac{10}{20} = \frac{1}{2}$ as a fraction   or   0.5 as a decimal   or   50% as a %

c) the probability of getting a number less than 6 (1, 2, 3, 4 or 5): is
   5 in 20   or $\frac{5}{20} = \frac{1}{4}$ as a fraction   or   0.25 as a decimal   or   25% as a %

d) the probability of getting 7 or more
   (7, 8, 9, 10, 11, 12, 13, 14, 15, 16, 17, 18, 19 or 20) is:
   14 in 20   or $\frac{14}{20} = \frac{7}{10}$ as a fraction   or   0.7 as a decimal   or   70% as a %

These probabilities can be shown and compared on a **probability line**:

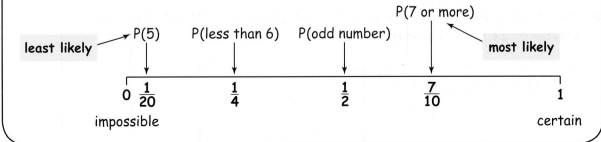

Practice

1 A set of cards is numbered from 1 to 10. Suppose you take a card at random.

  a) Find the probability of each event as a fraction, decimal and %.
  i)   getting a 3   **ii)**   getting an odd number
  iii) getting less than 7   **iv)**   getting 8 or more
  b) Show the results on a probability line.

2 a) A dice is thrown. What is the probability of getting:
  i) 5   **ii)** an even number   **iii)** 8   **iv)** a number less than 7?
  Give each answer as a fraction, a decimal and a %.
  b) Compare the answers with your results from Activity 3 on page 314.

3 This spinner is numbered 1 to 5.

  a) When it is spun, what is the probability of getting:
  i) 4   **ii)** an even number   **iii)** an odd number?
  Give each answer as a fraction, a decimal and a %.
  b) Show the results on a probability line.

4 A pack of cards is shuffled, then one is picked out at random.

  a) Find as fractions (in their simplest form) the probability of getting:
  i)   a king   **ii)** a club   **iii)** the king of clubs
  iv) a red card   **v)**   a red king   **iv)** a jack or queen.
  b) List the events in order of their probabilities, starting with the least likely.

5 Five 2p coins, two 10p coins and a £1 coin are put
into a bag.
A coin is taken out at random.

  a) Find as a fraction the probability that it is:
  i) 2p   **ii)** 10p   **iii)** £1.
  b) Show the results on a probability line.

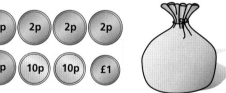

6 In a raffle the tickets are numbered from 1 to 1000.
All the tickets ending in 0 or 5 win a prize. If you buy a ticket, what is the probability that
you will get a prize? Give your answer as a decimal and a %.

7 A person says he was born in a leap year. What is the probability that he was born:

  a) on Christmas Day   **b)** in January   **c)** in June
  d) in February   **e)** before May 1st?

In some situations the possibilities are not equally likely.
Look at this spinner.
There are 3 numbers, but the probability of getting 1 is not $\frac{1}{3}$.
It is $\frac{1}{4}$ because the sector for 1 covers a quarter of the spinner.
If the spinner is spun 40 times, about 10 results should be 1s.

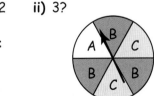

Sometimes you cannot use logic to find probabilities.
For example, suppose you toss a drawing pin. The possibilities for how it lands (point up or point down) are not equally likely. The only way of finding the probability is by doing an experiment like the one you did in Activity 2 on page 313.

Data that has been collected can also be used to find probabilities.

> ### Example
>
> In one year, 304 635 boys and 289 999 girls were born in the UK.
> The total number of babies was 594 634.
>
> This suggests that: the probability that a baby is a boy $= \dfrac{304\,635}{594\,634}$
>
> and: the probability that a baby is a girl $= \dfrac{289\,999}{594\,634}$
>
> Use your calculator to write these fractions as percentages.
> You should find that the probability of a boy is 51% and for a girl 49% (nearest %).
> These are typical values – usually near 50%, with slightly more boys than girls.

8 Look at the spinner shown above.

  a) What is the probability of getting: **i)** 2   **ii)** 3?
  b) Out of 40 spins, how many would you expect to be   **i)** 2   **ii)** 3?

9 **a)** If this spinner is spun, what is the probability of getting:
   i) A   **ii)** B   **iii)** C?
  b) Out of 60 spins, how many would you expect to be:
   i) A   **ii)** B   **iii)** C?

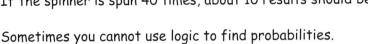

10 The table shows the number of sports accidents
  that needed hospital treatment during one year.
  a) Find the probability that a sports injury
  needing hospital treatment is from each
  category of sport.
  Give your answers as % to 1 dp.
  b) Interpret your answers to part **a)**.

Sport	No. of accidents
Ball sports	562 656
Combat sports	24 197
Wheel sports	38 757
Winter sports	23 410
Animal sports	22 625
Water sports	25 404
Other	43 020
Total	**740 069**

Identify possible outcomes of combined events

HD2/L2.1

Two events are **independent** if the first has no effect on the second.

For example, when a couple have 2 children, the gender of the first child has no effect on the gender of the second.

Using B to mean boy and G to mean girl, all the possibilities can be shown in a table like this ⟶

or

in a tree diagram like this.⟶

Both show that there are 4 possibilities:

BB　boy then another boy
BG　boy then a girl
GB　girl then a boy
GG　girl then another girl.

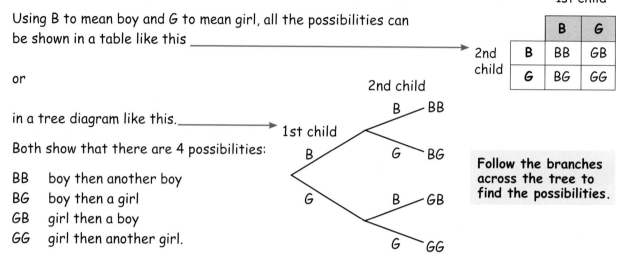

1st child

		B	G
2nd child	B	BB	GB
	G	BG	GG

Follow the branches across the tree to find the possibilities.

If the probability of a boy is about the same as a girl (see page 317), the 4 possibilities for 2 children are also approximately equally likely.

So about a quarter (1 out of 4) of all couples who have 2 children will have 2 boys, a quarter will have 2 girls, and a half (2 out of 4) will have one of each.

Practice

1 The table shows all the possibilities when two dice are thrown. The centre of the table shows some of the total scores:
$1+1=2, 2+1=3, 3+6=9, 6+2=8$

   a) Copy and complete the table.
   b) How many totals are there in the centre of the table?
   c) Write down the probability of each possible total, giving your answers as fractions.
   d) Which total is most likely?

Score on 1st dice

		1	2	3	4	5	6
Score on 2nd dice	1	2	3				
	2						8
	3						
	4						
	5						
	6			9			

2 Draw up another table like the one in question 1. This time multiply the two scores together, instead of adding them. (Multiplying gives the **product** of the scores).
What is the probability that the product is:　**a)** odd　**b)** even?

3 a) Draw a table to show all the possibilities when 2 coins are tossed.
  b) Draw a tree diagram to show all the possibilities when 2 coins are tossed.
  c) What is the probability of getting: **i)** 2 heads   **ii)** the same on each coin?

4 A tree diagram showing the possible genders
of 3 children has been started here.

Copy and complete it to show all the possible
combinations of genders in a set of triplets.

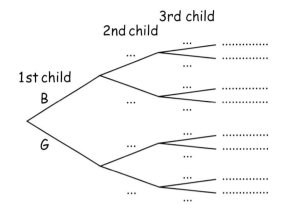

5 a) Draw a tree diagram to show all the possibilities when 3 coins are tossed.
  b) What is the probability of getting:
    i) 3 tails   **ii)** 2 tails and 1 head (in any order)?

6 A gardener has bought some white, purple and yellow
crocus bulbs, but she does not know which is which.

She plants 2 bulbs.

The tree diagram has been started to show all the
possible combinations of colours she could get when
her bulbs flower.

a) Copy and complete the diagram.
  b) Write a list of all the possible combinations of
    colours.

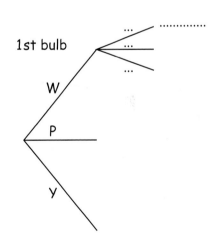

Activities

Use weather statistics to estimate the probability of different types of weather or
temperatures or hours of sunshine.

Find the insurance rates for different ages of car drivers or houses in different
areas – link these to accident and crime rates.

Find data about accident rates from government websites and deduce which
occupations or activities are the most likely to lead to accidents.

Answers

Number

1 Number Value

Numbers up to 1000

a Place value (p 1)

1. a) 165   b) 53   c) 212
   d) 403   e) 990   f) 68
   g) 820   h) 7   i) 217
   j) 902

2. a) Two hundred and thirty-six;
      2 hundreds, 3 tens, 6 units
   b) One hundred and five; 1 hundred, 5 units
   c) Six hundred and seventeen;
      6 hundreds, 1 ten, 7 units
   d) Eighty-two; 8 tens, 2 units
   e) Seven hundred and fifty;
      7 hundreds, 5 tens
   f) Five hundred; 5 hundreds
   g) One hundred and twenty-three;
      1 hundred, 2 tens, 3 units
   h) Ninety; 9 tens
   i) Nine hundred and thirty-eight;
      9 hundreds, 3 tens, 8 units
   j) Three hundred and forty-four;
      3 hundreds, 4 tens, 4 units

3.

Planet	Number of days	Planet	Approx number of years
Earth	365	Jupiter	12
Mercury	88	Neptune	165
Mars	687	Saturn	30
Venus	225	Uranus	84

4. Mercury, Venus, Earth, Mars, Jupiter, Saturn, Uranus, Neptune, Pluto

5. Ceres: five, Eris: five hundred and fifty-seven, Haumea: two hundred and eighty-five, Makemake: three hundred and ten, Pluto: two hundred and forty-eight.

b Count in 10s (p 3)

1. a) 41   91   121
   b) 461   501   531   571
   c) 403   443   483   513

2. a) 165   125   85
   b) 576   546   506   466
   c) 119   99   49   9

c Count in 100s (p 4)

1. a) 225   525   725   925
   b) 163   363   663   863

2. a) 782   582   282
   b) 807   607   307   7

d Odd and even numbers (p 4)

1. a)

$1 \times 2 =$	2	$6 \times 2 =$	12	$11 \times 2 =$	22	$16 \times 2 =$	32
$2 \times 2 =$	4	$7 \times 2 =$	14	$12 \times 2 =$	24	$17 \times 2 =$	34
$3 \times 2 =$	6	$8 \times 2 =$	16	$13 \times 2 =$	26	$18 \times 2 =$	36
$4 \times 2 =$	8	$9 \times 2 =$	18	$14 \times 2 =$	28	$19 \times 2 =$	38
$5 \times 2 =$	10	$10 \times 2 =$	20	$15 \times 2 =$	30	$20 \times 2 =$	40

   b) 0, 2, 4, 6 or 8

2. 8   80   180   280   172   754
   466   300   256   312   202

3. 3   209   77   99   403   127
   501   713   19   471   345

4. a) True. 32 and 180 are even.
   b) False. There cannot be as 435 is odd.
   c) True. 500 is even.

d) Maybe true. 1000 is even but each level may not necessarily have the same number of spaces.

Large numbers

a Place value (p 5)

1

	Thousands					
	M	H	T	U	H	T	U
a				2	0	0	0
b			3	2	0	0	0
c		1	6	5	2	0	0
d		1	0	9	8	7	0
e	4	2	0	0	0	0	0
f			9	0	7	5	0
g				8	0	2	5
h		1	1	9	0	0	0
i	6	2	9	9	0	0	0
j	9	0	0	0	3	0	6

2 a) Forty-five thousand

  b) Three hundred and seventy-five thousand

  c) Six million, seven hundred thousand

  d) Two hundred and five thousand

  e) Seventy thousand, seven hundred and fifty

  f) One million, two hundred thousand

  g) Five thousand and sixty-five

  h) Twenty thousand, three hundred and six

  i) Forty thousand, three hundred and fifty-two

  j) Nine million, six hundred and seventy thousand and eighty

  k) Five million, ninety thousand and four hundred

  l) Three million, five hundred and three thousand, two hundred and forty-one.

3 5 065, 20 306, 40 352, 45 000, 70 750, 205 000, 375 000 1 200 000, 3 503 241, 5 090 400, 6 700 000, 9 670 080

4 a) b)

Year	Estimated population
1100	25 000
1300	100 000
1500	50 000
1600	200 000
1801	1 117 000
1851	2 685 000
1939	8 700 000
2000	7 640 000

  c) i) 1300 and 1500
         1939 and 2000

      ii) 1939

b Greater than > and less than < (p 7)

1 a) $1 < 3$ **b)** $7 > 5$ **c)** $9 > 4$

  d) $0 < 1$

2 a) $2 < 8$ **b)** $6 > 1$ **c)** $18 > 13$

  d) $19 < 23$ **e)** $99 < 101$ **f)** $170 > 159$

3 a) $269 < 270$

  b) $15\,000 > 2500$

  c) $9\,900 > 7500$

  d) $50\,000 < 75\,000$

  e) $9 < 900$

  f) $10\,500 > 10\,499$

  g) $99\,900 < 100\,000$

  h) $4\,010\,000 > 4\,009\,999$

Positive and negative numbers in practical contexts (p 8)

1 a) −5°C, −2°C, 0°C, 1°C, 4°C, 8°C

  b) −10°C, −8°C, −7°C, 6°C, 9°C, 10°C

  c) −18°C, −15°C, −10°C, 0°C, 12°C, 20°C

2 a) 9°C, −1°C

  b) −16°C, −17°C, −14°C

  c) −18°C, −19°C, −17°C

3 a) M Patel   **b)** J Robinson

4 Yes because the account will be £225 overdrawn, £25 more than the agreed overdraft.

Compare numbers of any size (p 9)

1 a) 2 500 000   b) 2 560 000
   c) 1 700 000 000   d) 1 850 000 000
   e) 30 000 000   f) 300 000 000
   g) 1 300 000 000   h) 500 000 000
   i) 250 000

2 a) 2.5 million   b) 5.6 million
   c) 4.75 million   d) 3.8 billion
   e) 2.95 billion   f) 1.05 billion
   g) 20 million   h) 600 million
   i) 34.5 million   or 0.6 billion

3 £22 600 000,   £17 900 000
   £14 100 000,   £14 000 000

4 a) i)   Company B
       ii)   Company C
       iii) Company B
   b) i)   Companies C, E and G
       ii)   Company E

2 Addition and Subtraction

Three-digit numbers

a Add in columns (p 11)

1 a) 379   b) 779   c) 496
   d) 489   e) 780   f) 672
   g) 394   h) 787   i) 500
   j) 318   k) 946   l) 305
   m) 400   n) 931   o) 621
   p) 610

2 a) 989   b) 472   c) 241

3 341 employees
4 440 photocopies
5 350 miles
6 482 boxes
7 £795
8 £926
9 215 tickets

b Add in your head (p 12)

1 a) 57   b) 57   c) 69
   d) 43   e) 72   f) 63
   g) 85   h) 98

2 a) 241   b) 181   c) 398
   d) 387   e) 208   f) 185
   g) 406   h) 246   i) 362
   j) 301   k) 568   l) 538
   m) 574   n) 371   o) 501
   p) 493

c Subtract in columns (p 14)

1 a) 211   b) 321   c) 200
   d) 515   e) 309   f) 227
   g) 70   h) 31   i) 153
   j) 199   k) 179   l) 67
   m) 245   n) 69   o) 299
   p) 501   q) 329   r) 299

2 a) 387   b) 97   c) 698

3 325 envelopes
4 £324
5 127 more men
6 £106
7 163 miles

d Subtract in your head (p 16)

1 a) 23   b) 16   c) 28
   d) 56   e) 19   f) 42
   g) 37   h) 34

2 a) 149   b) 199   c) 149
   d) 622   e) 401   f) 205
   g) 206   h) 199   i) 326
   j) 490   k) 157   l) 699
   m) 199   n) 299   o) 709
   p) 294

e Use addition and subtraction (p 16)

1 £94
2 53p
3 16
4 61
5 £741
6 £385
7 352
8 a) 540
   b) 184
9 682
10 £77

Add and subtract large numbers (p 17)

1 a) 52 359   b) 39 505
   c) 28 595   d) 160 499
   e) 3 740 000   f) 6 527 500
   g) 312 089   h) 212 475
   i) 1 525 500   j) 895 276
   k) 363 250   l) 296 350

2 23 868 miles

3 £47 762

4 1 332 000 people

5 1 734 miles

6 £1 186 400

7 a) £65 150   b) £934 850

8 a) 1 068 181   b) 250 235

Add and subtract numbers of any size (p 19)

1 a) 4.9°C   b) 4.7°C
   c) 4.8°C   d) 0.7°C

2 a) 3.5 million   b) 2.6 million
   c) £1 113 500 million d) £6.5 billion
   e) £2.25 billion   f) £2.41 billion

3 £2.19 million

4 a)

Outlet	Total Profit (£ million)
A	4.7
B	−0.8
C	10.7
D	−4.4
Total	10.2

   b) i)   £0.4 million
       ii)   £0.8 million
       iii) £1.2 million
       iv)   £5.5 million
       v)   £15.1 million
       vi)   £3.6 million (D **lost** £3.6 million more
             than B.)

3 Multiplication and Division

Recall multiplication facts (p 20)

1 a) 15   b) 80   c) 14   d) 16
   e) 25   f) 18   g) 36   h) 42
   i) 81   j) 32   k) 48   l) 63
   m) 64   n) 28   o) 49   p) 54
   q) 48 eggs r) £36   s) 42 days

2 a) i)   2, 4, 6, 8, 0
       ii)   30, 50, 86, 162, 204, 318
   b) i)   5, 0
       ii)   65, 80, 100, 115, 125
   c) They add up to 3, 6 and 9
   d) The digitd in each number add up to 9.
       As the digit in the tens column increases by 1
       the digit in the units column decreases by 1.

3 a) 60   b) 200   c) 360
   d) 350   e) 270   f) 280
   g) 540   h) 720   i) 640
   j) 630   k) 480   l) 810

Multiplication methods (p 23)

1 a) 48   b) 93   c) 88   d) 126
   e) 105   f) 172   g) 260   h) 450
   i) 234   j) 448   k) 304   l) 534
   m) 616   n) 441   o) 485   p) 837

2 60 rings

3 75 plates

4 180 minutes

5 £333

6 161 rooms

7 378 pages

8 425 miles

9 600 grams

Division

a Different ways (p 24)

1 a) 5   b) 5   c) 4   d) 7
   e) 4   f) 5   g) 9   h) 7
   i) 6 r 1   j) 4 r 2   k) 6 r 1   l) 5 r 3
   m) 7 r 1   n) 6 r 3   o) 6 r 6   p) 8 r 3

2 £5 each

3 £8 per length

4 9p per pencil

5 8 tins

6 7 sweets (with 2 sweets left over)

7 6 cars (with 2 spare seats)

8 5 teams (with 4 players left over)

9 7 tables (with 2 spare seats)

10 2 trays left over (they get 6 each)

b Standard method (p 26)

1. a) 24   b) 33   c) 40
   d) 12   e) 15   f) 23
   g) 13   h) 19   i) 29
   j) 29 r 1   k) 18 r 2   l) 12 r 2
   m) 12 r 3   n) 17 r 2   o) 31 r 2
   p) 14 r 5
2. 18 doses
3. 12 patients
4. £17
5. £25
6. 14 metres
7. a) 12 trays   b) 3 plants
8. 14 adults
9. 2 pens

Multiply and divide whole numbers by 10, 100 and 1000

a Multiply by 10, 100 and 1000 in your head (p 27)

1. a) 130   b) 750   c) 1360
   d) 2020   e) 900   f) 1900
   g) 1200   h) 3700   i) 12 900
   j) 12 000   k) 30 300   l) 23 000
   m) 70 000   n) 904 000   o) 750 000
   p) 3 581 000
2. 500 pence
3. 200 ten pence coins
4. 660 millimetres
5. 1 500 centimetres
6. 2 500 envelopes
7. £450
8. 500 disks
9. £2 500
10. 12 000 millimetres
11. 250 000 leaflets
12. 78 000 grams

b Divide by 10, 100 and 1000 in your head (p 29)

1. a) 46   b) 460   c) 46   d) 46
   e) 53   f) 530   g) 53   h) 530
   i) 990   j) 75   k) 57   l) 38
   m) 64   n) 2.3   o) 96.2   p) 2.75

2. £120
3. 15p
4. 72 centimetres
5. 240 metres
6. 2750 boxes
7. 350 boxes
8. 176 calls
9. 15 people
10. a) 18 kilometres
    b) 3.5 kilometres
    c) 0.675 kilometres

Use multiplication and division (p 30)

1. a) 28   b) 30   c) 10   d) 6
   e) 4   f) 63   g) 9   h) 6
   i) 8   j) 8   k) 9   l) 48
2. 10 two pence coins
3. 9 boxes
4. £9
5. £63
6. 9 boxes
7. 8 weeks
8. 40 hours
9. £9 an hour

Recognise numerical relationships

a Multiples (p 31)

1. a) 12, 32, 40
   b) 21, 42, 70
   c) 12, 24
   d) 10, 30
2. a) 6, 12, 18, 24, 30, 36, 42, 48, 54, 60, 66, 72, 78, 84, 90, 96
   b) 8, 16, 24, 32, 40, 48, 56, 64, 72, 80, 88, 96
   c) 9, 18, 27, 36, 45, 54, 63, 72, 81, 90, 99
   d) 11, 22, 33, 44, 55, 66, 77, 88, 99
   e) 20, 40 , 60, 80
3. a) 50, 100, 150, 200, 250, 300, 350, 400, 450, 500, 550, 600, 650, 700, 800, 850, 900, 950
   b) 200, 400, 600, 800
   c) 250, 500, 750
4. Any 5 multiples of 1000 (e.g. 2000, 3000, 4000, 5000, 6000)

b Square numbers (p 32)

1 1, 4, 9, 16, 25, 36, 49, 64, 81, 100
   (diagonal line from top left to bottom right)

2 9, 25, 49, 81, 100

c Use mental arithmetic to multiply multiples of 10 (p 32)

1 a) 90   b) 240   c) 140   d) 450
   e) 480   f) 270   g) 1400   h) 2000
   i) 4200   j) 3500   k) 3600   l) 7200

2 £100

3 800 words

4 £1200

5 300 people

6 2400 people

7 2700 kilograms

8 1500 people

9 3600 seconds

Multiply and divide large numbers

a Multiply large numbers (p 35)

1 a) 1025   b) 1920   c) 2236
   d) 2790   e) 2128   f) 1247
   g) 4368   h) 2508   i) 2520
   j) 3015   k) 2430   l) 3087
   m) 4305   n) 29 704 o) 12 825
   p) 27 360 q) 32 256 r) 12 978
   s) 67 968 t) 27 504

2 £720

3 1680 tins

4 £1134

5 1116 children

6 4080 seats

7 7000 miles

8 9125 kilometres

9 31 025 days old

b Divide large numbers (p 36)

1 a) 985   b) 405   c) 881
   d) 3652   e) 4 210   f) 6 825
   g) 1813 r 4 h) 10 416 r 4
   i) 8375 r 2 j) 8158 r 2
   k) 5555 r 5 l) 20 056 r 1

2 250 people

3 £225

4 3221 tickets

5 4250 boxes

6 5200 samples

7 £22 500

8 £62 500

9 a) 42   b) 52   c) 45   d) 37
   e) 13   f) 25   g) 65   h) 136

10 a) £2040   b) £1250
   c) £2930   d) £3582

Efficient methods of calculating with numbers of any size

a Multiples, factors and prime numbers (p 38)

1 a) 98, 364, 720
   b) 165, 720, 945, 25 245
   c) 165, 243, 720, 945, 25 245
   d) 243, 720, 945, 25 245

2 a) 15   b) 12   c) 6   d) 7

3 a) 3, 6 and 12   b) 3 and 9

4 a) 5   b) 29

5 3

b Calculations with numbers of any size (p 39)

1 a) £300   b) £90   c) £36

2 42 000 statements

3 £150 000

4 128 000 people

5 £1 500 000

6 £300 000

7 £2 900 000

8 6.5 million

9 £375 000 or £0.375 million

10 £2.16 million

4 Rounding and Estimation

Approximation and estimation (numbers up to 1000)

a When to estimate (p 42)

1 a) Yes   b) No   c) Yes   d) No
   e) No   f) Yes   g) Yes   h) No
   i) Yes   j) No

b Round to the nearest 10 and 100 (p 43)

1 a) 40   b) 20   c) 20   d) 90
  e) 140   f) 220   g) 420   h) 990
  i) 410   j) 100   k) 700   l) 310
  m) 240   n) 180   o) 370   p) 710
  q) 370   r) 810   s) 900   t) 280
2 a) 200   b) 400   c) 600   d) 900
  e) 200   f) 400   g) 600   h) 900
  i) 400   j) 200   k) 800   l) 700
  m) 900   n) 1000   o) 200   p) 400
  q) 900   r) 800   s) 300   t) 100

c Use estimates to check calculations (p 45)

1 a) £70 + £30 = £100
     so answer is probably incorrect
  b) £160 − £70 = £90
     so answer is probably correct
  c) £70 + £60 = £130
     so answer is probably correct
  d) £90 ÷ £30 = 3
     so answer is probably incorrect
  e) £85 − £25 = £60 (or £90 − £30 = £60)
     so answer is probably correct
  f) 10 × £25 = £250
     so answer is probably correct
  g) £160 × 3 = £480
     so answer is probably incorrect
  h) £160 − £120 = £40
     so answer is probably incorrect
2 a) 889   b) 126
3 a) £338   b) £736   c) £858

Approximate and estimate with large numbers

a Round to the nearest 1000, 10000, 100000, 1000000 (p 47)

1 a) 5000   b) 7000
  c) 27 000   d) 31 000
  e) 256 000   f) 105 000
  g) 951 000   h) 1 366 000

2 a) 30 000   b) 50 000
  c) 150 000   d) 210 000
  e) 770 000   f) 230 000
  g) 680 000   h) 1 910 000
3 a) 500 000   b) 300 000
  c) 1 500 000   d) 4 600 000
  e) 1 700 000   f) 2 800 000
  g) 2 000 000   h) 12 800 000
4 a) 3 000 000   b) 3 000 000
  c) 7 000 000   d) 16 000 000
  e) 4 000 000   f) 2 000 000
5 a) 3 000 000 unemployed
  b) 7000 job losses
  c) 13 000 (or 12 800) people attend charity event
  d) £2 000 000 lottery win
  e) 4700 (or 5000) new jobs created
  f) 3 000 000 copies of record sold
  g) £21 000 000 profits for company
  h) 280 000 (or 300 000) people to benefit

b Use estimates to check calculations (p 49)
(Note: there are alternative methods.)

1 a) True
     500 000 − 200 000 = 300 000
  b) False
     100 000 × 2 doesn't equal 600 000
  c) True
     130 000 × 3 = 390 000
  d) False
     170 000 − 90 000 doesn't equal nearly 200 000
  e) True
     620 000 + 35 000 = 655 000
  f) True
     500 000 + 150 000 = 650 000
  g) False
     167 000 + 20 000 = 187 000

2 a)

Income	Total £
2976 tickets @ £25 each *Check: Round 2976 to 3000* *25 × 3 = 75 gives 75 000*	74 400
2204 tickets @ £12 each *Check: Round 2204 to 2000* *12 × 2 = 24 gives 24 000*	26 448
1008 progs. @ £3 each *Check: 3 × 1 000 = 3 000*	3024
592 mugs @ £8 each *Check: Round 592 to 600* *8 × 6 = 48 gives 4800*	4736
Total *Check: $\frac{74+26}{100}+3+5$* *= 108 thousand*	108 608

Expenditure	Total £
Hire of venue and insurance	17 750
Publicity and printing	4899
8 technicians @ £489 *Check: Round £489 to £500* *8 × 500 = 40 × 100 = 4000*	3912
50 security staff @ £175 each *Check: Round £175 to £200* *50 × 200 = 100 × 100* *= 10 000*	8750
Performers	0
Total *Check $\frac{18+5}{23}+4+9$* *= 36 thousand*	35 311

b) Income 108 608 −
Expenditure 35 311
Money Raised £73 297
Check: 110 − 35 = 75 thousand

Use estimates to check calculations with numbers of any size (p 50)

1 Between 1935 and 1945
1 585 − 912.3 = 672.7 million
= 670 700 000 increase
Check: 1600 million − 900 million
= 700 million

Between 1945 and 1955
1 585 − 1 181.8 = 403.2 million
= 403 200 000 decrease
Check: 1600 million − 1200 million
= 400 million

Between 1955 and 1965
1 181.8 − 326.8 = 855.0 million decrease
= 855 000 000 decrease
Check: 1200 million − 300 million
= 900 million

Between 1965 and 1975
326.8 − 116.3 = 210.5 million
= 210 500 000 decrease
Check: 300 million − 100 million
= 200 million

Between 1975 and 1985
116.3 − 72 = 44.3 million
= 44 300 000 decrease
Check: 120 million − 70 million
= 50 million

Between 1985 and 1995
114.6 − 72 = 42.6 million
= 42 600 000 increase
Check: 110 million − 70 million
= 40 million

Between 1995 and 2005
164.6 − 114.6 = 50 million
= 50 000 000 increase
Check: 160 million − 110 million
= 50 million

2 a) £12 678 328
b) 3 × £1000 000 + 20 × £60 000
+ 600 × £1000 + 30 000 × £50
+ 600 000 × £10
+ £12 300 000

5 Ratio and Proportion

Work with ratio and proportion

a Ratios (p 52)

1 a) 200 mℓ b) 400 mℓ c) 1000 mℓ = 1 ℓ
2 a) 1500 mℓ or 1.5 litres
  b) 3000 mℓ or 3 litres
  c) No. 800 + 4 800 = 5 600 mℓ

3

	Cement	Sand
Standard mortar	5 kg	25 kg
	10 kg	50 kg
	15 kg	75 kg
Strong mortar	3 kg	9 kg
	12 kg	36 kg
	20 kg	60 kg

b Direct proportion (p 53)

1 a) i)   300 mℓ milk   4 tsp caster sugar
              900 g plain flour   4 tsp dried yeast
              300 mℓ plain yogurt 1 tsp salt
              2 eggs   2 tsp baking powder
                                    4 tbls veg. oil

      ii) 75 mℓ milk   1 tsp caster sugar
              225 g plain flour   1 tsp dried yeast
              75 mℓ plain yogurt $\frac{1}{4}$ tsp salt
              1 small egg $\frac{1}{2}$ tsp baking powder
                                    1 tbls veg. oil

   b) 50 mℓ milk, 150 g plain flour
        50 mℓ plain yogurt

2 a) 8 teaspoons   b) 2 teaspoons
   c) 1 teaspoon

3 a) i)   500 mℓ (or 0.5 litres or $\frac{1}{2}$ litre)
      ii) 200 mℓ (or 0.2 litres or $\frac{1}{5}$ litre)
      iii) 100 mℓ (or 0.1 litres or $\frac{1}{10}$ litre)
   b) i)   2 litres water : 5000 g or 5 kg plaster
      ii) 4 litres water : 10 000 g or 10 kg
            plaster

Calculate with ratios

a Simplify ratios and find amounts (p 55)

1 a) 2 : 1   b) 1 : 5   c) 14 : 1
   d) 1 : 5   e) 3 : 1
2 a) 1 : 4   b) i) 12 litres
      ii) 100 millilitres
3 a) 5 : 1   b) i) 250 litres
      ii) 60 litres
4 a) 2 : 3   b) 2 : 9   c) 4 : 1 : 6
   d) 2 : 3 : 5   e) 33 : 18 : 10
5 a) i) 1 : 4   ii) 400 mℓ
   b) i) 2 : 3   ii) 600 mℓ

6 a) i)   50 : 450 : 500 = 1 : 9 : 10
      ii) 720 mℓ moss, 800 mℓ leaf green
      iii) 200 mℓ white, 2000 mℓ (2 ℓ) leaf green
   b) i)   200 : 300 : 500 = 2 : 3 : 5
      ii) 120 mℓ beige, 200 mℓ chocolate brown
      iii) 120 mℓ white, 300 mℓ chocolate brown
      iv) 300 mℓ white, 450 mℓ beige
7 a) 1 : 25   b) 4 : 1   c) 1 : 20
   d) 8 : 1   e) 1 : 50   f) 5 : 1
8 a) 1 : 5   b) 1 litre   c) 250 ml
9 a) 1 : 2   b) 400 g   c) $1\frac{1}{4}$ kg
10 a) 1 : 50   b) 12 mm
   c) 4500 mm = 4.5 m
11 a) 3 : 100   b) 5 : 4   c) 1 : 14
   d) 13 : 4   e) 1 : 50 000   f) 5 : 1
   g) 1 : 2 : 9   h) 4 : 2 : 2 : 3
12 a) 50 : 4 000 = 1 : 80
   b) 3200 mℓ or 3.2 ℓ   c) 30 mℓ
13 a) 1500 g : 400 g : 200 g = 15 : 4 : 2
   b) i)   150 g cereal, 40 g nuts
      ii) 160 g nuts, 80 g seeds
14 a) 500 m   b) 1 km

b Make a total amount (p 59)

1 15 black   30 white
2 a) 300 g oats   100 g fruit
   b) 750 g oats   250 g fruit
3 a) 600 g white   150 g brown
   b) 800 g white   200 g brown
4 a) 200 mℓ orange squash, 800 mℓ water
   b) 300 mℓ orange squash, 1200 mℓ water
5 a) 1 litre Tangerine, 1 litre White,
        3 litres Buttercup
   b) 200 mℓ Tangerine, 200 mℓ White,
        600 mℓ Buttercup
6 a) 12 : 5 : 3
   b) i)   480 g cereal, 200 g fruit, 120 g nuts
      ii) 1080 g cereal, 450 g fruit, 270 g nuts
7 a) 15 000 : 45 000 : 150 000
        1 : 3 : 10
   b) 30 Afro Caribbean, 90 Asian and 300 white
        people

8 a) $4000 : 10\,000 : 16\,000 = 2 : 5 : 8$

b) A £6000, B £15 000, C £24 000

c) £75 000

d) £187 500

e) A £41 000, B £102 500, C £164 000

c Use ratios to compare prices (p 61)

1 a) Pack of 9   b) 600 mℓ bottle

c) 375 g box   d) 100 envelopes

e) 24 tins

2 a) 0.5 tonne @ £27.50 = £55 per tonne

1 tonne @ £52.50 = £52.50 per tonne

5 tonnes @ £240 = £48 per tonne

25 kg @ £2.75 = £110 per tonne

40 kg @ £4.50 = £112.50 per tonne

5 kg @ £5.50 = £500 per tonne

b) £11   c) 24p

6 Using Algebra

Use formulae and number patterns (p 63)

1 a) multiples of 4

b) Cost = £4 × number of mugs

c) £24

d) Continue pattern (+ 4) giving £20, £24

2 a)

Number of friends in the group	2	3	4	5
Amount each person pays (£)	30	20	15	12

b) Amount each person pays = £60 ÷ number of friends in the group

c) £5

d) £5 × 12 = £60

3 a)

Length of fencing (metres)	3	4	5	6
Total cost of the fencing and gate (£)	66	78	90	102

b) Total cost = £12 × length of fence + £30

c) £150

d) Continue pattern (+ 12) giving £114, £126, £138, £150

4 a)

Number of weeks that Ian has been saving	1	2	3	4
Amount still to be saved (£)	240	200	160	120

b) Each number is 40 less than the previous number.

c) £0 Ian has now saved what he needs.

d) Continue pattern (−40) giving £80, £40, £0

5 a)

Number of hours taken	1	2	3	4
Total cost (£)	67	92	117	142

b) Each number is 25 more than the previous number

c) Total cost = £42 + £25 × number of hours

d) £242

e) Continue pattern (+ 25) giving £167, £192, £217, £242

Evaluate algebraic expressions and make substitutions in formulae (p 65)

1 a) i) Total cost = number of delegates × £15 + £75

ii) $T = 15d + 75$

b) i) £225   ii) £375   iii) £825

2 a) $C = 50w + 25$

b) i) 125 mins (or 2 hrs 5 mins)

ii) 150 mins (or $2\frac{1}{2}$ hours)

iii) 65 mins (or 1 hour 5 mins)

3 a) $F = 1.6m + 2$

b) i) £10   ii) £21.20   iii) £30

4 a) $P = 7.5a + 280$

b) i) £310   ii) £355   iii) £392.50

5 a) i) £40   ii) £80   iii) £430

b) £0.05 is the cost per extra card i.e. 5p

c) The overall cost per card decreases

6 a) 19   b) 0   c) 20   d) 16

e) 16   f) 6   g) 24   h) 22

i) 17

7 a) 50 metres   b) 27 metres

8 a) $P = 9t + 5n$

b) i) £87   ii) £160   iii) £420

9 a) £81 000   b) £ 114 000
  c) £195 000   d) £278 850
10 120 volts
11 a) 60 miles per hour
   b) 0.4 km per min
       (or 400 m per min or 24 km per hour)
12 a) $C = 10 + n(E + 28)$ or $C = n(E + 28) + 10$
   b) i) £109   ii) £185   iii) £298.75
13 a) 18 cm^2 b) 68 cm^2
   c) 10.5 cm^2 d) 285 cm^2
14 a) 100°C   b) 25°C

7 Mixed Operations and Calculator Practice

Solve problems (p 69)

1 576
2 £156
3 a) 786   b) 22
4 a) £468   b) £533
5 a) £895   b) £447.50
6 a) £153   b) £571
7 a) 270   b) 9
8 a) 5 × 5 litre tins and 1 × 2 litre tin
  b) 7 × 5 litre tins and 2 × 2 litre tins
  c) 25 × 5 litre tins, 1 × 2 litre tin and
      1 × 1 litre tin
9 a) 4 packs of 10
  b) 1 pack of 50 and 1 pack of 25
  c) 5 packs of 50 and 1 pack of 25
  d) 7 packs of 50 and 1 pack of 10
10 a) £51   b) £175
11 a) 3 packs of 2 (one pack is free) £6
   b) 4 packs of 2 (one pack is free) £9
   c) 1 pack of 10 £10
   d) 6 packs of 2 (2 packs are free) £12
12 a) £65   2 × £25 and 1 × £15
   b) £100   1 × £25 and 5 × £15
   c) £90   1 × £75 and 1 × £15
   d) £200   8 × £25
   e) £180   2 × £75 and 2 × £15
   f) £190   2 × £75, 1 × £25, 1 × £15

Calculate efficiently (p 72)

1 a) Part-time receptionist by £320 a year
  b) Warehouse assistant by £680 a year
2 a) i) £488   ii) £1324   iii) £1776
  b) £292
  c) £972
  d) i) 12 nights   ii) 6 nights

Calculate efficiently with numbers of any size (p 73)

1 55.6°C
2 £0.91 million
3 25th Feb £582.81 out
   26th Feb £817.11 out
   27th Feb £2778.86 in
   28th Feb £2097.68 out
4 1st quarter £20.05 in credit
   2nd quarter £18.65 in debit
   3rd quarter £16.75 in debit
   4th quarter £10.10 in credit
5 b)

Year	Profit £
2004	1 200 000
2005	700 000
2006	−300 000
2007	−550 000
2008	−100 000
2009	−1 000 000
2010	−500 000
2011	0
2012	500 000

  c) i) £2 200 000   ii) −£50 000

8 Fractions

Understand fractions

a Fractions in words, numbers and sketches (p 75)

1 a) $\frac{1}{3}$ b) $\frac{2}{5}$ c) $\frac{3}{8}$ d) $\frac{5}{12}$
2 a) one half   b) two thirds
  c) three quarters   d) three fifths
  e) five eighths   f) seven ninths

3 a) $\frac{1}{4}$, one quarter   **b)** $\frac{1}{5}$, one fifth

  c) $\frac{1}{9}$, one ninth   **d)** $\frac{5}{6}$, five sixths

  e) $\frac{2}{3}$, two thirds   **f)** $\frac{5}{8}$, five eighths

  g) $\frac{4}{7}$, four sevenths   **h)** $\frac{7}{12}$, seven twelfths

  i) $\frac{4}{9}$, four ninths

b Shade fractions (p 76)

1 b, c

2 b, d

3 a)   **b)**   **c)**
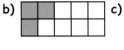

d)   **e)**   **f)**

or alternatives with the same number of sections shaded.

4 Fractions shown on students' own shapes

c Read about fractions (p 77)

1 half $\frac{1}{2}$, one third $\frac{1}{3}$

2 quarter $\frac{1}{4}$, three quarters $\frac{3}{4}$, one and a half $1\frac{1}{2}$, half $\frac{1}{2}$

3 two thirds $\frac{2}{3}$, half $\frac{1}{2}$

4 a third $\frac{1}{3}$, three quarters $\frac{3}{4}$, a fifth $\frac{1}{5}$, a quarter $\frac{1}{4}$, two thirds $\frac{2}{3}$, three and a half $3\frac{1}{2}$, two and a half $2\frac{1}{2}$

Equivalent fractions

a Use sketches (p 78)

1 a) $\frac{1}{2} = \frac{2}{4} = \frac{3}{6}$ **b)** $\frac{1}{5} = \frac{2}{10}$

  c) $\frac{3}{4} = \frac{6}{8} = \frac{9}{12}$ **d)** $\frac{3}{4} = \frac{12}{20}$

2 Students' own sketches

b Use numerators and denominators (p 79)

1 a) True   **b)** True   **c)** False   **d)** True

  e) True   **f)** False   **g)** False   **h)** True

2 a) 15   **b)** 24   **c)** 4   **d)** 8

  e) 10   **f)** 3   **g)** 2   **h)** 4

3 a) Yes   **b)** No, $\frac{1}{5}$

4 Any other 6 fractions equal to $\frac{1}{2}$ (eg $\frac{4}{8}$, $\frac{5}{10}$, $\frac{6}{12}$, $\frac{7}{14}$, $\frac{8}{16}$, $\frac{9}{18}$)

5 a) any 3 other fractions equal to $\frac{3}{8}$

  b) any 3 other fractions equal to $\frac{2}{5}$

  c) any 3 other fractions equal to $\frac{4}{11}$

  d) any 3 other fractions equal to $\frac{7}{9}$

6 Yes, $\frac{12}{30} = \frac{2}{5}$

7 2, 4, 5, 7, 11, 15, 20, 100

8 a) $\frac{6}{18}$, $\frac{4}{12}$, $\frac{5}{15}$, $\frac{2}{6}$, $\frac{3}{9}$ **b)** $\frac{6}{9}$, $\frac{8}{12}$, $\frac{4}{6}$, $\frac{12}{18}$, $\frac{10}{15}$

9 No, $\frac{15}{20} = \frac{3}{4}$ not $\frac{3}{5}$

10 a)   **b)**

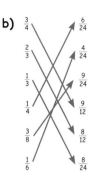

c Write quantities as fractions (p 81)

1 a) $\frac{1}{2}$ cm   **b)** $\frac{2}{5}$ cm   **c)** $\frac{3}{5}$ cm

2 a) $\frac{1}{2}$ kg   **b)** $\frac{3}{4}$ kg   **c)** $\frac{1}{10}$ kg

  d) $\frac{1}{5}$ kg   **e)** $\frac{3}{5}$ kg

3 a) $\frac{1}{200}$ km   **b)** $\frac{1}{20}$ km   **c)** $\frac{1}{2}$ km

  d) $\frac{1}{10}$ km   **e)** $\frac{3}{20}$ km

4 a) $\frac{1}{50}$ ℓ **b)** $\frac{1}{20}$ ℓ **c)** $\frac{1}{5}$ ℓ

  d) $\frac{1}{2}$ ℓ **e)** $\frac{2}{5}$ ℓ

5 a) $\frac{1}{5}$ m   **b)** $\frac{1}{2}$ m   **c)** $\frac{9}{10}$ m

  d) $\frac{4}{5}$ m   **e)** $\frac{1}{20}$ m   **f)** $\frac{1}{10}$ m

  g) $\frac{1}{5}$ m   **h)** $\frac{2}{5}$ m   **i)** $\frac{3}{5}$ m

  j) $\frac{1}{50}$ m

Compare fractions and mixed numbers

a Use fraction walls (p 82)

1 a) 2   **b)** 4   **c)** 8   **d)** 6

  e) 12   **f)** 6   **g)** 10   **h)** 14

2 a) $\frac{5}{16}$, $\frac{3}{8}$, $\frac{7}{16}$ **b)** $\frac{1}{2}$, $\frac{9}{16}$, $\frac{5}{8}$ **c)** $\frac{3}{4}$, $\frac{13}{16}$, $\frac{7}{8}$

3 a) 5, 10   **b)** 5   **c)** 2   **d)** 4

  e) 6   **f)** 10

4 $\frac{9}{10}$, $\frac{4}{5}$, $\frac{7}{10}$, $\frac{3}{5}$, $\frac{1}{2}$, $\frac{2}{5}$, $\frac{3}{10}$, $\frac{1}{5}$, $\frac{1}{10}$

b Compare fractions (p 83)

1 a) $\frac{1}{8}$ b) $\frac{1}{10}$ c) $\frac{1}{20}$ d) $\frac{1}{12}$

2 a) $\frac{1}{20}$ b) $\frac{1}{17}$ c) $\frac{1}{5}$ d) $\frac{1}{11}$

3 $\frac{1}{20}, \frac{1}{15}, \frac{1}{12}, \frac{1}{10}, \frac{1}{6}, \frac{1}{5}, \frac{1}{4}, \frac{1}{3}, \frac{1}{2}$

4 $\frac{1}{7}, \frac{1}{8}, \frac{1}{9}, \frac{1}{11}, \frac{1}{13}, \frac{1}{14}, \frac{1}{16}, \frac{1}{17}, \frac{1}{21}$

5 a) $\frac{3}{4}$ b) $\frac{1}{2}$ c) $\frac{2}{5}$ (with sketches)

6 a) $\frac{1}{2}$ b) $\frac{3}{5}$ c) $\frac{3}{4}$ (with sketches)

7 Meera is wrong: $\frac{3}{7}$ is less than $\frac{1}{2}$

c Improper fractions and mixed numbers (p 84)

1 a) $1\frac{1}{2} = \frac{3}{2}$ b) $1\frac{4}{5} = \frac{9}{5}$
   c) $2\frac{3}{10} = \frac{23}{10}$ d) $3\frac{5}{6} = \frac{23}{6}$
   e) $1\frac{7}{9} = \frac{16}{9}$ f) $2\frac{3}{8} = \frac{19}{8}$

2 Mixed numbers and improper fractions on students' own shapes

3 a) Mixed numbers and fractions on students' own shapes
   b) $\frac{5}{8}, \frac{3}{4}, \frac{7}{8}, 1\frac{1}{4}, 1\frac{3}{8}, 1\frac{1}{2}, 2\frac{1}{2}$

d Find or estimate fractions (p 85)

1 a) i) $\frac{1}{2}$ ii) $\frac{1}{2}$ b) i) $\frac{1}{3}$ ii) $\frac{2}{3}$
   c) i) $\frac{3}{4}$ ii) $\frac{1}{4}$

2 a) $\frac{4}{5}$ full   b) $\frac{3}{5}$ full   c) $\frac{2}{5}$ full

3 a) $\frac{1}{2}$ full   b) $\frac{2}{3}$ full   c) $\frac{3}{4}$ full

4 a) $\frac{3}{5}$ left   b) $\frac{1}{3}$ left   c) $\frac{1}{5}$ left

5 a) i) $\frac{1}{4}$ ii) $\frac{3}{4}$ b) i) $\frac{1}{3}$ ii) $\frac{2}{3}$
   c) i) $\frac{2}{3}$ ii) $\frac{1}{3}$ d) i) $\frac{5}{6}$ ii) $\frac{1}{6}$

Find a fraction of something

a By dividing when the numerator is 1 (p 87)

1 a) £28   b) 24 cm   c) £156
   d) £250   e) £204   f) £91
   g) 72 m   h) £286

2 12

3 a) £15   b) £30

4 a) 14   b) 56

5 15 litres

6 £85

7 a) 3500   b) 17 500

8 a) 125 g   b) 875 g

b Finding more than one part (p 88)

1 a) £630   b) £104   c) 30 m
   d) 150 g   e) 350 cm   f) £460
   g) £1800   h) 1460 kg

2 240

3 18

4 £33 750

5 31 050

6 165

7 £5750

8 360 000

c Using other methods (p 89)

1 a) £1140   b) 42 kg   c) 324 m

2 a) Students' own sketches
   b) i) £480   ii) £120   iii) £600

3 a) Students' own sketches
   b) i) 64 m   ii) 32 m   iii) 96 m

4 a) Students' own sketches
   b) i) £3600   ii) £2400   iii) £6000
   c) 1200

d Scale up and down (p 90)

1 a) 180 g potatoes, 300 g leeks, 500 mℓ milk, 100 mℓ stock
   b) 360 g potatoes, 600 g leeks, 1 ℓ or 1000 mℓ milk, 200 mℓ stock

2 a) $\frac{3}{5}$ b) 300 g nuts, 150 g breadcrumbs, 180 g tomatoes, 3 onions

3 a) $\frac{4}{5}$ b) 240 g flour, 128 g sugar, 112 g butter, 48 g ground almonds

4 a) 3 loaves, 600 g butter, 6 tomatoes, $1\frac{1}{2}$ cucumbers, 300 g pate, 225 g chicken, 375 g cheese, $\frac{3}{4}$ lettuce
   b) 5 loaves, 1 kg or 1000 g butter, 10 tomatoes, $2\frac{1}{2}$ cucumbers, 500 g pate, 375 g chicken, 625 g cheese, $1\frac{1}{4}$ lettuce

Put fractions in order (using a common denominator) (p 91)

1 a) $\frac{2}{3}$ b) $\frac{7}{9}$ c) $\frac{9}{10}$

2 a) $\frac{2}{5}, \frac{3}{5}, \frac{4}{5}$ b) $\frac{1}{7}, \frac{2}{7}, \frac{4}{7}, \frac{5}{7}$
   c) $\frac{1}{9}, \frac{2}{9}, \frac{4}{9}, \frac{5}{9}, \frac{8}{9}$

3 a) $\frac{1}{4}$ b) $\frac{3}{5}$ c) $\frac{3}{4}$ d) $\frac{4}{7}$

4 a) $\frac{3}{8}, \frac{1}{2}, \frac{5}{8}$ b) $\frac{1}{4}, \frac{3}{10}, \frac{2}{5}$
   c) $\frac{5}{16}, \frac{1}{2}, \frac{3}{4}, \frac{7}{8}$ d) $\frac{2}{3}, \frac{3}{4}, \frac{5}{6}, \frac{11}{12}$
   e) $\frac{3}{7}, \frac{1}{2}, \frac{5}{7}, \frac{3}{4}$ f) $\frac{1}{4}, \frac{1}{3}, \frac{3}{8}, \frac{5}{12}, \frac{1}{2}$

5 a) $\frac{4}{7}, \frac{1}{2}, \frac{2}{7}$ **b)** $\frac{5}{6}, \frac{2}{3}, \frac{5}{9}$
c) $\frac{3}{4}, \frac{2}{3}, \frac{1}{2}, \frac{4}{9}$ **d)** $\frac{9}{10}, \frac{13}{15}, \frac{4}{5}, \frac{2}{3}$
e) $\frac{5}{8}, \frac{7}{12}, \frac{5}{9}, \frac{4}{9}$ **f)** $\frac{13}{15}, \frac{11}{15}, \frac{7}{10}, \frac{2}{3}, \frac{3}{5}, \frac{2}{9}$

6 a) $\frac{3}{4}, 1\frac{1}{2}, 1\frac{3}{5}, 2\frac{1}{3}, 2\frac{2}{5}$
b) $\frac{4}{7}, \frac{5}{8}, 1\frac{7}{12}, 1\frac{3}{4}, 1\frac{5}{6}$
c) $\frac{4}{9}, \frac{5}{7}, 1\frac{3}{4}, 2\frac{3}{7}, 2\frac{1}{2}, 2\frac{2}{3}$
d) $\frac{1}{4}, 1\frac{5}{8}, 1\frac{1}{3}, 1\frac{5}{6}, 1\frac{11}{12}, 2\frac{1}{8}$

Write one number as a fraction of another

a Simplest form (p 93)

1 $\frac{2}{5}$ **2** $\frac{3}{4}$ **3** $\frac{1}{3}$ **4** $\frac{2}{5}$ **5** $\frac{4}{5}$
6 $\frac{2}{3}$ **7** $\frac{1}{2}$ **8** $\frac{3}{4}$ **9** $\frac{1}{3}$ **10** $\frac{3}{4}$
11 $\frac{1}{5}$ **12** $\frac{3}{10}$ **13** $\frac{2}{3}$ **14** $\frac{3}{7}$ **15** $\frac{3}{5}$

b Write one quantity as a fraction of another (p 94)

1 a) $\frac{1}{4}$h   **b)** $\frac{1}{3}$h   **c)** $\frac{4}{5}$h
2 a) $\frac{1}{10}$kg   **b)** $\frac{3}{4}$kg   **c)** $\frac{2}{5}$kg   **d)** $\frac{16}{25}$kg
3 a) $\frac{3}{10}$cℓ **b)** $\frac{1}{2}$cℓ **c)** $\frac{1}{5}$cℓ **d)** $\frac{2}{5}$cℓ **e)** $\frac{3}{5}$cℓ
4 a) $\frac{1}{20}$kg   **b)** $\frac{1}{2}$g   **c)** $\frac{1}{4}$kg   **d)** $\frac{4}{5}$g   **e)** $\frac{1}{5}$kg
5 a) $\frac{1}{2}$lb   **b)** $\frac{1}{4}$lb   **c)** $\frac{3}{4}$lb   **d)** $\frac{3}{8}$lb
6 $\frac{2}{3}$ **7** $\frac{1}{5}$ **8** $\frac{3}{5}$ **9** $\frac{9}{16}$

10 a)

	Men	Women	Total
Car	240	300	**540**
Bus	120	270	**390**
Train	80	100	**180**
Cycle	15	10	**25**
Walk	25	40	**65**
Total	**480**	**720**	**1200**

b) i) Men: Car $\frac{1}{2}$ Bus $\frac{1}{4}$ Train $\frac{1}{6}$
Cycle $\frac{1}{32}$ Walk $\frac{5}{96}$
ii) Women: Car $\frac{5}{12}$ Bus $\frac{3}{8}$ Train $\frac{5}{36}$
Cycle $\frac{1}{72}$ Walk $\frac{1}{18}$
c) i) $\frac{2}{5}$ **ii)** $\frac{3}{5}$

c Estimate one quantity as a fraction of another (p 96)

1 a) $\frac{1}{4}$ **b)** $\frac{3}{4}$
2 a) i) $\frac{5}{6}$ **ii)** $\frac{4}{5}$
b) both are 200, giving 1
3 a) i) $\frac{3}{4}$ **ii)** $\frac{2}{3}$
b) aii

4 a) i) $\frac{1}{10}$ **ii)** $\frac{1}{60}$ **iii)** $\frac{1}{30}$ **iv)** $\frac{7}{8}$
b) i) $\frac{5}{8}$ **ii)** $\frac{3}{8}$
5 a) ii) $\frac{3}{4}$ **b)** $\frac{1}{4}$
Note: Other answers are possible.
6 a) 237 100 000
b) i) $\frac{1}{6}$ **ii)** $\frac{1}{3}$ **iii)** $\frac{3}{8}$ **iv)** $\frac{1}{12}$
7 Paper $\frac{3}{10}$, Glass $\frac{1}{7}$, Compost $\frac{2}{7}$, Metal $\frac{1}{9}$
Textiles $\frac{1}{70}$, Cans $\frac{1}{100}$, Other $\frac{1}{10}$

Add and subtract fractions

a Add and subtract fractions with the same denominator (p 98)

1 $\frac{4}{5}$ **2** $\frac{2}{5}$ **3** 1   **4** $\frac{2}{3}$
5 $\frac{5}{7}$ **6** $\frac{3}{7}$ **7** $\frac{1}{2}$ **8** $\frac{2}{3}$
9 $\frac{4}{5}$ **10** 1   **11** $\frac{2}{5}$ **12** $\frac{1}{2}$
13 a) $\frac{3}{5}$ **b)** $\frac{2}{5}$ **14 a)** $\frac{3}{8}$ **b)** $\frac{1}{4}$

b Add and subtract fractions with different denominators (p 99)

1 $\frac{7}{10}$ **2** $\frac{1}{2}$ **3** $\frac{5}{6}$ **4** $\frac{1}{3}$
5 $\frac{11}{12}$ **6** $\frac{7}{20}$ **7** $\frac{1}{8}$ **8** $\frac{8}{9}$
9 $\frac{4}{15}$ **10** $\frac{19}{21}$ **11** $\frac{11}{15}$ **12** $\frac{3}{10}$
13 $\frac{3}{4}$ **14** $\frac{9}{40}$ **15** $\frac{29}{30}$ **16** $\frac{1}{18}$
17 $\frac{17}{36}$ **18** $\frac{3}{20}$ **19** 0   **20** $\frac{7}{8}$
21 a) $\frac{3}{4}$ **b)** $\frac{1}{4}$

c Improper fractions and mixed numbers (p 100)

1 $1\frac{4}{5}$ **2** $2\frac{3}{5}$ **3** $4\frac{1}{3}$ **4** $5\frac{2}{3}$
5 $3\frac{3}{4}$ **6** $4\frac{1}{2}$ **7** $6\frac{1}{2}$ **8** $1\frac{3}{7}$
9 $5\frac{1}{4}$ **10** $2\frac{5}{6}$ **11** $6\frac{1}{3}$ **12** $6\frac{3}{5}$
13 $2\frac{7}{8}$ **14** $3\frac{7}{10}$ **15** $4\frac{4}{9}$ **16** $1\frac{1}{3}$
17 $1\frac{1}{7}$ **18** $1\frac{5}{8}$ **19** $1\frac{1}{6}$ **20** $1\frac{1}{2}$
21 $1\frac{9}{28}$ **22** $1\frac{1}{5}$ **23** $2\frac{5}{24}$

d Add and subtract mixed numbers (p 101)

1 $\frac{3}{2}$ **2** $\frac{5}{2}$ **3** $\frac{7}{2}$ **4** $\frac{5}{3}$
5 $\frac{8}{3}$ **6** $\frac{11}{5}$ **7** $\frac{19}{5}$ **8** $\frac{13}{6}$
9 $\frac{11}{6}$ **10** $\frac{19}{8}$ **11** $\frac{17}{7}$ **12** $\frac{47}{8}$
13 $\frac{49}{10}$ **14** $\frac{31}{9}$ **15** $\frac{44}{7}$ **16** $2\frac{3}{4}$
17 $\frac{17}{20}$ **18** $5\frac{1}{3}$ **19** $1\frac{5}{12}$ **20** $8\frac{1}{8}$
21 $2\frac{5}{6}$ **22** $\frac{19}{20}$ **23** $6\frac{1}{10}$ **24** $3\frac{13}{15}$
25 $6\frac{1}{14}$ **26** $7\frac{1}{4}$ **27** $1\frac{1}{8}$
28 3 hours   **29** $4\frac{1}{4}$ pounds
30 a) $15\frac{1}{2}$ hours   **b)** $1\frac{1}{4}$ hours
c) $2\frac{3}{4}$ hours

Multiplying fractions (p 103)

1 $\frac{1}{10}$ 2 $\frac{1}{21}$ 3 $\frac{2}{9}$ 4 $\frac{4}{7}$

5 $\frac{6}{11}$ 6 $\frac{1}{10}$ 7 $\frac{1}{2}$ 8 $\frac{7}{10}$

9 Yes   10 $\frac{1}{3} \times \frac{1}{4} = \frac{1}{12}$ and $\frac{1}{2} \times \frac{1}{6} = \frac{1}{12}$

11 $1\frac{2}{3}$ 12 $1\frac{4}{5}$ 13 $1\frac{1}{4}$ 14 $\frac{2}{3}$

15 3   16 $4\frac{1}{2}$ 17 $10\frac{1}{2}$ 18 $5\frac{5}{7}$

19 $1\frac{2}{3}$ kg

20 a) $2\frac{2}{5}$ m b) $1\frac{3}{5}$ m   c) $\frac{4}{5}$ m $\frac{1}{6} \times \frac{24}{5} = \frac{24}{30} = \frac{4}{5}$

Dividing fractions (p 105)

1 3   2 $\frac{1}{2}$ 3 4   4 2

5 $\frac{3}{4}$ 6 $\frac{7}{10}$ 7 $\frac{5}{6}$ 8 $1\frac{11}{24}$

9 8

10 $\frac{6}{9} = \frac{2}{3}$ (cancelling by 3) and $\frac{2}{3} \div \frac{1}{9} = \frac{2}{3} \times \frac{9}{1} = \frac{18}{3} = 6$

11 10   12 $\frac{1}{2}$ 13 $3\frac{3}{4}$ 14 $\frac{5}{8}$

15 $1\frac{17}{18}$ 16 $1\frac{5}{7}$ 17 $\frac{2}{3}$ 18 $2\frac{19}{28}$

19 6   20 8

9 Decimals

Understand decimals in context (up to 2 decimal places) (p 107)

1 a) 0.4 cm   b) 2.5 cm   c) 4.1 cm

   d) 5.6 cm   e) 7.8 cm   f) 10.3 cm

   g) 12.5 cm   h) 15.2 cm   i) 17.9 cm

   j) 20.6 cm   k) 21.4 cm   l) 23.7 cm

2 a) 2.5 cm, 2.4 cm, 2.1 cm, 1.9 cm, 1.8 cm

   b) 4.5 cm, 4.3 cm, 4.1 cm, 3.9 cm, 3.8 cm, 3.4 cm

   c) 11.1 cm, 10.7 cm, 10.6 cm, 10.5 cm, 9.9 cm, 9.6 cm

   d) 6.4 m, 6.2 m, 6.1 m, 5.8 m, 5.7 m, 5.5 m, 5.3 m, 5.1 m

   e) 2.9 km, 2.3 km, 2.1 km, 1.8 km, 1.7 km, 1.4 km, 1.2 km, 0.9 km

3 a) 100p   b) 218p   c) 109p

   d) 407p   e) 62p   f) 824p

   g) 570p   h) 904p   i) 1017p

   j) 1640p

4 a) £2.13   b) £4.02   c) £0.85

   d) £3.50   e) £1.03   f) £6.00

   g) £7.25   h) £9.37   i) £10.00

   j) £24.09

5 a) £1.02   £1.19   £1.21   £1.37

     £1.46   £1.87   £2.07

   b) £47.20   £56.78   £63.02

     £63.21   £66.18   £74.02

   c) £10.07   £10.08   £10.70

     £11.01   £11.17   £11.91

6 a) 22 cm   b) 107 cm   c) 173 cm

   d) 74 cm   e) 133 cm   f) 112 cm

   g) 115 cm   h) 67 cm   i) 260 cm

   j) 540 cm

7 a) 1.54 m   b) 3.15 m   c) 0.94 m

   d) 1.2 m   e) 7.62 m   f) 4.02 m

   g) 0.7 m   h) 2.5 m   i) 11.34 m

   j) 22.3 m

8 a) 1.5 m   1.55 m   1.63 m   1.67 m

     1.72 m   1.74 m   1.8 m

   b) 1.26 m   1.28 m   1.3 m   1.31 m

     1.32 m   1.39 m   1.4 m

   c) 0.72 m   0.84 m   0.88 m   0.99 m

     1.01 m   1.02 m   1.03 m

9 a) i) 1.5 m   ii) 7.5 m   iii) 10.5 m

   b) i) $2\frac{1}{2}$ m   ii) $6\frac{1}{2}$ m   iii) $15\frac{1}{2}$ m

Write and compare decimals up to 3 decimal places

a Place value (p 111)

1 a)

Hundreds 100s	Tens 10s	Units 1s	.	Tenths 10ths	Hundredths 100ths	Thousandths 1000ths
1	1	6	.	7	4	
	3	7	.	5	4	
	1	2	.	0	6	
		4	.	5	2	
5	3	6	.	0	9	
5	0	6	.	9	2	6
	6	7	.	3	5	4
1	2	7	.	0	6	3
1	0	7	.	4	0	1
2	3	1	.	0	7	9
6	1	9	.	1	0	2
4	2	3	.	0	7	3

   b) i) $\frac{7}{10}$ ii) $\frac{4}{100}$ iii) 2

     iv) $\frac{5}{10}$ v) 500   vi) $\frac{6}{1000}$

     vii) $\frac{5}{100}$ viii) $\frac{3}{1000}$ ix) $\frac{4}{10}$

     x) $\frac{9}{1000}$ xi) $600\frac{1}{10}$ xii) $20\frac{3}{1000}$

2

Hundreds 100s	Tens 10s	Units 1s	.	Tenths 10ths	Hundredths 100ths	Thousandths 1000ths
	5	7	.	3	2	
		6	.	9		
	1	2	.	4	1	
1	0	0	.	3	5	
	9	6	.	0	7	
	1	2	.	5	0	6
4	9	7	.	6	4	3

3 a) $\frac{7}{10}$ b) $\frac{6}{100}$ c) $\frac{3}{1000}$ d) $\frac{4}{100}$
e) $3\frac{5}{100}$ f) $6\frac{4}{10}$ g) $15\frac{7}{1000}$ h) $8\frac{2}{100}$
i) $10\frac{6}{10}$ j) $240\frac{9}{1000}$

4 a) $\frac{1}{10} + \frac{7}{100}, \frac{17}{100}$ b) $\frac{4}{10} + \frac{9}{100}, \frac{49}{100}$
c) $\frac{3}{100} + \frac{1}{1000}, \frac{31}{1000}$ d) $\frac{3}{10} + \frac{5}{100} + \frac{7}{1000}, \frac{357}{1000}$
e) $6 + \frac{2}{10} + \frac{3}{100}, 6\frac{23}{100}$ f) $8 + \frac{4}{100} + \frac{3}{1000}, 8\frac{43}{1000}$
g) $42 + \frac{1}{10} + \frac{7}{1000}, 42\frac{107}{1000}$
h) $1 + \frac{2}{10} + \frac{4}{100} + \frac{1}{1000}, 1\frac{241}{1000}$
i) $502 + \frac{6}{10} + \frac{9}{100}, 502\frac{69}{100}$
j) $17 + \frac{6}{10} + \frac{5}{100} + \frac{3}{1000}, 17\frac{653}{1000}$

b Write decimals as common fractions (p 112)

1 $\frac{1}{2}$ **2** $\frac{1}{5}$ **3** $\frac{2}{5}$ **4** $\frac{3}{5}$ **5** $\frac{4}{5}$
6 $\frac{3}{20}$ **7** $\frac{7}{20}$ **8** $\frac{11}{20}$ **9** $\frac{1}{4}$ **10** $\frac{9}{20}$

c Put decimals in order of size (p 113)

1 a) 7.21   7.11   6.76   6.12   5.98
5.03   5.02   4.68   4.32
b) 14.26   14.21   13.1   13.03
12.97   12.5   12.43   12.07
c) 26.47   26.42   26.11   25.17
25.06   21.8   21.5   21.43
d) 7.84   7.21   6.96   6.48   5.74
4.06   3.5   3.21   3.19

2 a) 0.003   0.23   0.307   1.05
1.075   1.11   1.761
b) 12.005   12.05   12.121   12.429
12.5   12.502
c) 19.006   19.16   19.375   20.095
21.101   21.35
d) 412.438   428.003   483.562   581.098
581.879   604.002

3 a) 2.032 kg   2.009 kg   1.762 kg   1.104 kg
1.045 kg
b) 1.707 kg   1.7 kg   1.077 kg   1.07 kg
1.007 kg
c) 0.405 kg   0.318 kg   0.201 kg   0.058 kg
0.041 kg

4 a) 11.099   11.909   12.009   12.037
12.056   12.201
b) 18.302   18.411   18.567   18.591
19.002   19.041
c) 25.021   25.072   25.503   25.601
26.001   26.057

Use a calculator to solve problems (p 114)

1 a) 14.05   b) £4.76   c) 2 m
d) 10.7   e) £3.71   f) £8.81
g) 4.5 m   h) 21.86   i) 15.09
j) £1.53 or £1.54   k) 1 kg   l) £28.60
m) 2.15   n) £0.78   o) 20
p) 6.75
2 £11.70   **3** £5.37   **4** £28
5 a) 5.25 m   b) 0.3 m
6 £29.10   **7** £8.76   **8** £4.26

Multiply and divide decimals by 10, 100 and 1000 (p 115)

1 a) 174   b) 563   c) 17 200
d) 117.43   e) 16 010   f) 11 220
2 a) 5.622   b) 0.6304   c) 0.0144
d) 10.201   e) 0.0574   f) 0.106
3 a) 261.1   b) 0.577   c) 104
d) 1.21   e) 4 7200   f) 0.56602
4 a) 300 mm   b) 370 cm   c) 1.45 m
d) 4.5 cm   e) 345p   f) £7.89
g) 66 cm   h) 0.65 litres   i) 1.567 kg
j) 1.8 cm   k) 600 m   l) 7.58 km
5 a) 7500 g — 2.3 m   b) 0.5 m — 750 g
230 cm — 0.75 ℓ   1.2 ℓ — 5 mℓ
2300 g — 7.5 kg   0.75 kg — 50 cm
75 cℓ — 1.25 ℓ   1.2 cm — 120 cℓ
125 cℓ — 2.3 kg   0.5 cℓ — 12 mm

Round decimal numbers and approximate (p 117)

1 a) 15   b) 13   c) 7   d) 4
   e) 10   f) 6   g) 16   h) 12
   i) 10   j) 12   k) 15   l) 1
   m) 4   n) 6   o) 0   p) 27
   q) 250   r) 100

2 £15   **3** 6m   **4** £6   **5** 6kg   **6** £14.50

7 a) 5.68   b) 16.59   c) 11.33
   d) 21.90   e) 155.79   f) 16.22
   g) 131.08   h) 6.75   i) 121.90
   j) 1.89   k) 2.40   l) 16.71
   m) 15.92   n) 13.56   o) 1.01

8 a) £1.27   b) 2.73kg   c) £12.16
   d) 17.42m   e) 22.57km   f) 14.73g
   g) 16.19cm   h) £9.64   i) 0.47ℓ
   j) 15.10km

9 a) £15   b) 14kg   c) 5.11 litres
   d) £31.57   e) £7.17   f) 6.60m

10 a) 9.5   b) 12.4   c) 13.2
   d) 6.7   e) 121.8   f) 5.7
   g) 91.8   h) 151.9   i) 65.1
   j) 73.6

11 a) £7.36   b) 6.2kg   c) 4.75 m
   d) £123.69   e) 9.2 litres

12 a) 5.792   b) 16.938   c) 3.749
   d) 0.974   e) 4.032   f) 16.770
   g) 12.081   h) 27.901   i) 1.550
   j) 17.700

Add and subtract decimals (p 120)

1 a) 9.1kg   b) 2.8m   c) 23.36cm
   d) £22.67   e) £7.77   f) £30.54
   g) £127.16   h) 1.35m   i) 163.16km
   j) £1.05   k) 127.57kg   l) £156.54

2 £136.86   **3** £882.81   **4** £6.51

5 £86.35   **6** 150.25m   **7** £415.41

8 309m

9 a) Team A 91.13 seconds,
    Team B 90.78 seconds
   b) Team B by 0.35 seconds

10 £244.67

11 £1927.61

12 a) 10.491   b) 6.315   c) 4.129

13 16.116 seconds

Multiply decimals (p 121)

1 a) 2.4   b) 0.02   c) 0.35
   d) 0.12   e) 14.6   f) 59.1
   g) 5.61   h) 14.85   i) 32.56
   j) 5.14 (to 2 dp)   k) 10.52 (to 2 dp)
   l) 1.24 (to 2 dp)   m) 37.89 (to 2 dp)
   n) 8.02 (to 2 dp)   o) 66.25 (to 2 dp)
   p) 13.05 (to 2 dp)

2 15 m   **3** 4.5 litres   **4** £335.63

5

Order No	Item description	Price of item	Quantity	Price
PHF/1357	A4 paper	£4.99	5	£24.95
PHF/1359	A3 paper	£8.75	3	£26.25
ZL/129	Box black pens	£3.59	8	£28.72
FL/376	White board pens	£3.02	6	£18.12
GH/76P	Chrome stapler	£6.50	3	£19.50
GH/76L	Staples	£1.49	12	£17.88
		Total cost of all items		£135.42

6 1238 m^2

7 a) 29.8   b) 1.6776   c) 22.3987

8 99915 yen (to the nearest yen)

9 35.5m^3 (to 1 dp)

10 10.1m, 8.0m^2

Divide decimals (p 123)

1 a) 12.3   b) 11.26   c) 15.98
   d) 33.21   e) 15.32   f) 11.18
   g) 141.02   h) 37.6425

2 a) 4   b) 3   c) 20   d) 2
   e) 130   f) 34   g) 26.5   h) 360
   i) 62.5   j) 48.57 (to 2 dp)
   k) 1236   l) 36   m) 18.35 (to 2 dp)
   n) 17   o) 960   p) 140
   q) 5.23 (to 2 dp)   r) 15.19 (to 2 dp)
   s) 3.11 (to 2 dp)   t) 47

3 1.2mm per day   **4** £5.50

5 50   **6** 6060

7 a) 1940   b) 31.5   c) 19.2   d) 113.6

8 £14   **9** £125

10 a) 32.02   b) 7.20   c) 13.76
   d) 5.68   e) 41.17   f) 12.56
   g) 12.45   h) 3.91   i) 21.62

11 a) £97.84   b) £65.23   **12** 9 rolls

10 Percentages

Understand percentages (p 126)

1 black = 19%, orange = 21%,
  grey = 17%, white = 43%

2 a) 34 squares red   16 squares green
     21 squares blue   19 squares black
  b) 10% are not coloured

3 a) 20%   b) 45%   c) 16%   d) 85%

4 a) 17p   b) 99p   c) 22p   d) 51p

Equivalent fractions, decimals and percentages

a) Convert between percentages and decimals (p 127)

1 a) 0.25   b) 0.35   c) 0.4
  d) 0.3   e) 0.8   f) 0.05
  g) 0.1   h) 0.15   i) 0.04
  j) 0.08   k) 0.55   l) 0.85

2 a) 0.16   b) 0.32   c) 0.49
  d) 0.95   e) 0.64   f) 0.52
  g) 0.175   h) 0.625   i) 0.378
  j) 0.036   k) 0.025   l) 0.0125

3 a) 75%   b) 25%   c) 65%
  d) 60%   e) 6%   f) 10%
  g) 1%   h) 50%   i) 5%
  j) 90%   k) 9%   l) 99%

4 a) 29%   b) 57%   c) 43%
  d) 150%   e) 250%   f) 87.5%
  g) 32.5%   h) 6.5%   i) 0.5%
  j) 20.5%   k) 16.8%   l) 245%

b) Write percentages as fractions (p 128)

1 a) $\frac{1}{10}$ b) $\frac{1}{5}$ c) $\frac{2}{5}$ d) $\frac{4}{5}$
  e) $\frac{7}{10}$ f) $\frac{9}{10}$ g) $\frac{1}{2}$ h) $\frac{1}{4}$
  i) $\frac{1}{20}$ j) $\frac{3}{20}$ k) $\frac{9}{20}$ l) $\frac{3}{100}$

2 a) $\frac{19}{100}$ b) $\frac{16}{25}$ c) $\frac{19}{20}$ d) $\frac{2}{25}$
  e) $\frac{43}{50}$ f) $\frac{18}{25}$ g) $\frac{41}{100}$ h) $\frac{24}{25}$

3 a) $\frac{3}{200}$ b) $\frac{1}{40}$ c) $\frac{8}{125}$ d) $\frac{1}{8}$
  e) $\frac{17}{200}$ f) $\frac{5}{8}$ g) $\frac{3}{8}$ h) $\frac{21}{200}$
  i) $\frac{7}{80}$ j) $\frac{2}{3}$ k) $\frac{16}{125}$ l) $\frac{3}{80}$

c) Write fractions as decimals and percentages (p 130)

1 a) 50%   b) 30%   c) 25%
  d) 45%   e) 80%   f) 28%
  g) 34%   h) 72%

2 a) 0.1, 10%   b) 0.2, 20%   c) 0.4, 40%
  d) 0.05, 5%   e) 0.06, 6%   f) 0.04, 4%
  g) 0.35, 35%   h) 0.16, 16%

3

Fraction	Decimal	%
$\frac{1}{2}$	0.5	50%
$\frac{1}{4}$	0.25	25%
$\frac{3}{4}$	0.75	75%
$\frac{1}{10}$	0.1	10%
$\frac{1}{5}$	0.2	20%
$\frac{1}{3}$	0.33	$33\frac{1}{3}$%

Fraction	Decimal	%
$\frac{2}{5}$	0.4	40%
$\frac{7}{10}$	0.7	70%
$\frac{7}{20}$	0.35	35%
$\frac{3}{5}$	0.6	60%
$\frac{9}{10}$	0.9	90%
$\frac{3}{20}$	0.15	15%

4 a) 0.55, 55%   b) 0.625, 62.5%
  c) 0.875, 87.5%   d) 0.28, 28%
  e) 0.675, 67.5%   f) 0.667, 66.7%
  g) 0.0625, 6.25%   h) 0.1333, 13.33%

5 a) $\frac{3}{4}$, 0.76, 78%   b) 0.49, $\frac{1}{2}$, 51%
  c) 0.22, 24% $\frac{1}{4}$ d) 19%, $\frac{1}{5}$, 0.21
  e) 50%, $\frac{5}{9}$, 0.58   f) 88%, $\frac{8}{9}$, 0.9
  g) 0.56, $\frac{7}{12}$, 59%   h) $\frac{7}{16}$, 45%, 0.49

Find percentages of quantities

a) Finding 1% first (p 132)

1 a) 5   b) 6.3 km   c) 8.94 g
  d) 30p   e) £4.80   f) 26 p

2 a) 12   b) £16.20   c) 2.4 km
  d) £1.50   e) 1.05 kg   f) 51 p

3 a) 280   b) 18 m   c) 30 g
  d) 3.8 km   e) £14.40   f) £1.28

4 a) 35   b) 112 kg   c) 6.3 litres
  d) £1.75   e) 28.35 cm   f) £6.23

5 a) 180   b) £65.70   c) 10.8 m
  d) 6.3 kg   e) 6.75 litres f) £5.58

b) Finding 50%, 25% and 75% (p 134)

1 a) 125 g plain flour, 60 g butter, 45 g sugar, 140 g blackberries, 100 g red currants, 75 g strawberries

   b) 2 people

2 a) £80   b) 25 m   c) 48 km
   d) £22.50   e) 390 kg   f) 450 g

3 £60

4 45 m

5 a) £15   b) 23 m   c) 3.2 m
   d) £6.20   e) 192 m   f) 49 cm
   g) £37.50   h) £6.85   i) 3.85 kg
   j) 24.75 m   k) 11.25 km   l) £1.65

6 a) £135   b) 72 m   c) 270°
   d) 390 m   e) 132 kg   f) £27.30
   g) £9.60   h) 93 cm   i) 3.6 m
   j) 13.86 kg   k) 15.45 g   l) £55.65

7 Students' checks on 3 parts of question 6

8 75 tickets

9 45 seats

10 a) 80 men   b) 240 women

11 360 blue tiles

12 a) £467.50   b) £1402.50

13 £10.74

14 £2.4 million

c) Using 10% (p 136)

1 a) £45   b) 9 m   c) £240
   d) 5.4 cm   e) 6.2 kg   f) 32.5 m
   g) 42p   h) 80p   i) £2.37
   j) £10.50   k) 0.28 tonnes l) 0.07 km

2 a) £24   b) 51 kg   c) 336 g
   d) 384 m   e) 10.8 kg   f) £6
   g) £3.60   h) £114.80   i) £44.24
   j) 72.8 m   k) 12.78 kg   l) 11.1 km

3 a) 4 kg   b) £640   c) £2.25
   d) £3.12   e) 21.8 m   f) 3.64 litres
   g) £6.48   h) 0.475 km

4 £32 990

5 £1290

6 a) £180   b) £480   c) £36
   d) £72   e) £19.60   f) £9.20
   g) £5.16   h) £1.92   i) £4.48
   j) £19.36   k) £24.16   l) £5.50

7 £96

8 £13.74

9 a) £196   b) £49.80   c) £11.98
   d) £13.90   e) £2.99   f) £12.00

10 £78.75

11 £48.29

12 £67.35

13 £206.50

14 Tanya 25% = £160 so $12\frac{1}{2}\% = £160 \div 2 = £80$
    Jack 10% = £64, 5% = £32, $2\frac{1}{2}\% = £16$ so
    $12\frac{1}{2}\% = £64 + £16 = £80$

Percentage increase and decrease (p 138)

1 Breakfast £8.40, Lunch £16.80, Evening Meal £25.20
   Single Room £73.50, Double Room £126

2 Jan £260, Kate £286, Luke £332.80 Wes £306.80

3 £18.20

4 a) i) £128   ii) £400   iii) £1200
       iv) £1472   v) £8.48   vi) £713.60
   b) i) £120   ii) £375   iii) £1125
       iv) £1380   v) £7.95   vi) £669

5 a) i) £88   ii) £132   iii) £858
       iv) £70.40   v) £11.77   vi) £70.84
   b) i) £104   ii) £156   iii) £1014
       iv) £83.20   v) £13.91   vi) £83.72

6 £630

7 27.5 m

8 a) £52   b) £27.30

9 a) 675   b) 225

10 72

11 £855

12 £780

13 a) £11.16   b) £82.77   c) £550.56
    d) £1357.80   e) £14 229

14 a) £842.40   b) £2073.60   c) £91.26
    d) £22.95   e) £18.09

15 coffee maker £58.32, washer £478.80, radio £31.08, camera £203.99

16 £1242

Another way to increase or decrease by a percentage (p 141)

1   £10.30
2   chocolate bar 88p, tube sweets 55p,
    packet sweets £1.32, box chocs £8.25
3   £39.20
4   £10 800

Use a calculator to find percentages (p 142)

1   a) 14.3   b) £62.90   c) 4.92 m
    d) £426.40   e) 4.75 kg   f) 331.2 g
    g) 26.32   h) 5.06 cm   i) 2.24 litres
    j) 26.18 m   k) £483.84   l) £16.02
    m) 46.62 km   n) £6.75   o) £1.19
    p) 9 cm   q) £13.53   r) 8.84 m
    s) 1.75 m   t) £31.75
2   dress £43.19   stereo £209.99
    guitar £202.80   dinner service £107.99
    TV £299.99
3   Shop A: £181.30
    Shop B: £186.75   Shop A is best buy.
4   £19 665
5   £9620
6   See table

Year	Amount	Bayleys	Anchor	Rock Solid	Direct
APR		6.6	6.5	6.3	6.7
a)	£1000	£66	£65	£63	£67
b)	£2500	£165	£162.50	£157.50	£167.50
c)	£1750	£115.50	£113.75	£110.25	£117.25
d)	£3225	£212.85	£209.63	£203.18	£216.08
e)	£12 500	£825	£812.50	£787.50	£837.50
f)	€0.8 million	£52 800	£52 000	£50 400	£53 600

7   a) 4.5 inches × 3 inches
    b) 7. 5 inches × 5 inches
    c) 7.98 inches × 5.32 inches
    d) 8.46 inches × 5.64 inches
8   a) £32.76   b) 113.4 g   c) 690 m
    d) 34.02 m   e) 31.68 km   f) 93.28 kg
    g) 65.55 m   h) 16.25 litres i) £22.40
    j) £8.75
9   £98.29

Write one number as a percentage of another (p 144)

1   a) 60%   b) 40%
2   40%
3

Asst.	1	2	3	4
Mark	18	56	52	66
out of	30	70	80	120
%	60%	80%	65%	55%

4   a) 50%   b) 33.33%   c) 5%
    d) 25%   e) 5%   f) 33%
5   8%
6   8%
7   a) £1045   b) 4.5%
8   60%
9   12.5%
10 28%
11 a) 32%   b) 14%   c) 57.3%
12 a) i) 66.7%   ii) 33.3%
    b) Students' checks
13 46.3%
14 a) 28.6%   b) 7.0%
15 14.2%
16 a) 50%   b) 50%   c) 25%   d) 25%
17 a) i) 20% women   ii) 25% children
        iii) 55% men
18 a) £593.26
    b) i) 20%   ii) 10%   iii) 30%   iv) 40%
(Other estimated percentages are acceptable.)

Measures, Shape and Space

1 Money

Add and subtract sums of money

a Line up the £s and pence (p 148)

1 a) £6.31   b) £16.80   c) £2.67
   d) £5.40   e) £4.35   f) £3.45
   g) £5.05   h) £2.60   i) £1.01
   j) £3.51   k) £25.56   l) £64.06

2 £8

3 £2.65

4 a) £4.24   b) £15.76

b Use a calculator to add and subtract money (p 149)

1 a) £2.05   b) £2.50   c) £2.98
   d) £3.02   e) £3.39   f) 97p
   g) £2.83   h) £4.45   i) 59p
   j) £6.25   k) £12.98   l) £12.52

Round money to the nearest 10p or £

a Round to the nearest 10p (p 150)

1 a) 10p   b) 30p   c) 30p   d) 40p
   e) 90p   f) 60p   g) 30p   h) 60p
   i) 50p   j) 80p   k) 90p   l) 0p

2 a) £2.10   b) £4.40   c) £5.70
   d) £7.20   e) £6.90   f) £9.10
   g) £5.70   h) £4.20   i) £3.30
   j) £8.90   k) £18.00   l) £28.00

3 a) 40p + 30p + 40p = £1.10
   b) 30p + 30p + 50p = £1.10
   c) 40p + 30p + 50p = £1.20

b Round to the nearest £ (p 151)

1 a) £3   b) £3   c) £5   d) £5
   e) £7   f) £7   g) £8   h) £9
   i) £2   j) £1   k) £10   l) £20

2 a) £12   b) £34   c) £16   d) £27
   e) £337   f) £9   g) £26   h) £24
   i) £13   j) £229   k) £77   l) £76

3 a) i) £2 + £2 + £6 = £10
       ii) £2 + £3 + £1 = £6
       iii) £4 + £4 + £5 = £13
   b) i) £10.16   ii) £6.53   iii) £13.09

Add, subtract, multiply and divide sums of money

a Understand place value of £s and pence (p 152)

1 a) £30   b) 3p   c) £3   d) 30p
   e) £300

2 a) £90   b) 90p   c) £9   d) 9p
   e) £900

3 a) £110   b) £25.20   c) £12.50
   d) £30.05   e) £230.15   f) £413.25
   g) £303.03   h) £325

4 £731.34, £732.61, £736.02, £741.12

5 a) ten pounds, fifty pence
   b) five pounds, ninety-nine pence
   c) twelve pounds, fifty pence
   d) five hundred and five pounds
   e) two pounds, fifteen pence
   f) four hundred and ten pounds, four pence
   g) forty-five pounds, forty-nine pence
   h) two hundred and nine pounds, nine pence

6 £342.41   £342.14   £324.41   £324.14
   £234.41   £234.14

b Add, subtract, multiply and divide money (p 153)

1 a) £19.43   b) £27.80   c) £89.94
   d) £33.98   e) £13.01   f) £4.50
   g) £5.16   h) £12.21   i) £31.25
   j) £116.55   k) £2.10   l) 83p

2 a) £253.50   b) £282.10

3 £137.20

4 £29.98

Convert between currencies (p 154)

1 Mr Blank 1404 US dollars
   Mrs Smith 3795 zloty
   Miss Patten 15 324 rand
   Ms Chang 236 960 yen
   Mr Davies 13 128 rupees
   Miss Bailey 108 500 euros

2 Mr Wragg £447   Miss Yen £183

Mr Ennis £28   Ms Masters £62

Miss Dank £945   Mr Caine £2110

Ms Wright £27

3 £646

4 a) $1466 (US dollars)   **b)** £900

    c) £50

5 a) 5874 zloty   **b)** 106 zloty   **c)** £133

2 Time

Read and record dates (p 156)

1 a) 1st, 31   **b)** February, 28 or 29

   c) 3rd, 31   **d)** 4th, 30

   e) May, 31   **f)** June, 30

   g) July, 31   **h)** August, 31

   i) September, 30   **j)** 10th, 31

   k) November, 30   **l)** December, 31

2 Today's date – long/medium/short format

3 a) 2nd April 1998   **b)** 22/02/01

4 10/Mar/1876

5 25th July 1978

Read, measure and record time

a am and pm (p 158)

1 a) 7:00 am   **b)** 4:00 pm   **c)** 11:30 am

   d) 5:00 pm   **e)** 1:00 am   **f)** 8:45 am

2 a) iii)   **b)** i)   **c)** ii)

3 Any suitable suggestions

4 a) 12 hours   **b)** 13 hours   **c)** 17 hours

   d) $9\frac{1}{2}$ hours   **e)** 11 hours   **f)** $11\frac{1}{2}$ hours

b Read and record time to the nearest 5 minutes (p 159)

1 a) 20 minutes past 2, 2:20

   b) 40 minutes past 1, 1:40

   c) 25 minutes past 11, 11:25

   d) 50 minutes past 1, 1:50

2 a) 35 minutes past 10, 25 minutes to 11, 10:35

   b) 40 minutes past 2, 20 minutes to 3, 2:40

   c) 40 minutes past 12, 20 minutes to 1, 12:40

   d) 45 minutes past 3, 15 minutes to 4, 3:45

   e) 35 minutes past 2, 25 minutes to 3, 2:35

   f) 55 minutes past 7, 5 minutes to 8, 7:55

   g) 50 minutes past 11, 10 minutes to 12, 11:50

   h) 55 minutes past 5, 5 minutes to 6, 5:55

Use the 12 and 24 hour clock

a The 12 and 24 hour clock (p 161)

1 a) 15:50   **b)** 07:00   **c)** 09:15

   d) 22:26   **e)** 15:42   **f)** 13:30

   g) 06:45   **h)** 00:05   **i)** 19:40

   j) 18:59   **k)** 06:42   **l)** 20:25

2 a) 03:15   **b)** 16:45   **c)** 07:30

   d) 08:50   **e)** 23:10   **f)** 12:00

   g) 15:55   **h)** 07:45   **i)** 17:35

   j) 14:50   **k)** 16:05   **l)** 11:10

3 a) 16:06   **b)** 15:15   **c)** 08:11

   d) 09:45   **e)** 18:50   **f)** 23:35

   g) 10:55   **h)** 12:12   **i)** 21:20

   j) 18:30

4 a) 11:30 pm   **b)** 6:45 am   **c)** 7:12 am

   d) 4:45 pm   **e)** 12:10 pm   **f)** 1:10 am

   g) 12:16 pm   **h)** 2:10 pm   **i)** 12 am

   j) 6:05 am   **k)** 1:46 pm   **l)** 9:15 am

5 a) 15:52, 8 minutes to 4 in the afternoon, 3:52 pm

   b) 03:46, 14 minutes to 4 in the morning, 3:46 am

   c) 17:05, 5 minutes past 5 in the afternoon, 5:05 pm

   d) 23:54, 6 minutes to midnight, 11:54 pm

   e) 19:09, 9 minutes past 7 in the evening, 7:09 pm

   f) 16:32, 28 minutes to 5 in the afternoon, 4:32 pm

b Use timetables (p 163)

1 a) i) 0628   ii) 0703

   b) 0702

   c) i) 0620   ii) 0633

   d) 0640

   e) 0707

   f) i) 0637   ii) 0648

2 a) 8:45 am   **b)** 12:20 pm

   c) i) 2:10 pm   ii) 3 pm

   d) 8:53 am   **e)** 12:45 pm

   f) i) 9:15 am   ii) 12:30 pm   iii) 12:55 pm

   g) 2:03 pm

Units of time (p 164)

1 a) 30   b) 3   c) 30   d) 200
   e) 15   f) 1000   g) 6   h) 104
   i) 14   j) 26
2 a) i) 30   ii) 30   iii) 29
   b) i) 62   ii) 61
3 a) minutes   b) seconds
   c) days   d) days or weeks
   e) hours   f) weeks or months
   g) years or centuries
4 a) 210   b) 3   c) 270   d) 5
   e) 36   f) $2\frac{1}{2}$ g) 49   h) 5
   i) 450   j) 5   k) 3000   l) 10
5 2   6 £1250   7 4 weeks
8 150 seconds   9 105 min   10 13 weeks
11 a) £78   b) £312
12 a) August 4th   b) 14 nights

Calculate time

a Add and subtract time in hours, minutes and seconds (p 166)

1 a) 5 hours 20 minutes
   b) 8 minutes 57 seconds
   c) 3 minutes 35 seconds
   d) 6 minutes 15 seconds
   e) 9 hours 45 minutes
   f) 4 hours 5 minutes
2 a) 2 hours 20 minutes
   b) 1 hour 40 minutes
   c) 1 minute 20 seconds
   d) 52 seconds
   e) 3 hours 35 minutes
   f) 1 minute 21 seconds
3 a) 11 minutes
   b) 7 minutes 15 seconds
4 55 minutes

b Find the difference between times by adding on (p 166)

1 a) 2 hours 41 min   b) 2 hours 42 min
   c) 2 hours 45 min   d) 1 hour 33 min
   e) 3 hours 48 min   f) 3 hours 50 min
   g) 4 hours 15 min   h) 8 hours 36 min
   i) 3 hours 41 min   j) 5 hours 50 min
   k) 9 hours 45 min   l) 21 hours 45 min

2 a) 3 hours 30 min   b) 6 hours 15 min
   c) 5 hours 30 min   d) 4 hours 15 min
   e) 5 hours 30 min   f) 7 hours 45 min
3 a) 5 hours 21 min   b) 4 hours 18 min
   c) 4 hours 6 min   d) 4 hours 12 min
   e) 4 hours 35 min   f) 2 hours 13 min
   g) 2 hours 32 min   h) 2 hours 24 min
   i) 3 hours 25 min   j) 3 hours 9 min
   k) 4 hours 14 min   l) 2 hours 12 min

c Find the difference between times by subtracting (p 168)

1 a) 2 hours 27 min   b) 2 hours 29 min
   c) 1 hour 13 min   d) 1 hour 33 min
   e) 7 hours 26 min   f) 6 hours 11 min
   g) 1 hour 49 min   h) 3 hours 45 min
   i) 1 hour 27 min   j) 4 hours 17 min
   k) 1 hour 40 min   l) 10 hours 29 min
2 a) i) 3 hours 48 minutes
     ii) 3 hours 43 minutes
     iii) 2 hours 38 minutes
     iv) 4 hours 7 minutes
   b) 5 minutes

Use timers (p 169)

1 Analogue and digital clocks (as in question) showing:
   a) 10:30 pm   b) 6 am
   c) 11:15 pm   d) 8:30 am
2 a) Start cooking 17:30   b) Start cooking 15:15
      Cooking time   1:30   Cooking time   2:15
   c) Start cooking 18:35   d) Start cooking 17:15
      Cooking time   1:40   Cooking time   1:45
3 a) 1 minute, 16 seconds and 47 hundredths of
      a second, 1 min 16.47 s
   b) 1 minute, 21 seconds and 5 hundredths of a
      second, 1 min 21.05 s
   c) 1 minute, 16 seconds and 59 hundredths of
      a second, 1 min 16.59 s
   d) 1 minute, 22 seconds and 2 tenths of a
      second, 1 min 22.20 s
   e) 1 minute, 17 seconds and 8 hundredths of a
      second, 1 min 17.08 s
   f) 1 minute, 22 seconds and 4 tenths of a
      second, 1 min 22.40 s

4 a) 2.21 s   b) 0.56 s   c) 2.15 s

   d) 10.68 s   e) 13.52 s   f) 25.02 s

5 02:16.43, 02:34.05, 02:44.51, 02:49.30, 03:00.24, 03:01.09

   Difference: 17.62 s, 10.46 s, 4.79 s, 10.94 s, 0.85 s

3 Length

Understand distance: kilometres and miles (p 172)

1 a) A, D   b) B, C

2 Appropriate answers

3 a) 50 miles   b) 2 km   c) $\frac{1}{2}$ mile

   d) 1 mile   e) 10 miles   f) 1 km

   g) 80 miles (a and g could be interchanged)

Find the distance

a Use distances marked on a map (p 173)

1 a) 8 miles   b) 8 miles   c) 8 miles

   d) 17 miles   e) 16 miles   f) 18 miles

   g) 11 miles   h) 15 miles

2 23 miles

3 26 miles

b Use a mile/kilometre chart (p 174)

1 a) 261 miles   b) 284 miles   c) 95 miles

   d) 239 miles   e) 84 miles   f) 416 miles

2 a) 98 km   b) 106 miles   c) 45 miles

   d) 82 km   e) 48 miles   f) 11 miles

   g) 72 km   h) 255 km

Measure length in metric units

a Measure in millimetres and centimetres (p 176)

1 b) i) 3 cm   ii) 31 mm

   c) i) 5 cm   ii) 48 mm

   d) i) 7 cm   ii) 66 mm

   e) i) 7 cm   ii) 74 mm

   f) i) 9 cm   ii) 93 mm

   g) i) 11 cm   ii) 106 mm

   h) i) 11 cm   ii) 113 mm

2 a) 2 cm   b) 30 mm   c) 5 cm

   d) 70 mm   e) 6 cm   f) 80 mm

   g) 4 cm   h) 90 mm   i) 12 cm

3 a) 1.5 cm   b) 12 mm   c) 2.6 cm

   d) 13 mm   e) 1.4 cm   f) 21 mm

   g) 8.4 cm   h) 39 mm   i) 12.5 cm

4 a) 32 mm, 3 cm 2 mm, 3.2 cm

   b) 89 mm, 8 cm 9 mm, 8.9 cm

   c) 31 mm, 3 cm 1 mm, 3.1 cm

   d) 63 mm, 6 cm 3 mm, 6.3 cm

   e) 103 mm, 10 cm 3 mm, 10.3 cm

   f) 76 mm, 7 cm 6 mm, 7.6 cm

   g) 21 mm, 2 cm 1 mm, 2.1 cm

   h) 52 mm, 5 cm 2 mm, 5.2 cm

5 e b f d h a c g

6 a) 74 mm, 7 cm 4 mm, 7.4 cm

   b) 46 mm, 4 cm 6 mm, 4.6 cm

   c) 68 mm, 6 cm 8 mm, 6.8 cm

7 a) 26 mm, 2 cm 6 mm, 2.6 cm

   b) 52 mm, 5 cm 2 mm, 5.2 cm

   c) 33 mm, 3 cm 3 mm, 3.3 cm

   d) 44 mm, 4 cm 4 mm, 4.4 cm

   e) The safety pin (a) is easiest because it starts at the 0 mark

b Measure in metres, centimetres and millimetres (p 179)

1 c) 99 cm, 990 mm   d) 102 cm, 1020 mm

   e) 104 cm, 1040 mm   f) 107 cm, 1070 mm

   g) 110 cm, 1100 mm   h) 112 cm, 1120 mm

2 c) 0.99 m   d) 1.02 m   e) 1.04 m

   f) 1.07 m   g) 1.1 m   h) 1.12 m

3 c) i) 361 cm   ii) 3.61 m

   d) i) 370 cm   ii) 3.7 m

   e) i) 361 cm   ii) 3.61 m

   f) i) 363 cm   ii) 3.63 m

   g) i) 369 cm   ii) 3.69 m

   h) i) 367 cm   ii) 3.67 m

4 c) 360.7 cm, 3607 mm, 3 m 607 mm, 3.607 m

   d) 370.3 cm, 3703 mm, 3 m 703 mm, 3.703 m

   e) 361.3 cm, 3613 mm, 3 m 613 mm, 3.613 m

   f) 363.1 cm, 3631 mm, 3 m 631 mm, 3.631 m

   g) 369.3 cm, 3693 mm, 3 m 693 mm, 3.693 m

   h) 366.6 cm, 3666 mm, 3 m 666 mm, 3.666 m

Estimate and measure lengths (p 180)

1 a) accurate   b) round up   c) accurate
  d) round up   e) accurate
2 a) 1 m   b) 40 km   c) 2 m
  d) 3 cm   e) 30 cm   f) 14 cm
  g) 11 cm   h) 3 mm   i) 1 mm
  (Alternative answers are possible for some parts.)
3 a) A i) 10 cm   ii) 10.3 cm
     B i) 7 cm   ii) 6.7 cm
     C i) 4 cm   ii) 4.3 cm
     D i) 13 cm   ii) 12.6 cm
     E i) 9 cm   ii) 8.8 cm
     F i) 2 cm   ii) 2.3 cm
     G i) 10 cm   ii) 9.5 cm
  b) as a) ii)

Convert between metric lengths (p 183)

1 a) 23 mm   b) 43 mm
  c) 43 500 mm   d) 4500 mm
  e) 12 450 mm   f) 2 000 000 mm
2 a) 1300 cm   b) 4000 cm
  c) 960 cm   d) 1370 cm
  e) 6.5 cm   f) 250 000 cm
3 a) 4 m   b) 5000 m   c) 24 m
  d) 5840 m   e) 740 m   f) 3.534 m
  g) 5.3 m   h) 0.68 m   i) 0.96 m
  j) 13 750 m   k) 659 m   l) 50 m
4 a) 7 km   b) 9.4 km   c) 4.5 km
  d) 95 km   e) 2.5 km
5 a) 4000 m   b) 500 cm   c) 450 mm
  d) 5 km   e) 25 cm   f) 3.6 m
  g) 2.75 m   h) 0.275 km   i) 3600 m
  j) 270 cm   k) 122 mm   l) 7.563 km
6 6 mm, 2.6 cm, 200 mm, 0.5 m, 650 mm, 150 cm, 2 m
7 2.5 m   8 10 km
9 a) 60 cm   b) 0.6 m
10 a) 75 cm   b) 750 mm
11 a) 50 cm   b) 25 cm   c) 500 mm
   d) 250 mm   e) $\frac{2}{5}$ m   f) $\frac{3}{10}$ m

Calculate with metric lengths

a Metres and centimetres (p 185)

1 a) 5.3 m   b) 3.3 m   c) 4.5 m
  d) 4.25 m   e) 10.65 m   f) 3.64 m
  g) 2.25 m   h) 2.75 m   i) 3.8 m
2 a) $7\frac{1}{2}$ m b) $2\frac{1}{4}$ m c) $3\frac{3}{4}$ m d) $4\frac{3}{4}$ m
  e) $4\frac{1}{4}$ m f) $1\frac{1}{4}$ m g) $11\frac{1}{4}$ m h) $2\frac{1}{2}$ m

b Metres and millimetres (p 186)

1 a) 2.675 m   b) 3.025 m   c) 2.115 m
  d) 2.25 m   e) 7.7 m   f) 2.034 m

c Mixed calculations (p 186)

1 a) 2.625 m   b) 5.47 m   c) 3.045 m
2 No, the items won't fit because their total width is 3.2 m.
3 5.45 m   4 2.9 m
5 a) 22.2 m   b) 13.8 m
6 a) $1\frac{1}{4}$ m
  b) i) $5\frac{9}{10}$ m or 5.9 m   ii) $1\frac{3}{5}$ m or 1.6 m

Use a scale on a plan or map (p 187)

1

Room	Length on plan	Actual length	Width on plan	Actual width
Bedroom	3 cm	$3\times2=6$ m	2 cm	$2\times2=4$ m
Kitchen	2.5 cm	$2.5\times2=5$ m	1.5 cm	$1.5\times2=3$ m
Living room	4 cm	$4\times2=8$ m	3 cm	$3\times2=6$ m
Bathroom	1.5 cm	$1.5\times2=3$ m	1 cm	$1\times2=2$ m
Cupboard	1 cm	$1\times2=2$ m	0.5 cm	$0.5\times2=1$ m

2

	Measurements on the plan				Actual measurements			
	House length	House width	Plot length	Plot width	House length	House width	Plot length	Plot width
House A	11 mm	6 mm	25 mm	16 mm	11 m	6 m	25 m	16 m
House B	10 mm	9 mm	35 mm	16 mm	10 m	9 m	35 m	16 m
House C	13 mm	10 mm	28 mm	16 mm	13 m	10 m	28 m	16 m
House D	17 mm	6 mm	31 mm	16 mm	17 m	6 m	31 m	16 m

3 a) A 57 mm   B 74 mm   C 32 mm   D 51 mm

b) A 5.7 m   B 7.4 m   C 3.2 m   D 5.1 m

c)

		Measurement on drawing in mm	Actual measurement
i)	The length of the house	70 mm	7 m
ii)	The width of the downstairs window	22 mm	2.2 m
iii)	The width of the smaller upstairs window	22 mm	2.2 m
iv)	The width of the larger upstairs window	28 mm	2.8 m
v)	The height of the downstairs window	19 mm	1.9 m
vi)	The height of the upstairs windows	16 mm	1.6 m
viii)	The width of the front porch	28 mm	2.8 m

4 a) i) approx 15 km   **ii)** approx 10 km

iii) approx 30 km   **iv)** approx 13 km

b) i) approx 19 km   **ii)** approx 17 km

iii) approx 28 km   **iv)** approx 35 km

5

	Length on map	Actual distance
Tutor's home to A	7.6 cm	3.8 km
Tutor's home to B	5.4 cm	2.7 km
Tutor's home to C	2.6 cm	1.3 km
Tutor's home to D	3.8 cm	1.9 km

(Allow ±0.2 km)

Use ratios on scale drawings (p 190)

1 a) 50 m   **b)** 30 m   **c)** 25 m

d) 20 m   **e)** 10 m   **f)** 6 m

g) 16 m   **h)** 11.5 m

2 a)

Room		on plan	actual distance
Lounge	Length	30 mm	6 m
	Width	25 mm	5 m
Kitchen	Length	18 mm	3.6 m
	Width	14 mm	2.8 m
Bathroom	Length	15 mm	3 m
	Width	12 mm	2.4 m
Bedroom	Length	24 mm	4.8 m
	Width	16 mm	3.2 m

b) 2 m

3 a) A 41 mm × 19 mm   B 38 mm × 19 mm

C 31 mm × 16 mm   D 47 mm × 16 mm

b) A 8.2 m × 3.8 m   B 7.6 m × 3.8 m

C 6.2 m × 3.2 m   D 9.4 m × 3.2 m

4 a) 6.6 m   **b)** 3.3 m   **c)** 3 m

d) 2.4 m   **e)** 4.05 m   **f)** 2.7 m

g) 11.4 m   **h)** 10.8 m

5

Distance		On map	Actual
a)	The Minster – Hospital	2.4 cm	0.6 cm
b)	The Minster – Cemetery	5.6 cm	1.4 cm
c)	Snow Close Farm – Hospital	5.6 cm	1.4 cm
d)	College – The Minster	4 cm	1 km
e)	Snow Close Farm – The Minster	6 cm	1.5 cm
f)	Schools – Hospital	3.6 cm	0.9 km
g)	Studley – The Minster	5 cm	2.25 km

6 a) 1 : 100   **b)** 1 : 50

c) 1 : 20   **d)** 1 : 2000

e) 1 : 50 000   **f)** 1 : 250 000

g) 1 : 160   **h)** 1 : 125 000

i) 1 : 1000   **j)** 1 : 200 000

k) 1 : 5   **l)** 1 : 2500

7 Lines of length:

a) i) 6 cm   **ii)** 13 cm

b) i) 3.2 cm   **ii)** 4.8 cm

c) i) 13 cm   **ii)** 8.7 cm

d) i) 7 cm   **ii)** 4.3 cm

e) i) 4.5 cm   **ii)** 3.6 cm

f) i) 4.2 cm   **ii)** 9 cm

g) i) 7 cm   **ii)** 11.4 cm

8 Scale drawing should have the dimensions shown below:

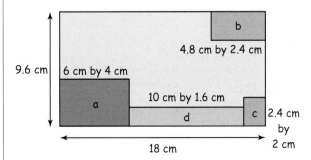

Measure lengths in imperial units (p 193)

1 a) $1\frac{1}{8}''$ b) $1\frac{13}{16}''$ c) $1\frac{3}{16}''$
   d) $1\frac{1}{8}''$ e) $\frac{3}{4}''$ f) $1\frac{5}{8}''$
   g) $1\frac{3}{8}''$ h) $1\frac{9}{16}''$ i) $\frac{11}{16}''$
   j) $1\frac{15}{16}''$ k) $1\frac{5}{8}''$ l) $\frac{3}{4}''$

Convert imperial lengths (p 194)

1 a) 180 in   b) 72 in   c) 336 in
   d) 192 in   e) 72 in   f) 432 in
   g) 113 in   h) 54 in
2 a) 6 ft   b) 240 ft   c) $2\frac{1}{2}$ ft
   d) 12 ft   e) 15 ft   f) 5 ft
   g) 4 ft   h) $2\frac{1}{3}$ ft
3 a) 2 yd   b) 5 yd
   c) 7040 yd   d) 4 yd
4 a) 3 miles   b) 9 miles
   c) 10 miles   d) 6 miles
5 a) 45 ft   b) 15 yd
6 Jan 4 ft 10 inches   Imran 5 ft 5 inches
   Carl 6 ft 1 inch

Convert between metric and imperial lengths

a Approximate metric/imperial conversions (p 195)

1

inches	centimetres
2 in	5 cm
6 in	15 cm
8 in	20 cm
3 in	7.5 cm
7 in	17.5 cm
10 in	25 cm
32 in	80 cm
18 in	45 cm
40 in	100 cm

yards/feet	centimetres	metres
2 ft	60 cm	0.6 m
$2\frac{1}{2}$ ft	75 cm	0.75 m
1 yd or 3 ft	90 cm	0.9 m
$3\frac{1}{2}$ ft or 1 yd $\frac{1}{2}$ ft	105 cm	1.05 m
$4\frac{1}{2}$ ft	135 cm	1.35 m
5 ft or 1 yd 2 ft	150 cm	1.5 m
7 ft or 2 yd 1 ft	210 cm	2.1 m
2 yd 2 ft	240 cm	2.4 m
3 yd	270 cm	2.7 m

2 a) i) 50 miles   ii) 130 miles
      iii) 18.75 miles
   b) i) 96 km   ii) 384 km
      iii) 48 km

3

		miles	km
a)	Paris to Boulogne	159	254
b)	Brussels to Calais	135	216
c)	Berlin to Cherbourg	847	1355
d)	Brussels to Dieppe	194	310
e)	Paris to Dunkerque	174	278
f)	Frankfurt to Amsterdam	283	453
g)	Munich to Boulogne	590	944
h)	Turin to Ostend	604	967
i)	Zurich to Calais	476	762
j)	Warsaw to Dieppe	1047	1675

b More accurate conversions (p 197)

1 a) 10.16 cm   b) 22.86 cm   c) 26.67 cm
   d) 30.48 cm   e) 45.72 cm
2 a) 9.14 m   b) 5.48 m   c) 4.11 m
   d) 2.13 m   e) 4.57 m
3 a) 5.91 in   b) 17.72 in   c) 9.76 in
   d) 98.43 in   e) 118.11 in
4 a) 29.8 miles   b) 9.9 miles   c) 24.8 miles
   d) 18.6 miles   e) 14.6 miles
5 a) 24.2 km   b) 40.3 km   c) 67.6 km
   d) 48.3 km   e) 15.3 km
6 a) 1030 km   b) 1755 km
7 99 miles
8 75 mph and 62 mph

4 Weight

Measure weight in metric units

a Know metric units of weight (p 198)

1 a) g   b) kg   c) g
   d) g   e) kg   f) mg

b Weigh on balance scales (p 199)

1 Any weights which match the given amounts
2 Any weights that add to 1 kg.
3 Any weights which match the given amounts

4 a) 7 kg   b) 4 kg
   c) 2 kg   d) 8 kg
   e) 4 kg 200 g   f) 7 kg 300 g
   g) 2 kg 600 g   h) 2 kg 870 g
   i) 5 kg 80 g   j) 5 kg 389 g
   k) 5 kg 20 g   l) 1 kg 5 g
5 a) 4000 g   b) 2000 g   c) 3000 g
   d) 1000 g   e) 1500 g   f) 1250 g
   g) 2500 g   h) 3750 g
6 a) 500 g   b) 700 g   c) 530 g
   d) 220 g   e) 750 g   f) 30 g
   g) 160 g   h) 420 g   i) 640 g
   j) 750 g or $\frac{3}{4}$ kg
   k) 500 g or $\frac{1}{2}$ kg
   l) 250 g or $\frac{1}{4}$ kg

c Measure in grams and kilograms (p 200)
1 a) 2   b) 500 g
   c) i) 500 g   ii) 2 kg 500 g   iii) 4 kg
      iv) 5 kg 500 g   v) 7 kg 500 g   vi) 9 kg
2 a) 6 kg   b) 10   c) 100 g
   d) i) 700 g   ii) 2 kg 300 g
      iii) 3 kg 400 g   iv) 4 kg 400 g
      v) 5 kg 200 g
   e) i) 0.7 kg   ii) 2.3 kg
      iii) 3.4 kg   iv) 4.4 kg
3 a) 500 g
   b) i) 50 g   ii) 150 g
      iii) 300 g   iv) 450 g
4 iii) 4 kg   iv) 5 kg   v) 7 kg   vi) 9 kg
5 a) i) 1 kg 200 g   ii) 2 kg 700 g
      iii) 3 kg 800 g   iv) 5 kg 200 g
      v) 6 kg 800 g
   b) i) 1.2 kg   ii) 2.7 kg
      iii) 3.8 kg   iv) 5.2 kg
      v) 6.8 kg
6 a) 50 g   b) 10 g
   c) i) 120 g   ii) 270 g
      iii) 490 g   iv) 640 g
      v) 830 g   vi) 940 g
7 a) A 6 kg   B 12.5 kg   C 18 kg
      D 27 kg   E 34.5 kg   F 42.5 kg
   b) i) 6.5 kg   ii) 9 kg   iii) 8 kg
8 a) 1 kg   b) 200 g

   c) A 2 kg 600 g   B 5 kg 200 g
      C 9 kg   D 12 kg 400 g
      E 16 kg 600 g   F 20 kg 800 g
   d) i) 2 kg 600 g
      ii) 3 kg 400 g
      iii) 4 kg 200 g

Estimate weights between marked divisions (p 202)
(*Allow any equivalent weights and some leeway.*)
1 a) A 250 g   B 650 g   C 975 g
      D 1525 g   E 1950 g   F 2350 g
   b) i) 400 g   ii) 550 g   iii) 400 g
      iv) 2100 g   v) 975 g   vi) 875 g
   c) A 1 kg   B 600 g   C 275 g
   d) D 975 g   E 550 g   F 150 g
   e) 2300 g

Choose units and instruments (p 203)
1 a) 410 g   b) 25 g   c) 5 kg
   d) 125 g   e) 1.5 kg   f) 50 g
   g) 250 g   h) 500 g   i) 1 kg
   j) 70 kg
   (*Allow reasonable alternatives.*)
2 a) kg   b) g   c) 10 g   d) 100 g   e) 10 g
3 a) P A   Q B   R A   S B   b) Q, S, R, P
   c) Scale B: there are extra divisions allowing you to weigh more accurately.

Convert between metric weights (p 204)
1 a) 4 kg   b) 6.5 kg   c) 5.485 kg
   d) 34.5 kg   e) 0.8 kg   f) 0.25 kg
   g) 300 kg   h) 0.75 kg   i) 4000 kg
   j) 3540 kg   k) 500 kg   l) 250 kg
2 a) 7000 g   b) 4500 g   c) 8268 g
   d) 16 400 g   e) 3050 g   f) 300 g
   g) 60 g   h) 1025 g   i) 2 g
   j) 7.5 g   k) 0.5 g   l) 1000 g
3 a) 44 000 mg   b) 30 500 mg
   c) 3300 mg   d) 830 mg
   e) 12 300 mg   f) 70 mg
   g) 50 000 mg   h) 3 750 000 mg
4 a) 503 tonnes   b) 42 tonnes
   c) 4.5 tonnes   d) 0.589 tonnes

5 a) 3000 kg   **b)** 500 mg

  c) 1500 g   **d)** 500 kg

  e) 7000 kg   **f)** 250 mg

  g) 0.75 tonnes   **h)** 5 tonnes

  i) 10 000 g

6 a) 0.4 kg 450 g   4 000 000 mg

    0.4 tonnes   450 kg

  b) 25 600 mg   0.25 kg   330 g   0.45 kg $\frac{1}{2}$ kg

Calculate using metric weights

a Kilograms and grams (p 206)

1 a) 5.7 kg   **b)** 3.075 kg   **c)** 3.15 kg

  d) 2.25 kg   **e)** 3.77 kg   **f)** 2.27 kg

  g) 4.7 kg   **h)** 1.6 kg   **i)** 3.94 kg

  j) 1.625 kg   **k)** 1.65 kg   **l)** 4.044 kg

  m) 2.345 kg   **n)** 0.996 kg or 996 g

2 a) 3.25 kg   **b)** 1.75 kg   **c)** 6.025 kg

3 6.05 kg   **4** 285 g   **5** 1.8 kg   **6** 960 g

7 850 g   **8 a)** 650 g   **b)** 3.25 kg

b Mixed calculations (p 207)

1 14.5 kg   **2** 10   **3** 3.61 kg

4 2.64 kg   **5** 2.1 kg   **6** 350 g

7 15   **8** 15 g   **9** 50

10 50

c Kilograms and tonnes (p 207)

1 16 days   **2** 400

3 a) 48 tonnes

  b) i) locomotive

     ii) 52 tonnes more than the cars

  c) 300 tonnes   **d)** 448 tonnes

Convert imperial weights (p 208)

1 a) 32 oz   **b)** 70 oz   **c)** 100 oz

  d) 192 oz   **e)** 72 oz   **f)** 55 oz

  g) 4 oz   **h)** 8 oz   **i)** 12 oz

  j) 40 oz

2 a) 1 lb   **b)** 6 lb   **c)** 3 lb

  d) 8 lb   **e)** 5 lb   **f)** 28 lb

  g) 56 lb   **h)** 35 lb   **i)** 4480 lb

  j) 3360 lb

3 a) 3 lb 8 oz   **b)** 5 lb 2 oz

  c) 2 lb 6 oz   **d)** 4 lb 9 oz

  e) 2 lb 12 oz   **f)** 1 lb 4 oz

  g) 3 lb 7 oz   **h)** 5 lb 1 oz

Calculate using imperial weights (p 209)

1 13 st 1 lb   **2** 8 st 3 lb

3 a) 68 lb   **b)** 4 st 12 lb

4 a) 149 lb   **b)** 10 st 9 lb

5 149 lb or 10 st 9 lb

6 a) See the chart below.

  b) 20 lb or 1 st 6 lb

  c) 1 st 2 lb or 16 lb

Week 1	Week 2	Week 3	Week 4	Week 5	Week 6
4 lb	4 lb	<u>4 lb</u>	3 lb	<u>3 lb</u>	2 lb
<u>12 st 11 lb</u>	<u>12 st 7 lb</u>	<u>12 st 3 lb</u>	<u>12 st</u>	11 st 11 lb	<u>11 st 9 lb</u>
179 lb	<u>175 lb</u>	171 lb	<u>168 lb</u>	<u>165 lb</u>	<u>163 lb</u>

Convert between metric and imperial weights

a Approximate metric/imperial conversions (p 210)

1 a) 100 g   **b)** more

2 a) 15 oz   **b)** 10 oz   **c)** 22 oz

  d) 11 oz   **e)** 19 oz   **f)** 29 oz

3 a) 350 g   **b)** 25 g   **c)** 125 g

  d) 200 g   **e)** 100 g   **f)** 90 g

  g) 65 g   **h)** 190 g

b More accurate conversions (p 211)

1 a) 8.8 lb   **b)** 10 kg   **c)** 19.8 lb

  d) 5.45 kg

2 a) 112 g   **b)** 5 oz   **c)** 420 g

  d) 2.5 oz

5 Capacity

Measure capacity and volume

a Know metric measures (p 213)

1 a) 2   **b)** 5   **c)** 10

2 a) 1000 ml   **b)** 500 ml   **c)** 250 ml

  d) 750 ml   **e)** 1500 ml

3 a) 3 l   **b)** 5 l   **c)** 9 l

  d) 12 l   **e)** 4$\frac{1}{2}$ l

b Measure in litres and millilitres (p 213)

1 a) i) 200 ml   **ii)** 600 ml

    iii) 1000 ml   **iv)** 1200 ml

    v) 1600 ml

  b) i) 0.2 litres   **ii)** 0.6 litres

    iii) 1 litre   **iv)** 1.2 litres

2 a) 1 litre   b) 100 mℓ   c) 50 mℓ
   d) i) 100 mℓ   ii) 250 mℓ
     iii) 500 mℓ   iv) 750 mℓ
     v) 950 mℓ
   e) 500 mℓ

3 a) 200 mℓ   b) 5   c) 20 mℓ
   d) 10 mℓ
   e) i) 30 mℓ   ii) 60 mℓ
     iii) 110 mℓ   iv) 130 mℓ
     v) 170 mℓ

Estimate between marked divisions (p 214)

1 i) 375 mℓ   ii) 275 mℓ   iii) 225 mℓ
   iv) 175 mℓ   v) 145 mℓ   vi) 60 mℓ

2 a) 200 mℓ
   b) i) 30 mℓ   ii) 60 mℓ   iii) 90 mℓ
     iv) 140 mℓ   v) 190 mℓ

Choose units and instruments (p 215)

1 a) litres   b) mℓ   c) mℓ
   d) litres   e) litres   f) mℓ

2 a) C   b) C   c) B   d) A   e) A

3 a) B   b) B   c) B   d) A
   e) A   f) A   g) B   h) A

4 a) 100 mℓ   b) 100 mℓ   c) mℓ
   d) 10 mℓ   e) litre
   f) litre or 100 mℓ   g) mℓ

Convert between millilitres, litres and centilitres (p 217)

1 a) 5000 mℓ   b) 2600 mℓ   c) 12 500 mℓ
   d) 3625 mℓ   e) 682 mℓ   f) 1850 mℓ
   g) 5063 mℓ   h) 750 mℓ   i) 83 mℓ
   j) 40 mℓ

2 a) 6 ℓ   b) 3.5 ℓ   c) 11 ℓ
   d) 1.1 ℓ   e) 65.4 ℓ   f) 0.33 ℓ
   g) 0.25 ℓ   h) 0.425 ℓ   i) 7.3 ℓ
   j) 5.32 ℓ

3 a) 3400 mℓ   b) 6250 mℓ
   c) 3630 mℓ   d) 8760 mℓ

4 a) i) 1 ℓ 200 mℓ   ii) 1.2 ℓ
   b) i) 3 ℓ 830 mℓ   ii) 3.83 ℓ
   c) i) 4 ℓ 935 mℓ   ii) 4.935 ℓ
   d) i) 5 ℓ 769 mℓ   ii) 5.769 ℓ

   e) i) 15 ℓ 800 mℓ   ii) 15.8 ℓ
   f) i) 5 ℓ 98 mℓ   ii) 5.098 ℓ

5 a) & g)   b) & k)   c) & l)
   d) & h)   e) & j)   f) & i)

6 C A D B

7 a) 3 ℓ   b) 2.5 ℓ   c) 30.5 ℓ
   d) 7.4 ℓ   e) 0.5 ℓ   f) 0.33 ℓ
   g) 0.75 ℓ   h) 32.1 ℓ

8 a) 3 cℓ   b) 50 cℓ   c) 25 cℓ
   d) 8.5 cℓ   e) 350 cℓ   f) 175 cℓ
   g) 50 cℓ   h) 46 cℓ

Calculate using metric units of capacity

a Litres and millilitres (p 219)

1 a) 3.7 ℓ   b) 3.425 ℓ   c) 9.075 ℓ
   d) 4.53 ℓ   e) 2.08 ℓ   f) 6.75 ℓ
   g) 205 mℓ   h) 5.96 ℓ   i) 2.66 ℓ
   j) 4.375 ℓ

2 6.5 ℓ   3 1.62 ℓ

4 a) 6.83 ℓ   b) 1.08 ℓ
   c) 4.57 ℓ   d) 8.99 ℓ

5 a) 500 mℓ   b) 400 mℓ   c) 400 mℓ
   d) 160 mℓ

6 a) 4.67 ℓ   b) 730 mℓ or 0.73 ℓ
   c) B and D   d) 800 mℓ   e) 5 ℓ

b Mixed calculations (p 220)

1 a) 20 mℓ   b) 7   2 20
3 4.8 ℓ   4 a) 9.6 ℓ   b) 7
5 9   6 a) 1.8 ℓ   b) 75 mℓ
7 4.25 ℓ   8 6 ℓ

9 a) 150 mℓ orange, 200 mℓ lemonade,
     75 mℓ rum, 50 mℓ tequila
   b) 475 mℓ

10 Answers depend on the number in the group.

Measure with metric and imperial units

Activity (p 221)

1 a) less   b) more

Convert between imperial units of capacity (p 222)

1 a) 3 pints   b) 2 pints
   c) 3.5 pints   d) 7.5 pints
   e) 32 pints   f) 12 pints
   g) 4 pints   h) 2 pints

2 a) 3 gallons   **b)** $\frac{1}{4}$ gallon
   c) 7.5 gallons   **d)** 17.5 gallons

3 a) 40 floz   **b)** 2 floz
   c) 10 floz   **d)** 5 floz

Calculate using imperial units of volume (p 222)

1 a) 20 floz lemonade   10 floz orange
   5 floz apple $2\frac{1}{2}$ floz lemon
   2 floz tequila
  b) $39\frac{1}{2}$ floz

2 $\frac{3}{4}$ gal   **3 a)** 40   **b)** £14.96

4 6 gallons

5 a) i) 80 pints   **ii)** 560 pints
   iii) 29 200 pints
  b) i) 10 gallons   **ii)** 70 gallons
   iii) 3650 gallons

Convert between metric and imperial measures

a Approximate metric/imperial conversions (p 223)

1 a) a litre   **b)** 1 pint

2 a) $3\frac{1}{2}$ pt   **b)** 2ℓ   **c)** 4 floz
  d) 10 gal   **e)** 18ℓ   **f)** 240 mℓ

3 $\approx \frac{2}{3}$ litre   **4 a)** 10ℓ   **b)** 6 floz

b More accurate conversions (p 224)

1 a) 1.14ℓ   **b)** 5 pt   **c)** 4.56ℓ
  d) 7.02 pt (to 2 dp)

2 a) 22.75ℓ   **b)** 110 gal (nearest gal)
  c) 40.95ℓ   **d)** 7.91 gal (to 2 dp)

3 a) 4.4 gallons   **b)** 54.5 litres (to 1 dp)

4 85.2 mℓ

5 a) 364 litres
  b) 3.85 gal   **c)** 21 weeks

Use conversion scales and tables (p 224)

1 a) i) 16 km   **ii)** 5 miles
   iii) 13 km   **iv)** 7.5 miles
  b) i) 68 km   **ii)** 42 or 43 miles
  c) i) 59.2 km   **ii)** 37 miles

2

Length in inches	1	2	$3\frac{1}{8}$	$3\frac{1}{2}$	$3\frac{7}{8}$	$4\frac{1}{2}$	$4\frac{3}{4}$	$5\frac{1}{4}$	$5\frac{3}{8}$	$5\frac{5}{8}$
Length in cm	2.5	5.1	8	8.9	10	11.5	12.1	13.3	13.7	14.3

3 a) 57 kg, 9 st   **b)** 66 kg, 10 st 6 lb
  c) 73 kg, 11 st 7 lb   **d)** 86 kg, 13 st 7 lb
  e) 94 kg, 14 st 11 lb   **f)** 300 mℓ, 10.6 floz
  g) 225 mℓ, 8 fl oz   **h)** 370 mℓ, 13 floz
  i) 510 mℓ, 18 fl oz

4 a) 3.107 miles   **b)** 96.56 km
  c) 77.25 km   **d)** 11.18 miles

6 Temperature

Measure temperature (p 226)

1 a) 40°C   **b)** 35°C
  c) i) 35.5°C   **ii)** 36.6°C   **iii)** 36.9°C
    iv) 37.2°C   **v)** 37.6°C   **vi)** 38°C

2 a) 50°C   **b)** 0°C
  c) i) 12°C   **ii)** 23°C   **iii)** 26°C
    iv) 35°C   **v)** 44°C

3 a) 50°C   **b)** 0°C   **c)** 2°C
  d) i) 3°C   **ii)** 16°C   **iii)** 27°C
    iv) 36°C   **v)** 44°C

4 a) 9°C   **b)** 7°C   **c)** 8°C
  d) 16°C   **e)** 24°C   **f)** 27°C

5 a) 9°C   **b)** 18°C   **c)** 7°C

6 a) 54°C   **b)** 11°C   **c)** 6 000°C
  d) 20°C   **e)** 0°C   **f)** 28°C

Temperatures below freezing point (p 227)

1 a) 50°C   **b)** −20°C
  c) i) 22°C   **ii)** 13°C   **iii)** 8°C
    iv) 2°C   **v)** −2°C   **vi)** −6°C
    vii) −12°C   **viii)** −17°C
  d) i) 22°C   **ii)** −17°C

2 a

3 a) 21°C   12°C   2°C   −2°C   −12°C   −21°C
  b) 23°C   14°C   4°C   0°C   −2°C   −4°C
  c) 7°C   5°C   3°C   1°C   −2°C   −7°C

4 a) 9°C   **b)** 14°C   **c)** 14°C
  d) 16°C   **e)** 22°C   **f)** 41°C
  g) 5°C   **h)** 17°C

5 a) 23°C   **b)** 26°C   **c)** 21°C
  d) 18°C   **e)** 26°C   **f)** 19°C
  g) 9°C   **h)** 14°C

6 a) i) 6°C   **ii)** −5°C
  b) days 4, 5, 6, 7, 8   **c)** 11°C

7

Sitting room	21°C	10°C	15°C	24°C
Dining room	20°C	9°C	14°C	23°C
Kitchen	16°C	5°C	10°C	19°C
Bedroom 1	19°C	8°C	13°C	22°C
Bedroom 2	16°C	5°C	10°C	19°C
Bedroom 3	13°C	2°C	7°C	16°C

Celsius and Fahrenheit (p 230)

1 a) i) 200°C   ii) 180°C   iii) 110°C
  b) i) 400°F   ii) 350°F   iii) 225°F
2 a) 350°F   b) 400°F
  c) dials showing:
    i) 180°C (meat), 350°F (meat)
    ii) 200°C (poultry), 400°F (poultry)

Use scales to convert temperatures (p 231)

1 32°F
2 a) 86°F   b) 113°F   c) 73°F
  d) 14°F   e) 5°F   f) 21°F
3 a) 27°C   b) 10°C   c) 6°C
  d) −8°C   e) 41°C   f) −20°C

Use formulae to convert temperatures (p 231)

1 a) 68°F   b) 86°F   c) 41°F
  d) 59°F   e) 50°F   f) 77°F
  g) 53.6°F   h) 57.2°F
2 a) 11°C   b) 19°C   c) 34°C
  d) 9°C   e) 27°C   f) 13°C
  g) 15°C

7 Perimeter, Area and Volume

Perimeter (p 232)

1 a) 10 cm   b) 12 cm   c) 10 cm
  d) 12.2 cm   e) 13.4 cm   f) 16.7 cm
2 a) 21 m   b) 21 m   c) 218 m   d) 18.8 m
3 5 m   4 3.8 m

5

Length	Width	Perimeter
6 cm	2 cm	16 cm
10 mm	6 mm	32 mm
20 cm	6.2 cm	52.4 cm
4.5 m	3 m	15 m
20 mm	15 mm	70 mm
2.7 m	2.25 m	9.9 m

6 Any dimensions which fit the given perimeters
7 1300 m   8 320 cm   9 7.8 m

Area

a Measuring in square units (p 235)

1 a) 5 cm^2 b) 7 cm^2 c) 8 cm^2
2 Any 4 shapes with an area of 12 cm^2

b Area of a rectangle (p 235)

1 a) 15 m^2 b) 26 m^2 c) 400 mm^2
  d) 15 cm^2 e) 1700 mm^2
  f) 9750 cm^2 or 0.975 m^2
2 Any dimensions which match the given areas
3 a) 54 m^2 b) 1080 g
4 a) £43.96   b) 4.32 m^2

Circles

a Circumference (p 237)

1 a) 10p diameter = 2.4 cm   £1 = 2.2 cm
  b) 10p circumference = 7.5 cm   £1 = 6.9 cm
2 a) 31.4 cm b) 44.0 cm   c) 4.7 m
  d) 5.0 m   e) 141.3 m or 141.4 m   f) 94.2 m

3

Radius	Diameter	Circumference
2 cm	4 cm	12.6 cm
15 mm	30 mm	94.2 mm
3 m	6 m	18.8 m
1.2 m	2.4 m	7.5 m
2.5 cm	5.0 cm	15.7 cm
12.0 mm	24.0 mm	75.4 mm

4 151 cm
5 251 cm

b Area of a circle (p 238)

1 a) 3.1 m^2 b) 28.3 cm^2
  c) 78.5 cm^2 d) 15.2 m^2
2 a) 153.9 cm^2 b) 12.6 m^2
  c) 201.0 mm^2 or 201.1 mm^2
  d) 452.2 mm^2 or 452.4 mm^2
  e) 706.5 cm^2 or 706.9 cm^2
  f) 6.2 m^2
3 a) 50.2 cm^2 or 50.3 cm^2
  b) 314 mm^2 or 314.2 mm^2 c) 95.0 cm^2
  d) 4.9 m^2 e) 153.9 mm^2
  f) 5024 cm^2 or 5026.5 cm^2

4 4420 cm^2

5 a) 31 cm

   b) 754.4 cm^2 or 754.8 cm^2

   c) 176.6 cm^2 or 176.7 cm^2

   d) 578 cm^2

Areas of composite shapes (p 239)

1 a) 21 m^2 **b)** 13 m^2 **c)** 21 m^2

   d) 18.32 m^2 **e)** 25.8 m^2

2 a) i) 5.76 m^2 **ii)** 10.32 m^2

   iii) 1.13 m^2 **iv)** 24.55 m^2

   b) 210 g   **c)** 1.23 kg or 1230 g

3 a) 10.8 m^2

   b) area of walls = 29.04 m^2

   area of door and window = 3.4 m^2

   walls minus door and window = 25.64 m^2

   c) No (3 litres are needed).

Use formulae to find areas (p 241)

1 a) 6 cm^2 **b)** 10.8 m^2 **c)** 94.5 cm^2

2 1.05 m^2 **3** 4800 cm^2 or 0.48 m^2

4 180 cm^2 **5** 2.2 m^2 **6** 4.7 m^2

Volume of a cuboid (p 242)

1 a) 6 cm^3 **b)** 4 cm^3 **c)** 18 cm^3

   d) 30 cm^3

2 a) 36 m^3 **b)** 8000 mm^3

   c) 32.4 cm^3 **d)** 28.8 cm^3

3

Length	Width	Height	Volume
2 cm	4 cm	5 cm	40 cm^3
3 m	3.4 m	3.5 m	35.7 m^3
4 mm	2.5 mm	6 mm	60 mm^3
9 cm	6.2 cm	8 cm	446.4 cm^3

4 0.5 m^3 **5** 4000 cm^3

6 Any dimensions which make a volume of 1 litre

7 112 500 cm^3 **8** 15 625 mm^3

9 a) 100 cm   **b)** 10 000 cm^2

   c) 1 000 000 cm^3

10 1st 0.384 m^3 2nd 0.224 m^3 3rd 0.14 m^3

   b) 0.748 m^3

11

Length	Width	Height	Volume
4 m	2 m	3 m	24 m^3
3 cm	2 cm	8 cm	48 cm^3
2 m	3 m	9 m	54 m^3
4 cm	4 cm	4 cm	64 cm^3

Volume of a cylinder (p 245)

1 a) 56.5 cm^3 **b)** 19.6 m^3

   c) 2.4 m^3 **d)** 17.0 m^3

   e) 6.8 cm^3

2 a) 9.0 m^3

   b) 1017.4 cm^3 or 1017.9 cm^3

   c) 62.2 m^3 **d)** 45.2 m^3

   e) 282.6 cm^3 or 282.7 cm^3

   f) 0.4 m^3

3 a) 3.77 m^3 **b)** 7.85 cm^3

   c) 2.71 cm^3 **d)** 0.92 m^3

   e) 0.11 m^3 **f)** 0.57 m^3

4 0.34 m^3

Use formulae to find volumes (p 246)

1 1200 cm^3

2 a) 75 000 cm^3 **b)** 2 bags

3 a) 60 cm^3 **b)** 416

4 27 litres

8 Shape

Angles (p 247)

1 a) 1   **b)** 2   **c)** 3

   d) 2   **e)** 4   **f)** 4

2 Any 10 objects with right angles

3 a) i) 90°   **ii)** 180°   **iii)** 270°

   b) 360°

4 a) 180°   **b)** 90°   **c)** 45°

   d) 90°   **e)** 45°   **f)** 135°

   g) 180°   **h)** 45°   **i)** 135°

Parallel lines (p 248)

1 B, D

2 Any examples of parallel lines

Describe positions (p 249)

1 a) bottom, right
   b) left, top
   c) bottom, between
2 There are lots of possible descriptions.
3 | conditioner | soap | shampoo |
  |---|---|---|
  | razor | toothbrush | toothpaste |

Give directions (p 250)

1 a) bank
   b) i) card shop   ii) right (east)
   c) left (south)
   d) east, right, south, Cross Street, left (east), Fir Tree Lane, right
2 Students' own questions and answers
3 a) Go straight on at the roundabout. The college is on the left.
   b) Turn left (south) at the roundabout. The college is on the left.
   c) Go straight on until the roundabout. Turn right (south) and the college is on the left.
   d) Turn right (east) at the first junction, follow the road past the school and round the left-hand bend. The college is on the right.
   e) Turn left (east) at the Town Hall then right (south) at the roundabout.
      The college is on the left.
   f) Turn left (east) at the traffic lights then right (south) at the roundabout.
      The college is on the left.
4 (Accept alternative routes.)
   a) Turn right (north) from the college, then left (west) at the roundabout.
      The post office is on the left.
   b) Turn left (west) out of the school. Turn right (north) at the 1st T-junction and left (west) at the 2nd T-junction. The petrol station is on the right.
   c) Turn left (west) out of the gym. Turn left (south) at the 1st T-junction and right (west) at the 2nd T-junction and right (north) at the 3rd T-junction.
      The fish and chip shop is on the right.
   d) Turn right (west) out of the cinema. Turn left (south) at the 1st T-junction, right (west) at the 2nd T-junction and right (north) at the 3rd T-junction, then first left. The supermarket is on the left.
   e,f) There is more than one good answer.
5 (Accept alternative answers.)
   a) **First route**
      On leaving the college, turn left (south). Follow the road round a right-hand bend and straight on to the T-junction.
      Turn right (north) at the T-junction, then take the second left turn.
      The swimming pool is on the left.
      Second route
      On leaving the college, turn right (north), then go left (west) at the roundabout (1st exit).
      Take the third left turn, then first right. The swimming pool is on the left.
   b) Either route with a valid reason, e.g. the first route because it has fewer road junctions.
6 Students' own questions and answers

Two-dimensional (2-D) shapes (p 252)

1

	Number of angles that are:		
	Right angles	Smaller	Larger
Square	4		
Rectangle	4		
Regular hexagon			6
Regular octagon			8
Regular pentagon			5
Parallelogram		2	2
Rhombus		2	2
Trapezium		2	2

2

	4 sides equal	4 right angles	Opposite sides equal	Opposite angles equal	No parallel lines	1 pair of parallel lines	2 pairs of parallel lines
Square	✔	✔	✔	✔			✔
Rectangle		✔	✔	✔			✔
Parallelogram			✔	✔			✔
Rhombus	✔		✔	✔			✔
Trapezium						✔	

3 a) 60°, 70°, 120°, 110°
  b) all 108°
  c) all 120°
  d) all 135°

Symmetry (p 254)

1 a) 1   **b)** 2   **c)** 1
  d) 1   **e)** 1   **f)** 2
  g) 1   **h)** 0   **i)** 1
  j) 0   **k)** 2   **l)** 1

2 H order 2, O order 2, Z order 2, S order 2, X order 4

3 a) parallelogram
  b) none
  c) rectangle, rhombus
  d) square, regular hexagon

4 square order 4, rectangle order 2, parallelogram order 2, rhombus order 2, regular pentagon order 5, regular octagon order 8

5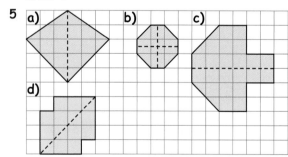

6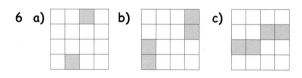

7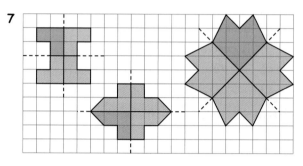

8 Students' diagrams

9 i) 4   **ii)** 0

10

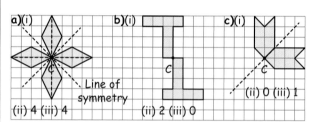

a)(i) (ii) 4 (iii) 4   Line of symmetry
b)(i) (ii) 2 (iii) 0
c)(i) (ii) 0 (iii) 1

11 Students' diagrams

Tessellations (p 256)

1 a) 8   **b)** 8   **c)** 28   **d)** 8
2 a) hexagon, triangle
  b) pentagon, 4 sided polygon (also called a quadrilateral)

Three-dimensional (3-D) shapes (p 257)

1

	Shape of faces	No. of faces	No. of edges	No. of vertices
Cube	Squares	6	12	8
Cuboid	Rectangles	6	12	8
Cylinder	Circles and curved rectangles	3	2	0
Square-based pyramid	Square and triangles	5	8	5
Triangular prism	Triangles and rectangles	5	9	6

Solve problems involving 2-D and 3-D shapes (p 258)

1 a) 88 cm b) 22 cm
2 a) length 1.6 m b) height 1.15 m
3 a) 45 tiles b) 180 tiles
4 a) i) 24 tiles ii) 112 tiles b) 136 tiles
5 360
6 a) 168 b) Net of box with dimensions
7 a) i) 295 cm ii) 310 cm iii) 350 cm
   b) 4-man c) 10 cm d) 4-man e) 9

Handling Data

1 Extracting and Interpreting Information

Information from lists, tables, pictograms and bar charts

a Find information from lists (p 261)

1 a) alphabetically b) easy to find items
  c) 791455 d) 787652
  e) Doctor's
2 a) by date b) Dinner date
  c) 20th April d) 17th March
  e) Between dinner date and wedding anniversary
3 a) at random b) 4
  c) none d) semi-skimmed
4 a) alphabetically b) by date
  c) by date d) alphabetically
  e) alphabetically in categories

b Find information from tables (p 264)

1 a) 11 b) 7 c) 11 d) 17
  e) Do-It-Yourself (19 students)
2 a) £11.00 b) £11.99 c) internet
  d) Horror e) £3.01
3 a) i) £207 ii) £147
  b) i) Nissan Qashqai ii) 3 days
  c) Toyota Aygo d) £68
  e) Mercedes C200 f) £42
4 a) 791/75S b) 11:10 c) 06:30
  d) Gatwick-Pula e) Saturday

5 a) Title and headings
  b) Redrawn table with suitable title and headings
  c) £1286 d) £500
  e) i) Altea or Hotel Esplendia ii) FB
  iii) 7 nights
  f) £105
6 a) £78.60 b) £225.50

c Find information from pictograms (p 267)

1 a) 450
  b) i) Music Box ii) George's Records
  c) 150 d) 2300
2 a) i) carpet ii) decorating
  b) i) curtains £80 bed linen £70 ii) £465
  carpet £155 furniture £104
  decorating £56 (allow ± £5)

d Find information from bar charts (p 268)

1 a)

Type of drink	Number of workers
Tea	25
Decaff tea	12
Coffee	30
Decaff coffee	13
Water	7
Fruit juice	15

  b) 13 c) 8 d) 25 e) 102

2 a) i) 38   **ii)** 65   **iii)** 39   **iv)** 21
  b) i) Soaps   **ii)** Music
  c) 17   **d)** 300

3 a) i) Germany   **ii)** Ireland
  b) UK 63 million, Spain 47 million,
    Netherlands 17 million, Ireland 5 million,
    Germany 82 million, Belgium 11 million
  c) i) 19 million   **ii)** 46 million
  d) 68 million

4 a)

Student	Height (m)
Jane	1.5
Debbie	1.6
David	1.9
Nasma	1.7
Tyla	1.8
Rae	1.7
Gary	1.8

  b) Jane   **c)** David
  d) Rae   **e)** Gary

5 a) Title, vertical axis label, (gridlines)
  b) Cheese   **c)** Baked Beans
  d)

Filling	No. bought
Baked Beans	16
Cheese	37
Chicken tikka	28
Coleslaw	19
Tuna	18

  e) 118   **f)** 19

6 a) Chart B
  b) i), ii) Students' own estimates
    (Exact values given below:)

Year	Chart A	Chart B
1	51	51
2	59	59
3	62	62
4	63	63
5	68	68

  iii) Chart B

7 a) Catering, Electrical installation
  b) i) 5   **ii)** 2   **iii)** 10
  c) i) 26   **ii)** 35   **iii)** 34
  d) i) 92   **ii)** 110

8 a) i) Pete   **ii)** £580
  b) i) Adrian   **ii)** £410
  c) Ahmed £560, Jim £490, Sam £530
  d) i) Sam   **ii)** £380
  e) i) Adrian and Jim   **ii)** £310
  f) i) Adrian £100, Ahmed £190, Jim £180,
      Pete £260, Sam £150
    ii) Pete, Ahmed, Jim, Sam, Adrian

9 Any 3 examples of each type of data

10

Magazine	a) nearest million	b) nearest 100 000
What's On TV?	3 million	3 300 000
TV Times	2 million	1 500 000
TV & Satellite	0 million	400 000
TV Choice	2 million	1 900 000
Total TV Guide	0 million	400 000
Radio Times	2 million	2 300 000

11 a) Maths   **b)** Media
  c) Allow ±2% on A & U, ±4% on others

Subject	A	B	C	D	E	U
Computing	8%	15%	22%	24%	20%	11%
Law	15%	19%	21%	20%	15%	10%
Maths	37%	19%	15%	11%	8%	10%
Media	12%	25%	32%	21%	8%	2%

Information from pie charts and line graphs

a Find information from pie charts (p 275)

1 a) work   **b)** 1.5 hours
  c) 24 hours – total time in a day

2 a) i) drive   **ii)** cycle
  b) i) $\frac{1}{4}$ **ii)** $\frac{1}{10}$ **iii)** $\frac{1}{5}$ **iv)** $\frac{2}{5}$ **v)** $\frac{1}{20}$
  c) walk 40, drive 160, bus 100,
    train 80, cycle 20

3 a) A $\frac{1}{8}$ B $\frac{1}{4}$ C $\frac{1}{2}$ D $\frac{1}{8}$
  b) A 6   B 12   C 24   D 6

4 a) i) T   **ii)** F   **iii)** T   **iv)** T
  b) Terraced, Flats, Semi-detached,
    Detached and Bungalows (approx. equal)

5 a)

Flatmate	Angle	Amount to pay
Bev	120°	120 × £0.50 = £60
Mary	40°	40 × £0.50 = £20
Jan	100°	100 × £0.50 = £50
Oliver	20°	20 × £0.50 = £10
Matt	80°	80 × £0.50 = £40
	Total	£180

6 a) Paint   **b)** more   **c)** true

  d) Wallpaper £1755, Paint £2340
  Decorating equipment £945,
  Plants £1440, Gardening tools £1260,
  Other £360 (Allow ±£22.50)

7 a) 1.5 workers

  b) 360 factory workers, 48 sales workers,
  42 managers

  c) 540

8 a) i), ii), iv), vi)

  b) i) 96°   **ii)** 2.5 people

  iii) & c) i)

	Fri	Sat
Men	240	250
Women	420	370
Children	240	580
Total	**900**	**1200**

  c) ii) as **a) i)**

b Find information from line graphs (p 279)
(Allow approximate answers.)

1 a) i) 1 hour   **ii)** 0.5°C

  b) every 2 hours

  c) i) 5.5°C   **ii)** 14:00

  d) i) −3.5°C   **ii)** 04:00

  e) i) 08:00   **ii)** 00:20, 18:40
                                      (±10 min)

  f) i) 04:00 to 14:00

    ii) 00:00 to 04:00 and 14:00 to 24:00

  g) 06:00 to 10:00

2 a) 5 mm

  b) i) November   **ii)** 150 mm

  c) i) September   **ii)** 35 mm

  d) 76 mm ± 5 mm   **e)** 1072 mm ± 20 mm

3 a) i) Graph B

    ii)

Year	Profits (£000s)
1	2130
2	2420
3	2390
4	2510
5	2680

  (allow ±10)

  b) Axis does not start at zero.

4 a) i) 10 gallons ≈ 45 litres

    ii) 5 gallons ≈ 23 litres

    iii) 18 gallons ≈ 82 litres

    iv) 7.5 gallons ≈ 34 litres

  b) i) 40 litres ≈ 9 gallons

    ii) 72 litres ≈ 16 gallons

    iii) 24 litres ≈ 5.5 gallons

    iv) 67 litres ≈ 14.5 gallons

  c) 68 litres   **d)** 18 gallons

5 a) i) 2°C   **ii)** 5°F

  b) i) 20°C ≈ 68°F   **ii)** 48°C ≈ 118°F

    iii) 75°C ≈ 167°F   **iv)** 27°C ≈ 81°F

    v) −10°C ≈ 14°F   **vi)** −7°C ≈ 19°F

  c) i) 50°F ≈ 10°C   **ii)** 140°F ≈ 60°C

    iii) 75°F ≈ 24°C   **iv)** 205°F ≈ 96°C

    v) 184°F ≈ 84°C   **vi)** 28°F ≈ −2°C

  d) freezes at 32°F, boils at 212°F

6 a) i) April   **ii)** May

  b) i) December

    ii) January and December

  c) i) 0.5 hour (or 30 minutes) per day

    ii) 2 hours per day

  d) i) 3.4 hours (or 3 hours 24 minutes) per day

    ii) 2.7 hours (or 2 hours 42 minutes) per day

  e) More sunshine in England than in Scotland.
  Least sunshine in winter, but more in spring
  than in summer.

7 a) 2009   **b)** 1972

  c) 42 million tonnes (of oil equivalent)

  d) 16 million tonnes

  e) 56 or 57 million tonnes

f) Solid fuel consumption fell sharply by about 8 million tonnes.
Petroleum consumption rose sharply by about 8 million tonnes.

g) Solid fuel consumption fell to just over a quarter of the original amount.
Gas consumption rose to about seven times as much.
Overall petroleum consumption fell by about half, but was fairly steady after 1999.
(Other wording and details are acceptable.)

2 Collecting and Illustrating Data

Record information using a tally chart (p 285)

1 a) Cat 19, Rabbit 5, Bird 7, Fish 6, Reptile 2, Other 11, None 13

b) Cat

c) Any type of pet not in the table, e.g. hamster

2 a)

Number of pets	Tally	Frequency
0	JHT JHT III	13
1	JHT JHT JHT JHT JHT I	26
2	JHT IIII	9
3	JHT II	7
4	II	2
5	I	1
6	II	2

b) i) 13   ii) 7   iii) 5

c) 60

3 a)

Grade	Tally	Frequency
Distinction	JHT II	7
Merit	JHT JHT IIII	14
Pass	JHT JHT JHT IIII	19
Fail	JHT JHT	10

b) 50   c) i) $\frac{7}{50}$ ii) $\frac{2}{25}$ d) 20%

4 a)

No. of Books	Tally	Frequency
1	JHT JHT JHT JHT JHT JHT I	31
2	JHT JHT JHT I	16
3	JHT JHT IIII	14
4	JHT JHT I	11
5	JHT III	8
6	IIII	4

b) 1   c) 84   d) 14% (nearest %)

5 a)

Time (minutes)	Tally	Frequency
6-10	III	3
11-15	JHT II	7
16-20	JHT III	8
21-25	JHT JHT	10
26-30	JHT	5

b) 21–25 minutes

c) 33

d) $\frac{1}{11}$

e) 45% (to nearest %)

Illustrate data using a pictogram or bar chart

a Using a pictogram (p 287)

1 Sunshine at a seaside resort in a week in August

Monday	☀ ☀ ☀
Tuesday	☀ ☀
Wednesday	☀ ◗
Thursday	☀ ☀ ◗
Friday	☀ ☀ ☀ ☀
Saturday	☀ ☀ ☀ ☀ ☀
Sunday	☀ ☀ ☀ ☀ ◗

Key: ☀ = 2 hours   ◗ = 1 hour

2 Number of houses sold by estate agent

January	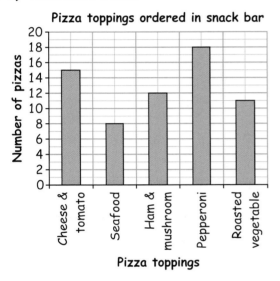
February	
March	
April	
May	
June	

Key: = 10 houses

3 Students' own design of pictogram

b Using a bar chart (p 288)

1 a) Allow other scales.

Pizza toppings ordered in snack bar

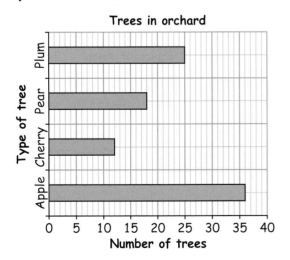

b) i) pepperoni   **ii)** seafood

2 a) Allow other scales.

Trees in orchard

b) i) apple   **ii)** 7

3 a) Students' own bar charts
   b) France
   c) i) 48   **ii)** 9

4 a) Students' own bar charts
   b) Adventure   **c)** Musical
   d) Adventure, Children's, Horror, Thriller, Comedy, Sci-fi, Musical

5 a) Students' own bar charts
   b) Most houses he supplies get 1 paper. As the number of papers increases, the number of houses decreases.

6 a) Students' own bar charts
   b) i) Sun   **ii)** Guardian
   c) Sun, Mail, Mirror, Star, Telegraph, Express, Times, Guardian

7 Allow other scales.

a)

Life expectancy

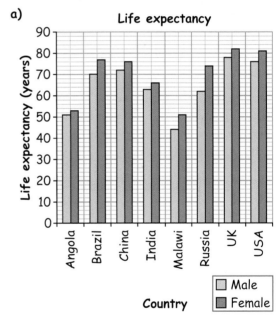

b) UK   **ii)** Malawi
   c) i) Life expectancy is greater for females than males in all countries.
   ii) Russia

d) The UK, closely followed by the USA, have the best life expectancy for both males and females.

Life expectancy is also reasonably good for both genders in China and Brazil.

Life expectancy in Russia is reasonably good for females but poorer for males.

Life expectancy for males is slightly better in India than Russia, but about 8 years worse for females.

Life expectancy is poor for both males and females in Angola and even worse in Malawi, especially for males.

8 a)

	Number of employees (thousands)		
Industry	Male	Female	Total
Construction	1905	236	2141
Education	865	2195	3060
Health and social work	818	3009	3827
Manufacturing	2139	689	2828
Wholesale and retail	2090	1901	3991

b) Allow other scales

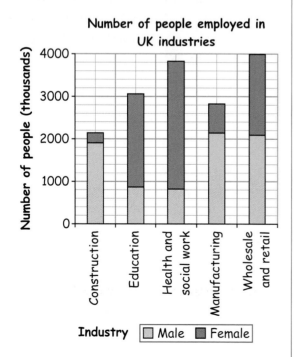

Number of people employed in UK industries

c) i) Wholesale and retail
ii) Manufacturing
iii) Health and social work

d) Many more men than women are employed in Construction and Manufacturing.

Many more women than men are employed in Education and Health and social work.

Roughly the same number of men and women are employed in Wholesale and retail.

Illustrate data using a pie chart (p 292)

1 a) Grades achieved by students on a course

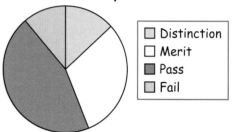

☐ Distinction
☐ Merit
■ Pass
▨ Fail

b) i) Fail   **ii)** Pass

2 a) 24   **b)** 15°

c)

Lunch	No. of students	Angle
Sandwich	6	$6 \times 15 = 90°$
Pasta	5	$5 \times 15 = 75°$
Baked potato	8	$8 \times 15 = 120°$
Salad	4	$4 \times 15 = 60°$
Pizza	1	$1 \times 15 = 15°$
Total	24	360°

d) Students' choice of lunch

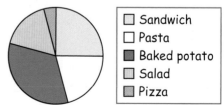

☐ Sandwich
☐ Pasta
■ Baked potato
▨ Salad
▨ Pizza

3 a) Angles: 0 cars 90°, 1 car 155°,
2 or more cars 115°

Car ownership by households in UK

☐	no car
☐	1 car
■	2 cars

b) 1   **c)** $\frac{1}{4}$

4 a) Angles: Europe 105°, N America 32°,
S America 21°, Asia 200°,
Other 2°

Production of cars

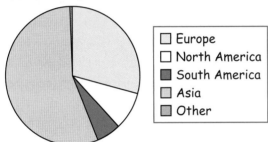

☐	Europe
☐	North America
■	South America
☐	Asia
▨	Other

b) Asia

5 a) Angles: Internet 114°, Cinema 4°,
Radio 9°, TV 90°, Press 79°,
Other 64°

Spending on advertising

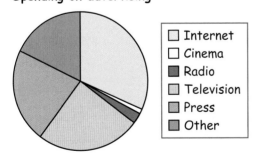

☐	Internet
☐	Cinema
■	Radio
▨	Television
▨	Press
■	Other

b) i) Internet   **ii)** more
iii) No (not true since 16.8 ÷ 3 = 5.6)
c) post, billboards, buses, etc.

6 a) Pop. angles   England 302°, NI 10°,
Scotland 30°, Wales 18°

Area angles   England 193°, NI 20°,
Scotland 116°, Wales 31°

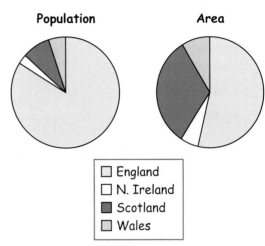

Population   **Area**

☐	England
☐	N. Ireland
■	Scotland
▨	Wales

b) England has over $\frac{3}{4}$ of the population but
just over $\frac{1}{2}$ of the land area. Scotland has
over $\frac{1}{4}$ of the land area but a much smaller
proportion of the population. Wales and
Northern Ireland also have a greater
proportion of the land area than their
population would suggest. This means that
the population density in England is greater
than in any of the other countries. Scotland
is more sparsely populated than the other
countries.

Draw graphs

a Draw a line graph (p 294)

1 a)

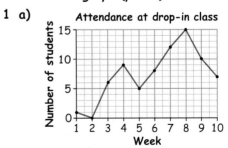

Attendance at drop-in class

b) i) Week 8   **ii)** Week 2
iii) Low at first, but increasing to 15 after
8 weeks.
Decreased at the end.

2 a) Allow other scales

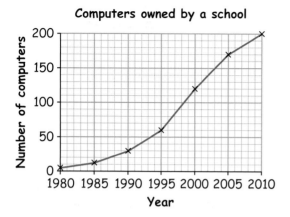

Computers owned by a school

b) i) 8   **ii)** 54   **iii)** 182
c) i) 1987   **ii)** 1998   **iii)** 2003

3 a) Allow other scales

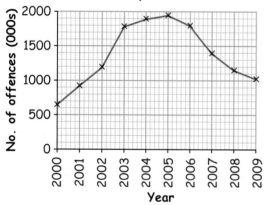

Motoring offences detected by camera

b) The number of motoring offences detected by camera increased every year between 2000 and 2005 with the greatest increase in 2003. After 2005 the number decreased each year with the greatest decrease in 2007.

c) The trend is downwards but the rate of decrease is slowing down. If this continues the number detected in 2010 would be about 900 000.

4 a)

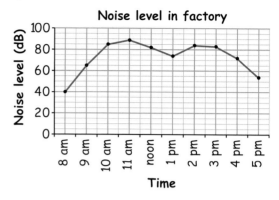

Noise level in factory

b) 9:45 am, 12:15 pm, 1:35 pm, 3:15 pm
c) 6 hours 50 minutes (allow ±10 min)
d) Starts low at 8 am (before most of work begins), builds to peak of almost 90 dB at 11 am (peak time for morning work), then falls to 74 dB at 1 pm (lunchtime – less work going on) before rising again to about 84 dB between 2 pm and 3 pm (peak period for afternoon work).
Finally falls to 54 dB at 5 pm (most of work finished for day).

5 a) i) Allow other scales

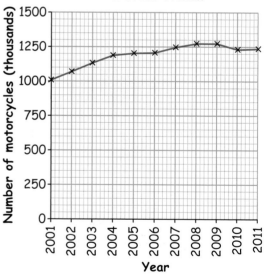

Number of motorcycles licensed in Great Britian

ii)   Allow other scales

Number of deaths and serious injuries to motorcyclists

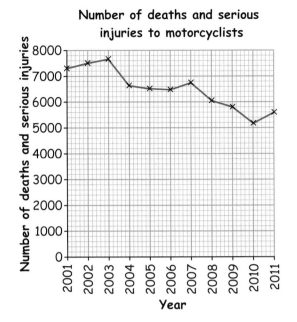

b)   The number of licensed motorcycles rose in most of the years between 2001 and 2011. The only exception was 2010 when the number fell slightly. Overall the trend was upwards. The number of deaths and serious injuries to motorcyclists rose between 2001 and 2003, but then fell in most of the years. The exceptions were 2007 and 2011 when the number rose. Overall the number of deaths and serious injuries fell.

6 a)

Clothing sales

b) i)   London Dec, Manchester Dec
    ii)   London Feb, Manchester Feb
    iii)   London – Jan, Nov, Dec
          Manchester – Dec

c) i)   Jan   ii) Jul

d)   Both sold a lot in January (sales) and in December (before Christmas). Both sold less in February (after the sales), but then sales rose during the spring with London reaching a peak in June and Manchester in July. Sales at both shops dipped slightly in August. From September onwards the sales in London increased, whereas in Manchester sales didn't rise significantly until after October.

7 a)

Average monthly temperatures

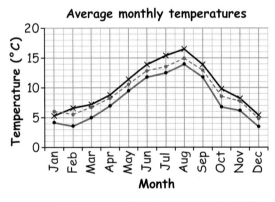

b)   E&W Jan, Dec
     Scot Jan, Feb, Mar, Dec
     NI Feb, Dec

c)   All low in winter months and high in summer, reaching a peak in August. Generally E&W has higher temperatures than NI, with Scotland the coolest. The only exception was in January when NI was warmer than E&W.

8 a)

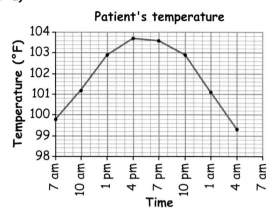

Patient's temperature

b) approx 19½ hours

c) 5:20 am (approx)

b Draw and use conversion graphs (p 298)

1 a) Horizontal 0.2 inches, Vertical 0.5 cm

c) i) 6 in ≈ 15.2 cm

ii) 9 in ≈ 22.9 cm

iii) 2.6 in ≈ 6.6 cm

iv) 4.3 in ≈ 10.9 cm

v) 10 cm ≈ 3.9 in

vi) 25 cm ≈ 9.8 in

vii) 18 cm ≈ 7.1 in

viii) 12.5 cm ≈ 4.9 in

Allow ±0.5 cm and ±0.2 in

2 a)

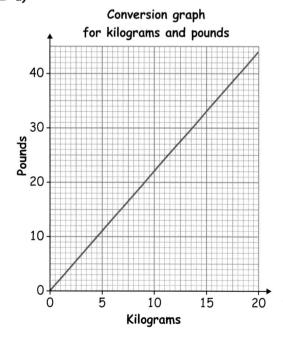

Conversion graph for kilograms and pounds

b) i) 15 kg ≈ 33 lb **ii)** 9 kg ≈ 19.8 lb

iii) 12.5 kg ≈ 27.5 lb **iv)** 30 lb ≈ 13.6 kg

v) 14 lb ≈ 6.4 kg **vi)** 38.5 lb ≈ 17.5 kg

Allow ±1 lb and ±0.5 kg

Choosing a statistical diagram (p 300)

A variety of statistical diagrams could be used.

The best types for each question are given below.

1 Line graph (or bar chart)

Cinema attendances each month in a year

The graph shows that the highest attendance was in August (perhaps because children were taken to the cinema in the school holidays).

The attendance was also quite high in July and February.

The attendance was low in the spring (March and April) and also September.

2 Pie chart (or bar chart)

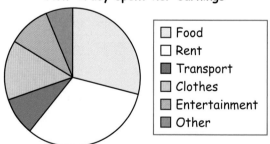

How Tracy spent her earnings

☐ Food
☐ Rent
■ Transport
▨ Clothes
▨ Entertainment
■ Other

Angles: Food 104°, Rent 115°, Transport 32°, Clothes 50°, Entertainment 36°, Other 22°

Tracy spent over a quarter of her earnings on rent (the largest amount) and over a quarter on food.

This left less than a half to be spent on transport, clothes, entertainment and other things.

3 Bar chart (comparative or component)

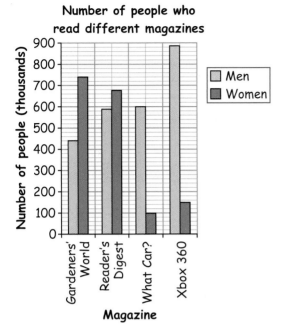

Number of people who read different magazines

Most of the readers of Gardeners' World and Reader's Digest are women. Most of the readers of What Car? and Xbox 360 are men – very few women read these magazines.

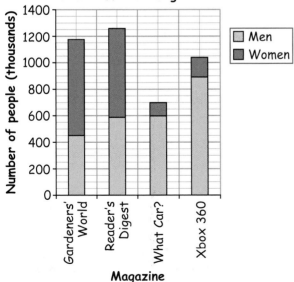

Number of people who read different magazines

The component bar chart shows more clearly that Reader's Digest has the most readers, followed by Gardener's World.
What Car? has the fewest readers.

4 Line graph (or bar chart)

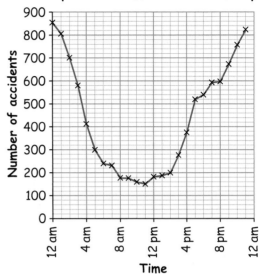

Number of drink drive accidents reported at different times of day

Many drink drive accidents occur at night between 5pm and 3am, with the largest number occurring around midnight.
Few drink drive accidents occur between 7am and 3pm, with the lowest number occurring around 11am.

5 Pie chart

Reasons for moving house

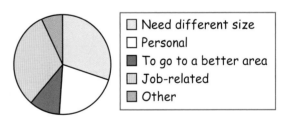

Angles: Different size 108°, Personal 76°, Better area 36°, Job 115°, Other 25°
The most common reason for moving house was related to jobs (roughly a third gave this reason). Over a quarter moved because they needed a different size of house.

6 Bar chart (comparative)

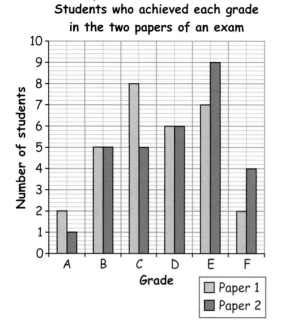

Students who achieved each grade in the two papers of an exam

The students had better results on Paper 1. The most common grade on Paper 1 was C, whilst on Paper 2 it was E. Also more students achieved grades A and C on Paper 1 whilst more got the low grades E and F on Paper 2. The same number of students achieved grade B and grade D on both papers.

7 Pie chart (or bar chart)

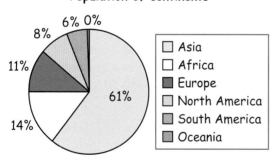

Population of continents

Angles: Asia 218°, Africa 52°, Europe 40°, North America 28°, South America 20°, Oceania 2°

Well over half of the world's population live in Asia.

Only a very small number live in Oceania.

8 Bar chart (comparative or component)

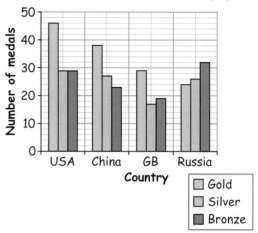

Medals won in the London 2012 Olympics

The comparative bar chart shows that the USA won more gold medals and more silver medals than the other countries and were only beaten by Russia in the number of bronze medals they won.

China was second in terms of gold medals won, with GB third and Russia fourth.

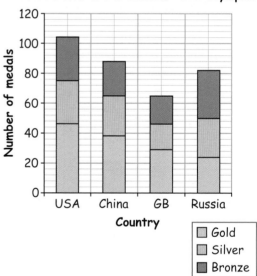

Medals won in the London 2012 Olympics

The component chart also shows clearly that the order in terms of gold medals was: USA, China, GB and Russia.

This chart also makes it easy to compare the total number of medals. The order in this case was: USA, China, Russia and GB.

It is harder to compare the number of silver medals and the number of bronze medals on this chart.

3 Averages and Range

Find the mean (p 302)

1 a) 1.8   b) 6   c) 6.6
   d) 4.25   e) 0.375 kg   f) £40

2 a) Andy 45, Imran 37, Mark 24, Stuart 28.75,
    Tim 39.75
   b) i) Andy   ii) Mark

3 a) £13.20   b) No – nobody gets £13.20
                   per hour

4 a) 303   b) 189.6

5 a) Jen £8.40, Kim £7.60, Ben £9.50,
    Nasma £8.40, Mike £7.20, Wes £7.80
   b) £8.15   c) Jen, Ben, Nasma

(Using a pictogram data): (pp 287–288)

1 6.4 hours per day (1 dp)

2 45 houses (nearest whole number)

(Using a bar chart data): (pp 288–290)

1 12.8 pizzas (1 dp)   2 22.8 trees (1 dp)

3 39.5 properties   4 190 films

6 2 665 000 readers (nearest thousand)

(Draw a line graph data): (pp 294–296)

1 7.3 students

3 1 380 000 offences (nearest thousand)

4 72.8 dB

5 Number of motorcycles per year:
   1190 thousand (nearest thousand)
   Number of motorcycle accidents per year:
   6 500 (nearest hundred)

6 London £256 000 (nearest £thousand)
   Manchester £202 000 (nearest £thousand)

Level 2 (p 305)

6 a) 20   b) 0.9 errors per page

7 a) 2.85 children per family   b) 10

8 a) 1.2 trains per day   b) 24%

9 a) 3   b) 18.1 years (1 dp)

10 a) i) 2 goals per match
       ii) 1.1 goals per match (1 dp)
   b) at home

Find the mode and median (p 307)

1 'Best' average underlined (matter of opinion)
   Mean given to 1 dp when not exact
   a) <u>mode = 1</u>, <u>median = 1</u>, mean = 1.6 (pints)
   b) <u>modes = 16</u>, <u>17</u>, median = 16.5, mean = 16.2
   c) mode = 18 years, <u>median = 19 years</u>,
       mean = 21.1 years
   d) <u>mode = 28 minutes</u>, <u>median = 29 minutes</u>,
       <u>mean = 29.6 minutes</u>
   e) no mode, <u>median = 1.2 m</u>, <u>mean = 1.21 m</u>

(Find the mean data): (pp 302–303)

1 a) mode = 2, median = 2
   b) no mode, median = 5
   c) no mode, median = 6 years
   d) modes = 4, 5, median = 4.5
   e) modes = 0.36 kg, 0.41 kg,
       median = 0.375 kg
   f) mode £10, median £10

2 No modes as all the scores are different.
   Medians: Andy 32.5, Imran 43, Mark 24,
   Stuart 29, Tim 39
   The mean uses **all** the scores.

(p 308)

2 a) i) 4 years   ii) 3 years   iii) 3.05 years
   b) Median

3 a) i) 14   ii) 13   iii) 12.95
   b) Mode

(Find the mean data): (pp 303–306)

Most appropriate averages given in brackets

3 mode, median both £9 per hour
   (mode , median)

6 mode = 0 errors, median = 1 error (median)

7 mode = 2 children, median = 2.5 children
   (mode)

8 mode = 1 train, median = 1 train (mode , median)

9 mode = 17 years, median = 18 years
   (mode, median)

10 home   mode = 1 goal, median = 1.5 goals
                    (mode)

     away   mode = 1 goal, median = 1 goal
                    (mode , median)

(p 309)

4 a) i) 1 attempt   ii) 2 attempts
  b) i) 2 attempts   ii) 2 attempts
  c) i) 1.85 attempts ii) 2.1 attempts
  d) men – lower mean and mode

> (*Using a bar chart data*): (p 289)
> 5   mean = 1.7 (to 1 dp), mode = 1, median = 1
> mode and median – possible number of
> newspapers

Find the range (p 310)

1 a) Apricots 18p, Grapefruit 12p, Peaches 33p,
    Pineapple 9p
  b) i) Pineapple   ii) Peaches

2 a)
Alex	mean = 43.3	range = 83
Cilla	mean = 44.5	range = 49
Kath	mean = 57	range = 75
Paul	mean = 76.9	range = 115
Rhona	mean = 66.5	range = 147
Paul	mean = 59.1	range = 88

  b) Paul, highest mean
  c) Cilla, lowest range

> (*Find the mean data*): (pp 302–303)
> 1 a) 4 students   b) 12 calls
>   c) 11 years   d) 8 letters
>   e) 0.09 kg   f) £240
> 2 Andy 109 runs, Imran 68 runs, Mark
>   26 runs, Stuart 23 runs, Tim 27 runs
> 3 £21 per hour
> 4 198 tickets
> 5 £2.30
>
> (*Find the mode and median data*): (p 307)
> 1 a) 6 pints   b) 6 students
>   c) 16 years   d) 11 minutes
>   e) 0.09 m

(p 311)

3 a) 77.9 minutes (1 dp)   b) 120 minutes
4 a) 5 visits   b) 3 visits
5 a) i) 7 years   ii) 5 years
  b) i) 10 years   ii) 11 years
  c) The boys' ages are more spread out than
    the girls' ages. On average the girls are
    older.

> (*Find the mean data*): (pp 305–306)
> 6 4 errors   7 5 children
> 8 4 late trains   9 6 years
> 10 Home 6 goals, Away 4 goals
>
> (*Find the mode and median data*):
> (pp 308–309)
> 2 3 years   3 12
> 4 Men 4 attempts, Women 4 attempts

6 a) i)   3 points   ii) 5 points
    iii) 5.1 points (1 dp)   iv) 6 points
  b) i)   3 points   ii) 3 points
    iii) 3.8 points (1 dp)   iv) 5 points
  c) On average the women were awarded
    fewer points and the points awarded to
    women were less variable.

4 Probability

Show that some events are more likely than others (p 313)

1 a), e) depend on students
  b) unlikely   c) impossible   d) certain
2 a) 2   b) 1 in 2
3 a) i)   1 in 5   ii) 2 in 5   iii) 3 in 5
  b) odd number
4 a) 52
  b) i)   4 in 52 (or 1 in 13)
    ii) 13 in 52 (or 1 in 4)   iii) 1 in 52
5 a) 26
  b) i)   1 in 26   ii) 5 in 26   iii) 21 in 26

Use fractions, decimals and % to measure probability (p 316)

1 a) i) $\frac{1}{10}$, 0.1, 10%   ii) $\frac{1}{2}$, 0.5, 50%
    iii) $\frac{3}{5}$, 0.6, 60%   iv) $\frac{3}{10}$, 0.3, 30%
  b)

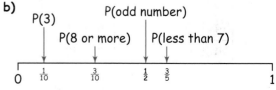

2 a) i) $\frac{1}{6}$, 0.17 (2 dp), 16.7% (1 dp)
    ii) $\frac{1}{2}$, 0.5, 50%   iii) 0, 0, 0%
    iv) 1, 1, 100%

3 a) i) $\frac{1}{5}$, 0.2, 20%   **ii)** $\frac{2}{5}$, 0.4, 40%

   iii) $\frac{3}{5}$, 0.6, 60%

b)

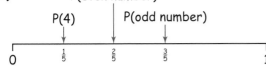

4 a) i) $\frac{1}{13}$ **ii)** $\frac{1}{4}$ **iii)** $\frac{1}{52}$ **iv)** $\frac{1}{2}$ **v)** $\frac{1}{26}$ **vi)** $\frac{2}{13}$

   b) the king of clubs, a red king, a king, a jack
   or queen, a club, a red card

5 a) i) $\frac{5}{8}$ **ii)** $\frac{1}{4}$ **iii)** $\frac{1}{8}$

   b)

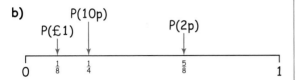

6 0.2, 20%

7 a) $\frac{1}{366}$ **b)** $\frac{31}{366}$ **c)** $\frac{5}{61}$ **d)** $\frac{29}{366}$ **e)** $\frac{121}{366}$

8 a) i) $\frac{1}{4}$ **ii)** $\frac{1}{2}$

   b) i) 10   **ii)** 20

9 a) i) $\frac{1}{6}$ **ii)** $\frac{1}{2}$ **iii)** $\frac{1}{3}$

   b) i) 10   **ii)** 30   **iii)** 20

10 a) Ball sports 76.0%, Combat sports 3.3%,
   Wheel sports 5.2%, Winter sports 3.2%,
   Animal sports 3.1%, Water sports 3.4%,
   Other 5.8%

   b) Far more hospital treatments are for
   injuries from ball sports than from any
   other type of sport.

Identify possible outcomes of combined events (p 318)

1 a)

Score on 1st dice

	1	2	3	4	5	6
1	2	3	4	5	6	7
2	3	4	5	6	7	8
3	4	5	6	7	8	9
4	5	6	7	8	9	10
5	6	7	8	9	10	11
6	7	8	9	10	11	12

Score on 2nd dice

   b) 36

c) $P(2) = \frac{1}{36}$, $P(3) = \frac{1}{18}$, $P(4) = \frac{1}{12}$, $P(5) = \frac{1}{9}$,
   $P(6) = \frac{5}{36}$, $P(7) = \frac{1}{6}$, $P(8) = \frac{5}{36}$, $P(9) = \frac{1}{9}$,
   $P(10) = \frac{1}{12}$, $P(11) = \frac{1}{18}$, $P(12) = \frac{1}{36}$

d) 7

2

Score on 1st dice

	1	2	3	4	5	6
1	1	2	3	4	5	6
2	2	4	6	8	10	12
3	3	6	9	12	15	18
4	4	8	12	16	20	24
5	5	10	15	20	25	30
6	6	12	18	24	30	36

Score on 2nd dice

   a) $\frac{1}{4}$ **b)** $\frac{3}{4}$

3 a)

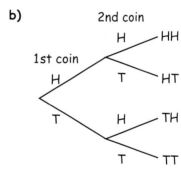

   b)

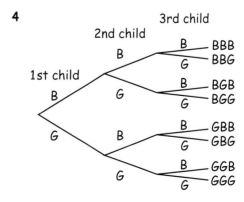

   c) i) $\frac{1}{4}$ **ii)** $\frac{1}{2}$

4

3rd child
2nd child

1st child

B — B — B — BBB
       G — BBG
     G — B — BGB
       G — BGG
G — B — B — GBB
       G — GBG
     G — B — GGB
       G — GGG

5 a)

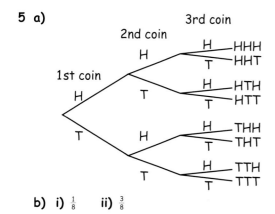

b) i) $\frac{1}{8}$ **ii)** $\frac{3}{8}$

6 a)

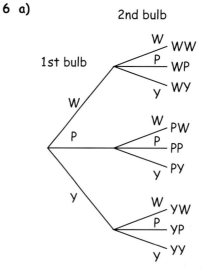

b) 2 white, 2 purple, 2 yellow,
1 white and 1 purple, 1 white and 1 yellow,
1 purple and 1 yellow

Index

Acknowledgements

The authors and the publisher would like to thank the following for permission to reproduce material:

Text Permissions
pp 50, Annual UK cinema admissions, reproduced with kind permission of the Cinema Exhibitors Association website with data originally sourced from EDI Rentrak; pp 49, Census 2001 Key Statistics for health areas in England & Wales; pp 96, energy consumed in the UK each year, Office for UK National Statistics; pp 97, Department for Environment, Food & Rural Affairs, the Welsh Assembly Government; pp 262, Google search 'Kirk Michael'; pp 264, various car models including Ford Focus, Mercedes C200, Nissan Qashqai, Toyota Aygo, VW Golf, and VW Tiguan; pp 273, National Readership survey of magazines, reproduced with kind permission of the National Readership Survey; pp 274, Percentage of A Level passes A-U in Computing, Law, Maths and Media, sourced from the Qualifications and Curriculum Development Agency statistics on UK; pp 280, Rainfall in England and Wales across a year, information sourced from the Met Office; pp 283, Data number of hours per day of sunshine over the course of a year, information sourced from the Met Office; pp 290, National Readership survey of newspapers 2011/2012, reproduced with kind permission of the National Readership Survey; pp 292, Number of cars owned by UK households, information sourced from the National Travel Survey; pp 293, number of cars produced in different world continents, International Organisation of Motor Vehicle Manufacturers; pp 293, type of advertising, reproduced with kind permission of the Advertising Association; pp 293, population and area of each country in the UK, sourced from the office for UK National Statistics; pp 295, licenced motorcycles in Great Britain between 2001 and 2011, Department for Transport; pp 296, average temperature in England, Wales, Scotland and Northern Ireland, information sourced from the Met Office; pp 300, cinema attendances, data sourced from the British Film Institute Statistical Yearbook with data suppliers CAA and Rentrak; pp 300, number of drink drive accidents reported in one year, Department of Transport.

Images
pp 2, ChristianAnthony / iStockphoto; pp 17, Felix140800 / iStockphoto; pp58, karlbarrett / iStockphoto; pp72, xavdlp / iStockphoto; pp 256 Smaglov / iStockphoto;

Every effort has been made to trace the copyright holders but if any have been inadvertently overlooked the publisher will be pleased to make the necessary arrangements at the first opportunity.